Green Chemistry
and Processes

Green Chemistry and Processes

Mukesh Doble
Professor, Department of Biotechnology,
India Institute of Technology, Madras, India

Anil Kumar Kruthiventi
Associate Professor, Department of Chemistry
Sri Sathya Sai University, India

AMSTERDAM • BOSTON • HEIDELBERG • LONDON
NEW YORK • OXFORD • PARIS • SAN DIEGO
SAN FRANCISCO • SINGAPORE • SYDNEY • TOKYO
Academic Press is an imprint of Elsevier

Academic Press is an imprint of Elsevier
30 Corporate Drive, Suite 400, Burlington, MA 01803, USA
525 B Street, Suite 1900, San Diego, California 92101-4495, USA
84 Theobald's Road, London WC1X 8RR, UK

This book is printed on acid-free paper.

Library of Congress Cataloging-in-Publication Data
Application submitted

British Library Cataloguing-in-Publication Data
A catalogue record for this book is available from the British Library.

ISBN: 978-0-12-372532-5

For information on all Academic Press publications visit our
Web site at www.books.elsevier.com

Printed in The United States of America
07 08 09 10   9 8 7 6 5 4 3 2 1

Working together to grow
libraries in developing countries

www.elsevier.com | www.bookaid.org | www.sabre.org

ELSEVIER   BOOK AID   Sabre Foundation
             International

Dedications

I would like to dedicate this book to Geetha, Deepak, and Niharika.

—Mukesh

I would like to dedicate this book to Bhagawan Sri Sathya Sai Baba.

—Anil Kumar

Contents

Preface

After the Second World War, industrialization took place at a tremendous pace without giving any thought to its effects on the environment, flora and fauna, and peoples' safety and heath. This led to increased global warming, depletion of ozone protective cover from harmful UV radiation, contamination of land and water ways due to release of toxic chemicals by industries, reduction in nonrenewable resources such as petroleum, destructions of forest cover due to acid rains, increased health problems, and industrial accidents resulting in loss of life and property.

The Kyoto Protocol is an international treaty bringing almost 180 nations of the world together in an effort to limit greenhouse gas emissions and reduce the effects of global warming. The reduction is expected to be achieved through industrial green house gas reduction (42%), use of renewable energy (22%), methane reduction from live stock and other sources (22%), energy efficiency (12%), and fuel switch from oil to natural gas (2%). The United States and Australia rejected the treaty and had refused to sign it even as late as November 2006. The EU-15 countries had agreed to cut, by 2012, 8% of the 1990 greenhouse gas emissions values, but the data collected by the European Commission (Oct. 2006) predicted that the values would be only 0.6%. Worse still, many European countries may even exceed their individual limits.

Industries are interested in green chemistry because reduction in energy, improvement in yield, and use of cheaper raw materials lead to reduction in working capital and an increase in profits. Industries are also interested in improving the safety, health, and working conditions of their workforce, because they are worried about litigations and pressures from NGO and other employee unions. Incremental improvements in processes are easily implementable, while major changes in processes require changes in equipment and hardware; therefore the manufacturing industries hesitate to undertake such a huge capital expenditure. Process intensification also requires scrapping the old equipment and using new equipment, which industries are hesitant to undertake due to expenditure. Alternate energy sources are still not as competitive as the petroleum-based energy. Although a few applications are found in the use of renewable energy, the latter has not made major inroads. Waste treatment is seen as a wasteful activity and industries are beginning to comply with government pollution bodies. On the other hand, preventing or reducing waste at the source through design of innovative processes is much more profitable for them. While industries weigh all actions with respect to profit and cost reduction, it is the duty of the government to take the larger view and weigh the long-term benefits to society. Unless the governments are committed nothing can be achieved.

Green chemistry involves reduction and/or elimination of the use of hazardous substances from a chemical process. This includes feedstock, reagents, solvents, products, and byproducts. It also includes the use of sustainable raw material and energy sources for this manufacturing process. This book is an interdisciplinary treatise dealing with chemistry and technology of green chemistry. Measuring "greenness" has always been a challenge and there are different approaches discussed here. Comparing various processes based on a set of common standards is discussed. Novel and innovative synthetic techniques that lead to reduction of raw material/solvent usage, milder operating conditions, less wasteful side products, etc., are discussed in Chapter 2. Heterogeneous catalysts provide mild and efficient environment. They are superior to methods that use stoichiometric amounts of chemical reagents or homogeneous catalysts since they produce minimum waste and side products. Examples of processes that use biocatalysts are discussed in detail in Chapter 4. Biocatalysts simplify reactions, telescope multiple steps into one and make it environmentally clean and friendly. Even reactions which cannot be performed with chemical synthesis can be carried out with ease using microbes. New solvents such as ionic liquids are becoming very popular since they are benign, environmentally friendly, and do not contribute toward voc. This book does not discuss the research that is in progress to identify new reagents which are mild and those that do not produce waste.

Miniaturization and innovative reactor designs have led to an increase in mass and heat transfer coefficients by several orders of magnitude, reduced reaction time dramatically, eliminated waste, and improved safety. Chapter 6 covers the approaches followed by manufacturers to achieve process miniaturization and intensification, decrease multistep reaction to single step, and combine reaction with unit operation. Use of membranes and other novel downstream processing techniques are also discussed.

Hydrocarbon fuel is a non-renewable source of energy and it is well accepted that it would not last long. Research using alternate energy sources is being pursued seriously in many research labs worldwide. Use of renewable raw material includes ethanol, biodiesel, etc. Biomass appears to be the future energy source in addition to solar, wind, and wave.

Processes that are milder (i.e., use less toxic solvents, do not produce dangerous intermediates, recalcitrant waste, or side product chemicals, etc.) are inherently safe. Such a philosophy could have prevented Bhopal. The principle of green chemistry believes in designing inherently safe processes. Industries that are seriously taking an all-out effort to decrease waste, improve efficiency, and decrease energy are discussed in Chapter 9. The final chapter deals with the future trends and frontier research in this area. Process improvements will never end. Biotechnology will play a crucial part in achieving the goals of green chemistry. Unless the politicians and administrators in countries of the world realize the importance of green chemistry and the urgency with which it has to be approached, the world will deteriorate into an uninhabitable mass of land.

I would like to thank Ms. N. Aparnaa for her secretarial help during the preparation of the manuscript.

Mukesh Doble
December 2006

About the Authors . . .

MUKESH DOBLE is a Professor at the Department of Biotechnology, Indian Institute of Technology, Madras, India. He has authored 110 technical papers, four books, filed three patents, and is a member of the American and Indian Institute of Chemical Engineers. The recipient of the Herdillia Award from the Indian Institute of Chemical Engineers, he received his B.Tech. and M.Tech. degrees in chemical engineering from the Indian Institute of Technology, Madras, and a Ph.D. degree from the University of Aston, Birmingham, England. He has postdoctoral experience from the University of Cambridge, England and Texas A&M University, USA. He has worked in the research centers of ICI India and GE India for 20 years.

ANIL KUMAR KRUTHIVENTI is an Assistant Professor, Department of Chemistry, Sri Sathya Sai University, India. The author of 40 technical papers, he has filed two patents in microbial mediated o-dealkylations of aryl alkyl ethers and is a member of the Indian Science Congress. He is the recipient of the Young Teachers Career Award (1996) from the All India Council for Technical Education, New Delhi, India. He received his M.Sc. and Ph.D. degrees from Sri Sathya Sai University, Prashanti Nilayam, A.P. India.

CHAPTER 1

Introduction

The chemical industry accounts for 7% of global income and 9% of global trade, adding up to US$1.5 trillion in sales in 1998, with 80% of the world's output produced by 16 countries. Production is projected to increase 85% by 2020 compared to the 1995 levels. This will be in pace with GDP growth in the United States, but at twice the per capita intensity. There will be strong market penetration by countries other than these 16, especially in commodity chemicals (OECD, 2001). Over the past half-century, the largest growth in volume of any category of materials has been in petrochemical-based plastics; and in terms of revenue it was pharmaceuticals. The latter, in the past two decades, has become number one. Overall production has shifted from predominantly commodity chemicals to fine and specialty chemicals, and now it is the life sciences. In the United States, the chemical industry contributes 5% of GDP and adds 12% of the value to GDP by all U.S. manufacturing industries, and it is also the nation's top exporter (Lenz and Lafrance, 1996). This information speaks volumes about the importance of chemical industries in our day-to-day life and in supporting the nation's economy. But it is plagued with several problems, such as running out of petrochemical feedstock, environmental issues, toxic discharge, depletion of nonrenewable resources, short-term and long-term health problems due to exposure of the public to chemicals and solvents, and safety concerns, among others.

About 7.1 billion pounds of more than 650 toxic chemicals were released to the environment in 2000 by the United States

1

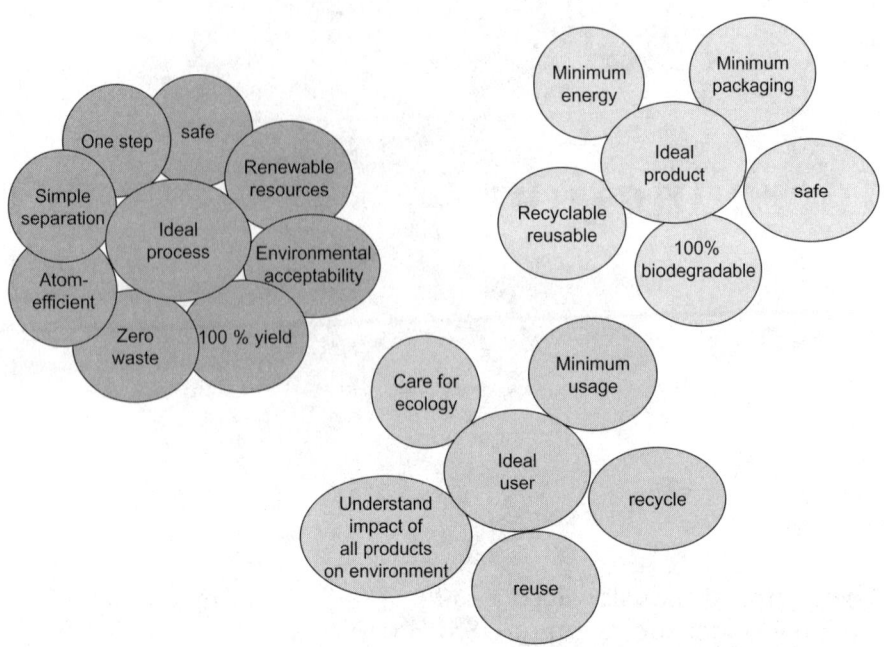

FIGURE 1.1. Criteria for ideal product, process of manufacture and user.

alone (Environmental Protection Agency, 2002, www.epa.gov). This inventory represents only a small fraction of the approximately 75,000 chemicals in commercial use in the United States. The health and environmental effects of many chemicals are not known completely, even though some have been in use for several decades. The U.S. industry spends about $10 billion per year on environmental R&D. An ideal manufacturing process and an ideal product should have certain criteria, which are depicted in Fig. 1.1. An *ideal process* is simple, requires one step, is safe, uses renewable resources, is environmentally acceptable, has total yield, produces zero waste, is atom-efficient, and consists of simple separation steps. An *ideal product* requires minimum energy and minimum packaging, is safe and 100% biodegradable, and is recyclable. Generally, the public focuses on the process and product, paying very little attention to the "ideal user." Figure 1.1 also lists an ideal user. An *ideal user* cares for the environment, uses minimal amounts, recycles, reuses, and understands a product's environmental impact. In addition, an ideal user encourages "green" initiatives.

Definition of Green Chemistry

Green chemistry involves a reduction in, or elimination of, the use of hazardous substances in a chemical process or the generation of hazardous or toxic intermediates or products. This includes feedstock, reagents, solvents, products, and byproducts. It also includes the use of sustainable raw material and energy sources for this manufacturing process (Anastas and Warner, 1998; Anastas and Lankey, 2000, 2002; Anastas et al., 2001). A responsible user is also required to achieve the goals of green chemistry. The U.S. Presidential Green Chemistry Challenge, March 1995, defines green chemistry as,

> the use of chemistry for source reduction or pollution prevention, the highest tier of the risk management hierarchy as described in the Pollution Prevention Act of 1990. More specifically, green chemistry is the design of chemical products and processes that are more environmentally benign.

Green and sustainable chemistry, a new concept that arose in the early 1990s, gained wider interest and support only at the turn of the millennium. Green and sustainable chemistry concerns the development of processes and technologies that result in more efficient chemical reactions that generate little waste and fewer environmental emissions than "traditional" chemical reactions do. Green chemistry encompasses all aspects and types of chemical processes that reduce negative impacts to human health and the environment relative to the current state-of-the-art practices (Graedel, 2001). By reducing or eliminating the use or generation of hazardous substances associated with a particular synthesis or process, chemists can greatly reduce risks to both human health and the environment.

Twelve Principles of Green Chemistry

The 12 principles of green chemistry are listed below (Clark and Macquarrie, 2002). Of course, over the years additional principles have been added to these original 12, but those could be derived from these 12 principles.

1. Prevention. It is better to prevent waste than to treat or clean it up after it has been generated in a process. This is based on the concept of "stop the pollutant at the source."

2. Atom economy. Synthetic steps or reactions should be designed to maximize the incorporation of all raw materials used in the process into the final product, instead of generating unwanted side or wasteful products (Trost, 1991, 1995).
3. Less hazardous chemical use. Synthetic methods should be designed to use and generate substances that possess little or no toxicity to the environment and public at large.
4. Design for safer chemicals. Chemical products should be designed so that they not only perform their designed function but are also less toxic in the short and long terms.
5. Safer solvents and auxiliaries. The use of auxiliary substances such as solvents or separation agents should not be used whenever possible. If their use cannot be avoided, they should be used as mildly or innocuously as possible.
6. Design for energy efficiency. Energy requirements of chemical processes should be recognized for their environmental and economic impacts and should be minimized. If possible, all reactions should be conducted at mild temperature and pressure.
7. Use of renewable feedstock. A raw material or feedstock should be renewable rather than depleting whenever technically and economically practicable. For example, oil, gas, and coal are dwindling resources that cannot be replenished.
8. Reduction of derivatives. Use of blocking groups, protection/de-protection, and temporary modification of physical/chemical processes is known as *derivatization*, which is normally practiced during chemical synthesis. Unnecessary derivatization should be minimized or avoided. Such steps require additional reagents and energy and can generate waste.
9. Catalysis. Catalytic reagents are superior to stoichiometric reagents. The use of heterogeneous catalysts has several advantages over the use of homogeneous or liquid catalysts. Use of oxidation catalysts and air is better than using stoichiometric quantities of oxidizing agents.
10. Design for degradation. Chemical products should be designed so that at the end of their function they break down into innocuous degradation products and do not persist in the environment. A life-cycle analysis (beginning to end) will help in understanding its persistence in nature.
11. Real-time analysis for pollution prevention (Wrisberg et al., 2002). Analytical methodologies need to be improved to allow for real-time, in-process monitoring and control prior to the formation of hazardous substances.

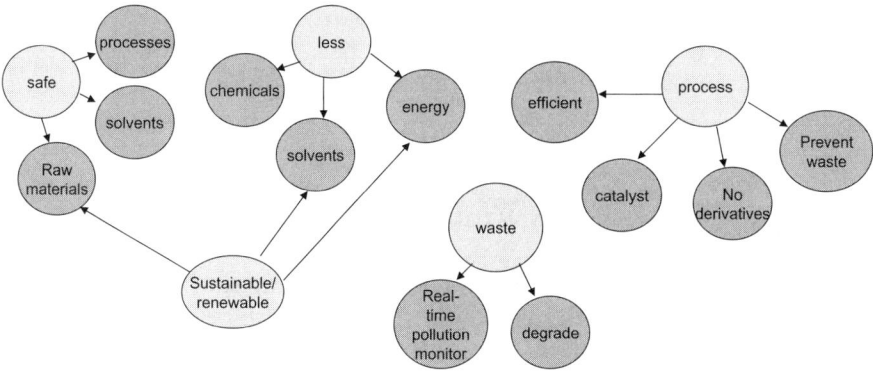

FIGURE 1.2. Key words in 12 principles of green chemistry.

12. Inherently safer chemistry for accident prevention. Substances and the form of a substance used in a chemical process should be chosen to minimize the potential for chemical accidents, including releases, storage of toxic chemicals, explosions, and fires.

Five main foci emerge from these 12 principles (see Fig. 1.2), namely

1. less,
2. safe,
3. process-oriented,
4. waste-reducing,
5. sustainable.

All five key words could be grouped around this, as shown in Fig. 1.2:

1. Uses fewer chemicals, solvents, and energy.
2. Has safe raw materials, processes, and solvents.
3. Process should be efficient, without waste, without derivatization, and should use catalysts.
4. Waste generated should be monitored in real time and should degrade.
5. All chemicals, raw materials, solvents, and energy should be renewable or sustainable.

As can be seen from the figure, the key words are interrelated at the lower level. By focusing on the five key words during process development, one could achieve the philosophy of green chemistry.

Initiatives Taken Up by Countries Around the World

In the United States, President Clinton announced the Presidential Green Chemistry Challenge in March 1995. The program included research grants, educational activities, and annual green chemistry awards to highlight the topic and encourage companies and researchers to focus on green chemistry. Later, in the United Kingdom, the Green Chemistry Network was established. In Italy, Germany, and Australia, similar activities were initiated. Many international organizations, such as the Organization for Economic Cooperation and Development (OECD), the International Union of Pure and Applied Chemistry (IUPAC), the European Chemical Industry Council (CEFIC), and the Federation of European Chemical Societies (FECS), also adopted green and sustainable chemistry as part of their agenda. The European Environmental Agency established a European Green and Sustainable Chemistry Award.

Green chemistry is being pushed by the EPA (USA) because it meets the agency's mission of protecting human health and the environment. It is also being promoted by the National Science Foundation (USA) as part of its mission to support and promote innovative, basic scientific research. The Department of Energy (DOE, USA) is involved in green chemistry programs because it expects that the country will benefit from increased energy efficiency and reductions in global warming pollutants. India's Department of Science and Technology (DST) and the Science and Engineering Research Council (SERC) actively sponsor and fund research by academia and industries in the area of green chemistry and processes, especially the academic–industry collaboration (IChemE, 2003).

The Maastricht Treaty states that those who are responsible for environmental pollution, resource depletion, and social cost should pay the full cost of their activities. If these costs are ultimately passed on to the consumer, then there is an incentive to reduce the levels of environmental damage and resource depletion. For example, discharge of industrial waste from a manufacturing operation into a water course results in a downstream user's incurring extra costs to treat and purify it. Unless there is government pressure, the upstream manufacturer does not have to pay the full cost of its operation, and it will get away with creating the damage. Problems created by industries several decades ago are surfacing now, creating serious health problems to the public. In such cases the government has to bear the cost of cleanup (e.g., "super funds").

Green chemistry has introduced several new terms and new research frontiers, including "eco-efficiency," "sustainable chemistry," "atom efficiency or economy," "process intensification and integration," "inherent safety," "product life-cycle analysis," "ionic liquids," "alternate feedstock," and "renewable energy sources."

Eco-efficiency is the efficiency with which ecological resources are used to meet human needs. It includes three scientific focus areas needed to achieve eco-efficiency. The three focuses, which are the same in the United States and Australia, are the use of alternative synthetic pathways, the use of alternative reaction conditions, and the design of chemicals that are less toxic than current alternatives or the design of inherently safer chemicals with regard to accident potential.

Great Britain's Royal Society of Chemistry publishes an international scientific journal entitled *Green Chemistry* bimonthly; the first issue was published in February 1999. The majority of the synthesis and process journals focus on papers related to this topic.

The Green Chemistry Expert System

The Green Chemistry Expert System (GCES) developed by the EPA (http://www.epa.gov/greenchemistry/tools/gces.html) allows users to build a green chemical process, design a green chemical, or survey the field of green chemistry. The system is useful for designing new processes based on user-defined input. The various features of GCES are contained in five modules:

1. The "Synthetic methodology assessment for reduction techniques" (SMART) module quantifies and categorizes the hazardous substances used in or generated by a chemical reaction.
2. The "Green synthetic reactions" module provides technical information on green synthetic methods.
3. The "Designing safer chemicals" module includes guidance on how chemical substances can be modified to make them safer.

4. The "Green solvents/reaction conditions" module contains technical information on green alternatives to traditional solvent systems. Users can search for green substitute solvents based on physicochemical properties.
5. The "Green chemistry references" module allows users to obtain toxicity and other data about a large number of chemicals and solvents.

The interest in the field of green chemistry is growing dramatically, motivated as much as by economic as by environmental concerns. Faced with rising environmental costs, litigation,, and pressures from environmental groups, institutions now recognize that more efficient processes leading to bottom-line energy and environmental savings are necessary. Legal and societal pressures are increasing as a result of today's increased scrutiny of the chemical businesses, increased medical bills due to long-term exposure to toxic chemicals, and the environmental impacts these businesses cause (see Fig. 1.3). For example, the chemical industry's response to the Montreal Protocol in quickly innovating benign alternatives to ozone-depleting compounds demonstrates its ability to respond quickly to environmental challenges. Many companies have already replaced their ozone-depleting CFCs with milder chemicals.

Although exposure reduction may reduce risk at a cost, it does nothing to address consequences should there be a failure in the controls. Green chemistry, based in hazard reduction, reduces or eliminates risk at the source—and hence reduces the consequences. Green chemistry can address toxicological and waste generation

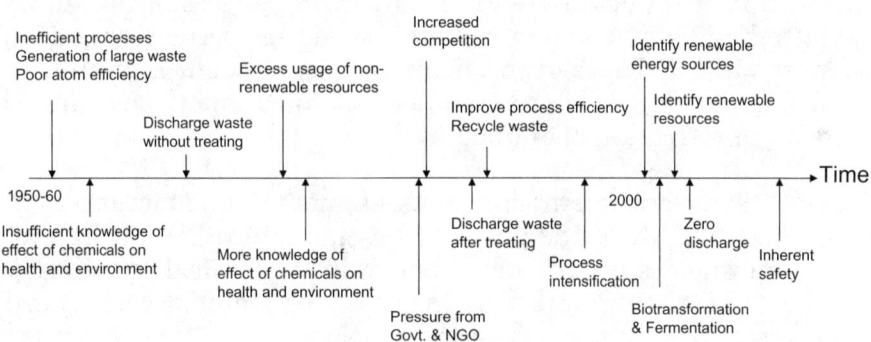

FIGURE 1.3. Changes taking place in the philosophy of process development by industries.

concerns as well. The molecular nature of products can be designed to maximize efficiency of their function while minimizing characteristics that increase physical and toxicological hazards. Green chemistry also provides governmental organizations with an effective way to address environmental concerns.

How Green Chemistry Is Being Addressed

One nation's air emissions problems may be addressed through the increased use of clean fuel as an energy source (for example, Brazil uses ethanol as an add-on to car fuel), while another country may solve its air emissions problems by selecting alternatives to volatile organic solvents (VOCs) (Marteel et al., 2003). A third country may use effective exhaust oxidation catalysts. In New Delhi, India, vehicles now use compressed natural gas (CNG) to reduce particulate, CO, and SO_2 emissions.

Cross Interactions from Green Chemistry

3M's CEO and chairman, Livio DeSimone, and Dow Chemical's chairman of the board, Frank Popoff, together with the World Business Council for Sustainable Development in Eco-Efficiency advocate that

> corporate leaders should invest in eco-efficient technological innovations and move toward sustainable business practices. Governments should establish a framework that encourages long-term progress without harming private-sector competition. Society should demand and establish a viable market for eco-efficient products and processes. (DeSimone and Popoff, 1997)

They propose the establishment of a new contract among society, government, and industry to address the rising costs and inefficiencies of traditional processes and establish an ecoefficiency philosophy throughout the business community.

Green chemistry links the design of chemical products and processes with their impacts on human health as well as on the environment, which includes toxicity, explosiveness, and other hazards. Green chemistry is highly cross-disciplinary/interdisciplinary, involving large government laboratory initiatives or collaborations among geographically dispersed organizations that are performing research to achieve common goals (Curzons et al., 1998). The various branches of science and engineering that green chemistry encompasses include microbiology, biotechnology, chemical engineering, synthetic organic chemistry, enzyme technology, scale-up, toxicology, analytical chemistry, catalysis, environmental chemistry, engineering design, and mechanical engineering. Knowledge of biotransformations, safety, hazards, and toxicity of chemicals is also essential.

Green chemistry is being practiced at four levels: (1) basic academic research; (2) industry-specific research and development; (3) government–industry–academic collaborations; and (4) government national laboratories worldwide. For example, programs have been started at the Center for Process Analytical Chemistry at the University of Washington in Seattle, the National Environmental Technology Institute at the University of Massachusetts in Amherst, the Gulf Coast Hazardous Substances Research Center at Lamar University in Beaumont, Texas, the Center for Clean Industrial and Treatment Technologies at Michigan Technological University in Houghton, and the Emission Reduction Research Center at the New Jersey Institute of Technology in Newark. Recent collaborative initiatives include the Kenan Center for Utilization of CO_2 in Manufacturing at North Carolina State University in Raleigh, green oxidation catalysis at Carnegie-Mellon University in Pittsburgh, the Center for Green Manufacturing at the University of Alabama in Tuscaloosa, Biocatalysis at Michigan State University in East Lansing, Polymers and Green Chemistry at the University of Massachusetts at Amherst, and Green Chemistry Research and Education at the University of Massachusetts at Boston. The Green Chemistry Institute (GCI) was founded by several participant organizations, namely the EPA's Office of Pollution Prevention and Toxics, the LANL, the University of North Carolina, and industrial organizations including Hughes Environmental (now Raytheon Environmental) and Praxair. The goals of this institute are to promote green chemistry through information dissemination, education, and research, outreach through conferences, symposia, and workshops, and GCI's e-mail lists.

The Nobel Prize in Chemistry

Yves Chauvin (France), Robert Grubbs (USA), and Richard Schrock (USA) shared the 2005 Nobel Prize in chemistry for their development of metathesis. Apart from its applications in the polymer industry, the technique can reduce hazardous waste generation through efficient production and manufacturing. The discovery of olefin metathesis dates back to the 1950s, when formation of olefins was observed when gaseous alkenes were passed over the Mo catalyst. Chauvin understood the mechanism of the process. He realized that olefin metathesis is initiated by the formation of a metal carbene. Grubbs played a major role in the development of the right catalysts for metathesis, which revolutionized its industrial applications. The scientists were rewarded for their efforts to develop environmentally friendly chemical processes. This Noble Prize will definitely push researchers and scientists who work in the area of green chemistry.

The Patent Scene

Over 3200 green chemistry patents were granted in the United States between 1983 and 2001, with most of them assigned to chemical industries and government sectors. Global emphasis on green chemistry technology relative to chemical, plastic, rubber, and polymer technologies has increased since 1988. The number of granted patents was fairly constant from 1983 to 1988 (averaging 71 patents per year) but increased to 251 patents granted per year by 1994. From 1995 to 2001, an average of 267 green chemistry patents was granted each year. The United States is the largest inventor, with 65% of all green chemistry patents, followed by Europe (24%) and then Japan (8%). The Procter & Gamble Company (USA) was granted the most green chemistry patents (74) followed by Bayer (66) between 1983 and 2001. Table 1.1 shows the top 10 companies who were busy in green chemistry and processes during the years 1983 to 2001. Sixty-one percent of the green chemistry patents belong to chemical technology, namely the chemical, plastics, polymer, and rubber technology areas (Nameroff et al., 2004).

The Measure of Greenness

How does one "measure greenness"? Is it based on the raw material, process, or product? There are no clear quantitative metrics that connect the raw materials to various environmental issues at the individual, society, global, and ecological levels. Quantifying the true costs of waste and the potential savings offered by green chemical technology (GCT) is difficult. The impact of certain

TABLE 1.1
Top 10 Companies Busy in Green Chemistry and Processes Between 1983 and 2001

Rank	Organization	Sector	Total Patents	1997–2001
1	The Procter & Gamble Co.	Chemicals	74	34
2	Bayer AG.	Chemicals	66	32
3	BASF Group	Chemicals	61	32
4	Exxon Mobil Co.	Energy	50	22
5	Eon AG.	Chemicals	49	26
6	GlaxoSmithKline	Pharma.	43	43
7	Minnesota Mining & Manufacturing Co.	Instruments	43	23
8	General Electric	Electrical	41	13
9	Henkel KGaA	Chemicals	37	13
10	Dow Chemicals	Chemicals	35	0

chemicals on human health or the environment may be felt much later in time, which further complicates the cost calculations.

Rafiqul Gani, Sten Bay Jørgensen, and Niels Jensen developed a systematic methodology for the generation of sustainable process alternatives with respect to new processes while attempting to retrofit design. The generated process alternatives are evaluated using (1) sustainability metrics, (2) environmental impact factors, and (3) inherent safety indices (see Fig. 1.4). The alternatives for new process design as well as retrofit design are generated through a systematic method based on the path-flow analysis approach. In this approach, a set of indicators is calculated; these indicators include energy and material waste costs, potential design targets that may improve the process design, operation cost, sustainability metrics, environmental impact factors, and inherent safety indices. Steady-state design data and a database with properties of compounds, including data related to environmental impact factors and safety factors, also have to be provided.

Uerdingen et al. (2003) introduced mass and energy indicators that use information about accumulation, mass, and energy being circulated in closed and open loops within a process with respect to mass and energy entering and/or leaving the process (or loop). These indicators provide important information about the process

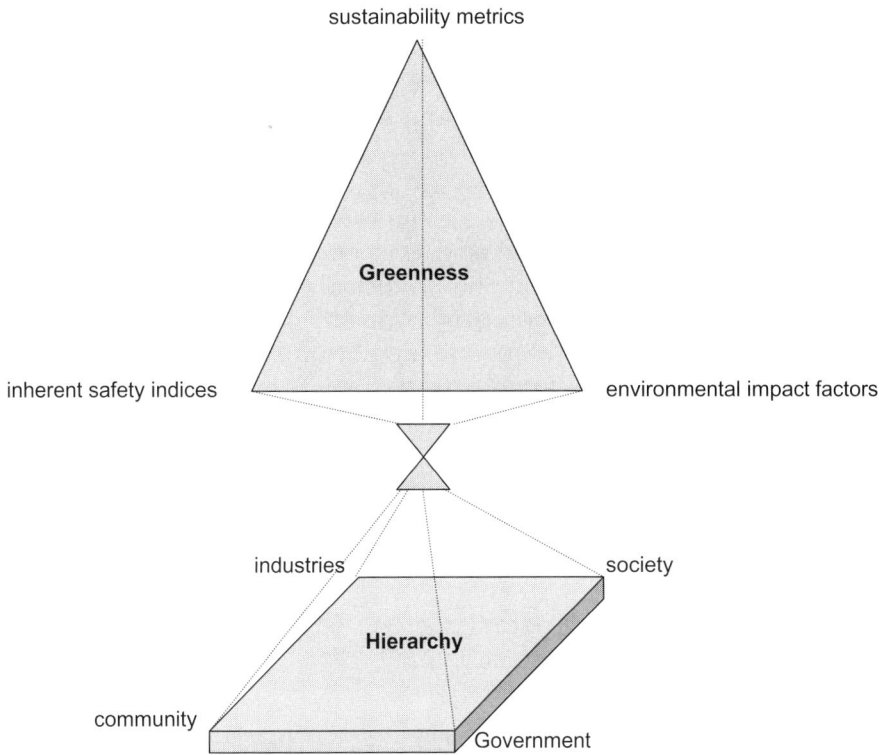

FIGURE 1.4. Impact of green chemistry on the stakeholders.

in terms of its operation, efficiency, and process flowsheet and its comparative cost versus other processes (Uerdingen, 2002). A process that has a higher recycle rate needs more hardware and utilities and hence is less economical than one whose recycles are at a minimum. In addition, the energy cost also increases linearly with the amount of recycle.

"Sustainable development" means providing for human needs without compromising the ability of future generations to meet their needs. The American Institute of Chemical Engineers and Britain's Institution of Chemical Engineers for the chemical process industries have defined the sustainability metrics that will help engineers to address the issue of sustainable development. The sustainability metrics, as proposed by the IChemE (Tallis, 2002), include 49 submetrics covering the

1. environment (related to resource usage, emissions, effluents, and waste),
2. economy (related to profit, value, tax, and investments),
3. society (related to workplace and society).

Environmental Factors

Young et al. (2000) developed environmental impact factors calculated using the Waste Reduction (WAR) algorithm. These factors encompass (1) physical potential impacts (acidification of soil, greenhouse enhancement, ozone depletion, and photochemical oxidant depletion), (2) human toxicity effects (air, water, and soil), and (3) eco-toxicity effects (aquatic and terrestrial). The important parameters are as follows:

PI (human toxicity potential by ingestion),
TPE (human toxicity potential by exposure both dermal and inhalation),
TTP (terrestrial toxicity potential),
ATP (aquatic toxicity potential),
GWP (global warming potential),
ODP (ozone depletion potential),
PCOP (photochemical oxidation potential),
AP (acidification potential).

As the preceding names imply, detailed information is needed about the impact potential of chemicals to estimate these parameters. Currently such detailed information is not available for most of the chemicals used in the chemical and allied industries.

Several tools and methodologies are available for measuring environmental impact, including (Deuito and Garrett, 1996; Dunn et al., 2004).

life-cycle assessment (LCA) (Dewulf and Van Langenhove, 2001a and 2001b),
material input per unit service (MIPS),
environmental risk assessment (ERA),
materials flow accounting (MFA),
cumulative energy requirements analysis (CERA),
material input per delivered function analysis (MFA),
environmental input–output analysis (Department of Environment, 1997),

eco design,
life-cycle costing,
total cost accounting,
cost-benefit analysis.

Long-term effects of chemicals on the ecology and fauna are very difficult to obtain (Chenery, 1953). Thus, we have to make several assumptions and approximations while using these impact factors. In addition, the line connecting the indicators to the environment and the economic growth and/or social development needs to be developed. These indicators should also be relevant to diverse groups such as industries, public, NGOs, government agencies, etc. There are several unifying or common factors, such as the availability of resources, energy, land, water, clean air, and environmental diversity. Examples of indicators based on this principle include MIPS (Schmidt-Bleck, 1993), MFA, the energy-based indices such as CERA, and the energy-based sustainability index developed by Dewulf et al. (2001a and b; 2000). The area of land and that of water are limiting factors that can be used to develop a sustainable process index. For example, the electronic and semiconductor industries require large quantities of clean water, and the choice of the location will depend on that water's uninterrupted availability. Depletion of groundwater could be a disadvantage for setting up such manufacturing plants. The petrochemical and oil industries require large areas of land, preferably near the coastal area. Atomic power plants are located near the coast, so that the sea water is used for cooling the reactor core. Industries would like to evaluate technologies using money-based indices such as total cost accounting, cost-benefit analysis, shareholders' value, and value addition. Countries below the poverty line would not mind compromising on many of these resources to provide enough food, clothing, and shelter to its residents. That is one of the reasons why several hazardous manufacturing industries have been relocated in Third World countries in 1980s.

Safety and Risk Indices

The best-known measure for safety is risk. Risk is defined as the potential for loss or the probability of a specified undesired event's occurring in a particular period of time and its consequences. An inherently safe chemical process is one that avoids hazards instead of controlling them. Heikkilä (1999) developed a method for

TABLE 1.2
Inherent Safety Indices

Total Inherent Safety Index (ISI) Chemical Inherent Safety Index (ICI)	Score	Process Inherent Safety Index (IPI)	Score
Subindices for reaction hazards:		Subindices for process conditions:	
Heat of reaction for the main reaction	0–4	Inventory	0–5
Heat of the reaction for the side reactions	0–4	Temperature	0–4
Chemical interactions	0–4	Pressure	0–4
Subindices for hazardous substances:		Subindices for process equipment:	
Explosiveness	0–4	IIsbl	0–4
Toxicity	0–6	IOsbl	0–3
Corrosivity	0–2	Process structure	0–5
Maximum ICI score	28	Maximum IPI score	25
Maximum ISI score	53		

measuring the intrinsic safety of a process based on two major classifications: (1) process equipment and (2) properties of the chemical substances present in the process. Heikkilä's method requires the evaluation of a number of interrelated factors and subindices, as given in Table 1.2. They are the heat of the main reaction and side reactions, the chemicals inventory, the reaction temperature and pressure, the chemical interactions, the explosiveness, corrosivity, and toxicity of the chemicals handled, the equipment, and the process structure.

Mass and Energy Indices

The mass- and energy-based indicators help to identify process alternatives for which sustainability metrics, environmental impact factors, as well as inherent safety indices can be estimated. The goal here is to optimize the process so that there is improvement in all these metrics with respect to the base case design. The application of the methodology requires a combination of tools ranging from databases to process simulation software (such as Aspen®, Hypro®, Sim Sci®, etc.), to computational routines for various types of indicators and process synthesis/design tools.

Process economics is a well-developed subject, while analysis of the potential gains in environmental performance, obtained through the introduction of novel product or process, is difficult to judge since complex interactions exist among feedstock, effluents, intermediates, and energy streams, etc. In addition, one needs to consider the system boundaries as well as the complex nature of the interaction of chemicals with the natural environment (in both the short and long terms). The system boundary is not just the area surrounding the site where the product is manufactured; it extends to the regions where the product is being used and creates an impact and also to the place where it is finally disposed. The assembly of these factors results in an ambiguous picture of benefits and drawbacks of alternative and new technologies, making the goal of establishing a compelling case for a particular chemical technology as being "green" or "not so green" difficult.

There is a fundamental need for a system of scientifically sound measures and indicators based on physical principles that are interpretable and relevant to various stakeholders. Many assessment methodologies now being developed include not only the product or process but also the entire supply chain and disposal, i.e., the cradle-to-grave approach (Krotscheck and Narodoslawsky, 1996; DeSimone and Popoff, 1997; Hoffmann et al., 2001; Tallis, 2002; Sikder, 2003). Apart from LCA and life-cycle costing, the energy-based sustainability index, the area of land-based sustainable process index, MIPS, and MFA are also cradle-to-grave approaches. The main differences among all these indices lie in the way they normalize the environmental impacts. There is still no readily available simple, efficient, and unambiguous methodology suitable for comparing alternative chemical technologies in such a way that they faithfully represent the benefits of implementation of green chemical technologies. In addition, certain technologies may also affect the upstream raw materials, which also need to be considered while one estimates the benefits. For example, the introduction of a new photocopier design also affects the type of toner used in the machine.

The Hierarchical Approach

Lapkin et al. (2004) has come up with four vertical hierarchy levels: (1) product and process, (2) company, (3) infrastructure, and (4) society, with each level's corresponding stakeholders (refer back to Fig. 1.4). Looking back at Fig. 1.4, in order from bottom to top, we see that the first level of the hierarchy is the products and processes. The main stakeholders at this level are scientists and engineers who have developed the products and processes. These stakeholders would develop processes that are the most efficient, consume less energy, and generate the minimum amount of waste. The next level is the company; the stakeholders at this level are a company's business managers. The system's boundary depends on how far the company's responsibility extends down the product's life cycle. It could be a cradle-to-gate boundary or a cradle-to-grave boundary depending upon the product and the way it is used by the public. For example, a company making surfactants may say its responsibility ends as soon as it sells the product. It would not care how the product is used or what the fate of its product is after its use. In contrast, a responsible company wants to develop surfactants that would degrade but not remain in the environment and cause harm to the flora and fauna. The third hierarchical level relates to infrastructure. The stakeholders here are the local bodies and local governments. The local bodies want rounded development in their region and prosperity to their locality. But at the same time they do not want toxic waste causing risks and health hazards to the local community or long-term damage to the local environment. The final level is the society, whose stakeholders include the public at large, NGOs, state, and country. The choice of appropriate indicators depends on the specifics of the industry sector and even on the types of products. The indicators should reflect specific byproducts, wastes, and emissions that are characteristic of the process or the product. In addition, the resources needed for the delivery of service, the operation of a process, or the manufacture of a product need to be considered as well.

Green chemistry imposes additional constraints on the product and process characteristics. The issues are different for different product sectors and industries. Thus, the specialty chemicals and the pharmaceuticals industries use many different types of synthetic organic reactions. The product volume may be small, but the purities expected are very high. They are very dependent on solvents, as both reactants and products are often solids, and produced molecules may possess toxicity and have other effects on the environment. The most important issues are atom efficiency of individual reactions, solvent recovery and reuse, use of benign solvents, product toxicity, and product end-of-life. Such data for all their raw materials and products may not be available.

In the case of drugs and pharmaceuticals, toxicity of the products is a major issue. Given that a significant number of products have been the subject of litigation over recent years, a possible indicator of how good a product is could be the amount spent on litigation and compensation claims per product per year. In the pharmaceutical business a "blockbuster drug" may suddenly get bogged down with hundreds of litigations, even leading to its withdrawal from the global market. A typical example is Vioxx, a COX2 inhibitor made by Pfizer (USA), which, after earning billions of dollars, was withdrawn in 2002 in the United States due to long-term cardiac side effects not noticed during its drug trials. On the other hand, in the bulk chemical sector, the volumes are large and the chemistry is simple. Here reaction selectivity and energy usage are much more significant. Hence, companies have made many attempts to develop environmental metrics for specific products and processes, such as the organic synthesis-oriented indicators used by GlaxoSmithKline (GSK; USA) for external assessment. Because the only stakeholders at this level are technical personnel, technical indicators, which are potentially meaningless to the wider audience or even to company management, are used here. For example, GSK uses reaction mass efficiency (RME), which is calculated as the ratio of the mass of the product to the sum of the mass of reactants. The RME indicator characterizes the material input efficiency of a synthesis reaction. For a service industry, the normalization is performed not on the unit product but on the "service delivered." For example, if a company is providing a VOC recovery facility, then the delivered service is the volume of treated air or the amount of recovered solvent with minimum resources (Shonnard and Hiew, 2000). A cryogenic condensation technology may recover the VOC from the air efficiently, but it may be very energy-intensive. On the contrary, a biofilter may perform the same task at ambient conditions, but may not be 100% efficient (Place et al., 2002).

The product or process level indicators are relevant to the specific manufacturing industry. These indicators include the

1. process's efficiency,
2. product's performance,
3. process's energy usage,
4. product's water usage,
5. waste generated.

All these indicators have factory gate-to-gate boundaries. Increasing productivity and the recovery of products, reducing waste and energy consumption as well as the expenditure of valuable raw materials, and making firm requirements related to process safety and process controllability are well understood by the manufacturing chemical and mechanical engineers. They have practiced them for a long time due to economic considerations as well as the need for bottom-line improvements.

At the level of company, the indicators could be

1. How expensive is the delivery of a product or a service in terms of money and energy?
2. How much money is being spent on remediation of waste?
3. Do the emissions and waste discharged by the company cause any employee health problems?
4. Are there any risks or unsafe incidents causing damage to company property or employees?

The boundaries for the company are beyond gate-to-gate and reach gate-to-user or customer. They also reach the public, who may be affected by the emissions and waste discharged.

At the infrastructure level, the cumulative indexes are based on the

1. mass and the area of land used by the industry (its footprint),
2. amount of water consumed.

The system boundary in this case should be cradle-to-grave.

Finally, at the level of society, indicators could correspond to the most important issues relating to the

1. potential use of renewable energy,
2. total greenhouse gas emissions,
3. chronic illness due to certain chemical usage.

Such indicators could be used to compare various alternative technologies delivering the same product.

As mentioned, several groups of process indicators could be used for measuring the sustainability or greenness of a process. There is no single set of universally accepted indicators. Table 1.3 lists another set of process indicators that are appropriate to use when comparing various technologies. These indicators not only include "within the boundary" factors but also consider factors that affect the environment as well as factors that affect the company. These indicators can take up different values if they are considered for the short, medium, or long term and also will depend on the type of stakeholders.

TABLE 1.3
Various Process Indicators for Measuring the Greenness of a Process

Atom efficiency:
Amount of greenhouse gases (GHG) produced per kg of product
Amount of ozone-depleting gases (ODG) produced per kg of product
Amount of freshwater used per kg of product
Amount of solvent used per kg of product
Amount of solvent lost per kg of product
Amount of energy required (fresh energy input–energy generated) per kg of product
Amount of waste (solid + liquid and gaseous) produced per kg of product
Amount of nonbiodegradable material produced per kg of product
Amount of cytotoxic material produced per kg of product
Amount aquatoxic material produced per kg of product
Amount of ecotoxic material produced per kg of product
Amount VOC produced per kg of product
Amount of consumption of nonrenewable energy per total energy consumed
Expected monetary compensation that must be paid due to toxic release per kg product
Expected cleanup cost per kg release
Environmental impact (in monetary terms) per kg release
Nonusability of land or water body due to release per kg release (relates to land value)

The Sustainable Process Index

The Sustainable Process Index (SPI) is a cumulative index based on the principle that the area of land suitable for feedstock generation, habitat, production, and dissipation of effluents is a limiting resource (Institute of Chemical Engineers, 2002; Lange, 2002). It is calculated as the ratio of total land area required to sustainably manufacture a product or provide a service to the average available land area per individual, specific to the location of the production facility. The SPI is based on the following four principles aimed at minimizing the influence of technology on the environment:

1. The flow of anthropogenic material into the environment should not exceed the rate of its assimilation. If that happens, there will be a steady increase in the total amount of anthropogenic material in that locality.

2. Anthropogenic materials released into the environment should not affect global material cycles.
3. Renewable resources should only be used at a rate not exceeding the natural local production rate of the renewable resource. If this principle is not followed, there will be a slow erosion of the renewable resource.
4. The natural variety of species and landscapes should be sustained (maintain biodiversity).

The total land area is calculated as the specific area per unit of product or service provided. For example, for an energy generation facility, the total energy requirement is expressed in $m^2/kW/$ year. For a manufacturing industry, it will be m^2 land area required/ kg product manufactured. Five factors contribute to the total land area:

1. the area required to produce the raw materials,
2. the area necessary to provide process energy,
3. the area for installation of process equipments,
4. the area required for the staff, including habitat,
5. the area required to accommodate the products and byproducts.

The area required to store the waste and effluents as well as to treat them will also be included in the third factor just listed.

SPI is the only index that specifically accounts for local conditions, such as the density of population, mode of energy generation, as well as ecological and climatic conditions responsible for the dissipation of manmade effluents. SPI includes the social dimension through the number of employees and the area of land required to accommodate the employees in a specific location of the production facilities.

Conclusions

Green chemistry has become the important philosophy for the 21st century and beyond. Chemical and allied industries have taken this philosophy very seriously due to societal and governmental pressures with respect to environmental issues. In addition, depletion in fossil fuel and increased global competition have forced industries to look at biotechnology and green routes for achieving efficient manufacturing processes. Measuring the greenness of a process is a very difficult task. Several indicators are now being defined to measure the process characteristics (Mac Gillivrary, 1995; Warhurst et al., 2001). There are several levels in the green chemistry hierarchy, and each level is interested in certain indicators. Both developing a line of sight between the process and the environmental impact at a larger dimension and assigning a cost to this impact are probably impossible. It is easy to compare similar processes, but it is very difficult to compare diverse technologies. In addition, all the physical and chemical data, the short- and long-term impacts of chemicals on the environment, and the relationship between people and the ecology have not been fully studied. On hindsight people realize the mistakes they have made, but by that time it may be too late. In addition to considering an ideal process and product, one needs to include an ideal user as well. The user could also help support the green chemistry initiative by being responsible in product selection, usage, and disposal. The user may be able to bring pressure on manufacturing organizations to adapt to sustainable development. A product recycle could lead to five times more employment than a remediation operation. Of the four Rs—Reduce, Recycle, Reuse, and Remediate—the first two are part of the green chemistry principles; the third one describes a responsible user; and the fourth R should be the last option. This book deals with the latest developments in green chemistry and green process technologies, with relevant industrial examples.

References

Anastas, P. T. and Lankey, R. L., Sustainability through green chemistry and engineering, *ACS Symp. Series*, **823**: 1–11, 2002.

Anastas, P. T., Kirchhoff, M. M., and Williamson, T. C., Catalysis as a foundational pillar of green chemistry. *Appl. Catal. A: Gen.*, **221**: 3–13, 2001.

Anastas, P. T. and Williamson, T. C., *Green Chemistry: Frontiers in Benign Chemical Syntheses and Processes*, Oxford University Press, Oxford, 1998.

Anastas, P. T. and Lankey, R. T., *Green Chem.*, **2**: 289, 2000.

Anastas, T. T. and Warner, J. C., *Green Chemistry: Theory and Practice*, Oxford University Press, New York, 1998.

Chenery, H. B., Process and production functions from engineering data. In *Studies in Structure in the American Economy*, W. W. Leontieff, ed., Oxford University Press, New York, 1953, p. 299.

Clark, J. H. and Macquarrie, D. J., *Handbook of Green Chemistry & Technology*, Blackwell, Oxford, 2002.

Curzons, A. D., Constable, D. J. C., Mortimer, D. N., and Cunnigham, V. L., So you think your process is green, how do you know? Using principles of sustainability to determine what is green—a corporate perspective, *Green Chem.*, **3**: 1–6, 2001.

Dewulf, J., Van Langenhove, H., Mulder, J., van den Berg, M. M. D., van der Kooi, H. J., and de Swaan Arons, J., Illustrations towards quantifying the sustainability of technology, *Green Chem.*, **2**: 108–114, 2000.

Dewulf, J., Van Langenhove, H., and Dirckx, J., Energy analysis in the assessment of the sustainability of waste gas treatment systems, *Sci. Total Environ.*, **273**: 41–52, 2001a.

Dewulf, J. and Van Langenhove, H., Assessment of the sustainability of technology by means of a thermodynamically based life cycle analysis, *Environ. Sci. Pollut. Res.*, **8**: 1–7, 2001b.

Department of the Environment, Secretary of State's Guidance—Printworks. PG6/16(97), The British Government, Stationery Office, 1997.

DeSimone, L. D. and Popoff, F., *World Business Council for Sustainble Development, Eco-Efficiency: The Business Link to Sustainable Development*; MIT Press, Cambridge, MA, 1997.

DeSimone, L. D. and Popoff, F., *Eco-Efficiency: The Business Link to Sustainable Development*, MIT Press, Cambridge, MA, 1997.

DeVito, S. C. and Garrett, R. L., eds., *Designing Safer Chemicals: Green Chemistry for Pollution Prevention*, American Chemical Society Symposium Series 640, American Chemical Society, Washington, DC, 1996.

Dunn, P. J., Galvin, S., and Hettenbach, K., The development of an environmentally benign synthesis of sildenafil citrate (Viagra) and its assessment by green chemistry metrics, *Green Chem.*, **6**: 43–48, 2004.

Gani, R., Jorgensen, S. B., and Jensen, N., Design of Sustainable Processes: Systematic Generation & Evaluation of Alternatives, Proceedings of 7th World Congress of Chemical Engineering, Glasgow, Scotland, 2005.

Graedel, T. E., Green chemistry as systems science, *Pure Appl. Sci.*, **73**(8): 1243–1246, 2001.

Heikkilä, A.-M., Inherent safety in process plant design—an index-based approach, Ph.D. thesis, VTT Automation, Espoo, Finland, 1999.

Hoffmann, V. H., Hungerbuühler, K., and McRae, G. J., Multiobjective screening and evaluation of chemical process technologies, *Ind. Eng. Chem. Res.*, **40**: 4513–4524, 2001.

IChemE, Facilitating the uptake of green chemical technologies, Report of Crystal Faraday Partnership (www.crystalfaraday.org), Crystal Faraday Partnership, Davis Building, 165–189 Railway Terrace, Rugby CV21 3HQ, UK, 2003.

IChemE, The sustainability metrics. Sustainable development progress metrics, Institution of Chemical Engineers, Rugby, UK, 2002.

Incinerator Sector Guidance Note IPPC S5.01; Environment Agency, UK, 2003.

Krotscheck, C. and Narodoslawsky, M., The sustainable process index. A new dimension in ecological evaluation, *Ecol. Eng.*, **6**: 241–258, 1996.

Lange, J. P., Sustainable development: Efficiency and recycling in chemical manufacturing, *Green Chem.*, **4**(6): 546–550, 2002.

Lapkin, A., Metrics of green chemical technology, commissioned by Research Development and Technology Transfer steering group of Crystal Faraday Partnership (www.crystalfaraday.org), 2002.

Lapkin, A., Joyce, L., and Crittenden, B., Framework for evaluating the "greenness" of chemical processes: Case studies for a novel VOC recovery technology, *Environ. Sci. Tech.*, **38**: 5815–5823, 2004.

Lenz, A. J. and Lafrance, J., Meeting the Challenge: US Industry Faces the 21st Century—The Chemical Industry; US Department of Commerce, Office of Technology Policy, 1996.

MacGillivray, A., Indicators, indicators everywhere. In *Accounting for Change; Papers from an International Seminar*, New Economics Foundation, London, 11–14, 1995.

Marteel, A. E., Davies, J. A., Olson, W. W., and Abraham, M. A., Green chemistry and engineering: Drivers, metrics, and reduction to practice, *Ann. Rev. Environ. Resour.*, **28**: 401–428, 2003.

Nameroff, T. J., Garant, R. J., Albert, M. B. Adoption of green chemistry: An analysis based on US patents, *Research Policy*, **33**: 959–974, 2004.

Place, R. N., Blackburn, A. J., Tennison, S. R., Rawlinson, A. P., and Crittenden, B. D., Method and equipment for removing volatile compounds from air,WO02/072240, filed Mar. 13, 2002.

Schmidt-Bleek, F., MIPS re-visited, *Fresenius Environ. Bull.*, **2**: 407–412, 1993.

Shonnard, D. R. and Hiew, D. S., Comparative environmental assessment of VOC recovery and recycle design alternatives for a gaseous waste stream, *Environ. Sci. Tech.*, **34**: 5222–5228, 2000.

Sikdar, S. K., Sustainable development and sustainability metrics, *AIChE J.*, **49**: 1928–1932, 2003.

Tallis, B., Sustainable development progress metrics, IChemE Sustainable Development Working Group, IChemE, Rugby, UK, 2002.

Trost, B. M., Atom economy—a challenge for organic synthesis: Homogeneous catalysis leads the way, *Angew. Chem., Intl. Ed. Engl.*, **34**: 259–281, 1995.

Trost, B. M., The atom economy—a search for synthetic efficiency, *Science*, **254**: 1471–1477, 1991.

Uerdingen, E., Retrofit design of continuous chemical processes for the improvement of production cost-efficiency, PhD thesis, ETH-Zürich, Switzerland, 2002.

Uerdingen, E., Gani, R., Fisher, U., and Hungerbuhler, K., A new screening methodology for the identification of economically beneficial retrofit options for chemical processes, *AIChE J.*, **49**(9): 2400–2418, 2003.

Warhurst, A. et al., Environmental and Social Performance Indicators. Methodology and Field Handbook. A Report by Mining Environment Research Network; University of Bath, 2001.

Wrisberg, N., Udo de Haes, H. A., Triebswetter, U., Eder, P., and Clift, R., eds., *Analytical Tools for Environmental Design and Management in a Systems Perspective*, Kluwer Academic Publishers, New York, 2002.

Young, D., Scharp, R., and Cabezas, H., The waste reduction (WAR) algorithm: Environmental impacts, energy consumption, and engineering economics, *Waste Management*, **20**: 605–615, 2000.

CHAPTER 2

Newer Synthetic Methods

Introduction

Waste prevention and environmental protection are major requirements in an overcrowded world of increasing demands. Synthetic chemistry continues to develop various techniques for obtaining better products with less damaging environmental impacts. The control of reactivity and selectivity is always the central subject in the development of a new methodology of organic synthesis. Novel, highly selective reagents appear every month. New reactions or modifications of old reactions have been devised to meet the ever-increasing demands of selectivity in modern synthesis. Periodic review articles and books appear in the literature on these newer reagents. The scope of this chapter is to focus on newer techniques (experimental) for improving the yield and reducing the duration of the reactions and also to discuss the need for a good synthetic design. In other words, newer methods of kinetic activation, which minimize the energy input by optimizing reaction conditions, will be discussed along with the need for an elegant synthetic design.

In most reactions, the reaction vessel provides three components (as shown in Fig. 2.1):

- solvent,
- reagent/catalyst,
- energy input.

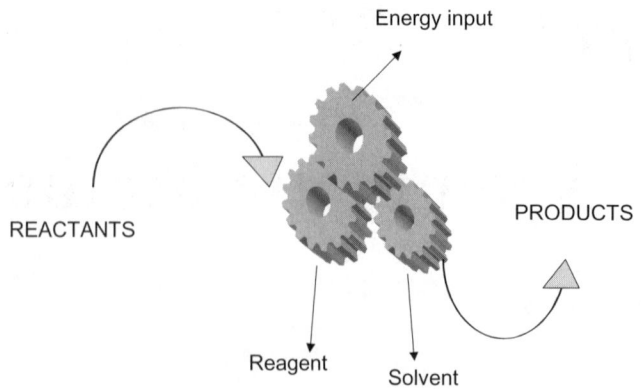

Figure 2.1. Components of chemical reaction.

Hence, efforts to green chemical reactions focus predominantly on "greening" these three components. By "greening," we mean to

- Use benign solvents or completely dispense with the solvent.
- Use alternate, more efficient and effective reagents/catalysts.
- Optimize the reaction conditions by using cost-effective, eco-friendly alternative processes.

The role of alternate reagents, solvents, and catalysts in greening chemical reactions is discussed in other chapters.

In this chapter we shall see newer methods of kinetic activation of molecules in chemical reactions. Pressure and temperature are important parameters in reaction processes in chemical systems. However, it is a less well-known fact that other than thermally initiated reactions can also lead to sustainable results. The basic requirement is to capture the energy required by a reaction. The energy required for synthesis as well as that required for cooling are of interest here.

In order to minimize energy and control reactions with a view to *green* chemistry, attempts are being made to make the energy input in chemical systems as efficient as possible. Approaches are being taken and possibilities investigated to use until now scarcely used forms of energy, so-called nonclassical energy forms, in order to optimize the duration and product yield and avoid undesired side products. Teams working in this area are also interested in the energetic aspects of the preparation of starting substances and products and the conditioning of reaction systems (e.g., surface activation, emulsification, homogenization, degassing, etc.).

We now have six well-documented methods of activating molecules in chemical reactions, which can be grouped as follows: the *classical methods*, including

- thermal,
- photochemical,
- electrochemical,

and the *nonclassical methods*, which include

- sonication,
- mechanical,
- microwave.

Each of these methods has its advantages and niche areas of applications, alongside its inherent limitations. A comparative study of these techniques is given in Table 2.1.

What do we mean by classical and nonclassical energy forms? In classical processes, energy is added to the system by heat transfer; by electromagnetic radiation in the ultraviolet (UV), visible, or infrared (IR) range; or in the form of electrical energy. On the other hand, microwave radiation, ultrasound, and the direct application of mechanical energy are among the nonclassical forms.

Sonochemical Processes

Ultrasound, an efficient and virtually innocuous means of activation in synthetic chemistry, has been employed for decades with varied success. Not only can this high-energy input enhance mechanical effects in heterogeneous processes, but it is also known to induce new reactions, leading to the formation of unexpected chemical species. What makes sonochemistry unique is the remarkable phenomenon of *cavitation*, currently the subject of intense research, which has already yielded thought-provoking results.

The majority of today's practitioners accept a rationale based on "hotspot" interpretation, provided this expression is not taken literally, but rather as "a high-energy state in a small volume." One should also recall that only a small part (10^{-3}) of the acoustic energy absorbed by the system is used to produce a chemical activity (Margulis and Mal'tsev, 1968). High-power, low-frequency

TABLE 2.1
Classical and Nonclassical Ways of Kinetic Activation of Chemical Reactions

	Thermal	Photochemical	Electrochemical	Sonication	Microwave	Mechanical
Mode of activation	Convection currents	Electronic excitation	Electron transfer at the electrode	Cavitation	Dipole fluctuation	Mechanical
Applicability: Solid reactants	Not appropriate	Not appropriate	Not suited	Suited	Most suited	Most suited
Liquids	Most suited	Most suited	Most suited	Most suited	Not appropriate	Not appropriate
Paste	Not appropriate	Suitable	Not suitable	Suitable	Ideally suited	Not suitable
Comparative duration of the reaction	Long	Long	Medium	Short	Very short	Medium
Equipment for industrial use	Very well developed	Very well developed	Needs improvement	Needs improvement	Needs improvement	Well developed
Solvents usable	All	Selected solvents transparent to required light	Mostly dry ether solvents	Other than halogenated solvents	Mostly solventless conditions	No solvent
Average yield of the product	Medium	Medium to low	Medium	Good	Very good	Medium

(16–100 Hz) waves are often associated with better mechanical treatment and less importantly with chemical effects. With high-frequency ultrasound, the chemistry produced displays characteristics similar to high-energy radiation (more radicals are created). One of the most striking features in sonochemistry is that there is often an optimum value for the reaction temperature. In contrast to classical chemistry, most of the time it is not necessary to go to higher temperatures to accelerate a process. Each solvent has a unique fingerprint.

Sonochemistry in heterogeneous systems is the result of a combination of chemical and mechanical effects of cavitation, and it is very difficult to ascribe sonochemistry to any single global origin, other than the overriding source of activity, namely, cavitation.

The real benefit of using ultrasound lies in its unique selectivity and reactivity enhancement. The heterogeneity of the reaction phase would be particularly significant. In fact, heterogeneous reactions are those in which ultrasound is likely to play the most important role by selective accelerations between potentially competitive pathways.

Apart from the use of ultrasound in enhancing the reactivity in organic reactions, ultrasound has varied uses in industry, such as welding, cutting, emulsification, solvent degassing, powder dispersion, cell disruption, and atomization. It was reported that the sonochemical decomposition of volatile organometallic precursors was shown to produce nanostructured materials in various forms with high catalytic activities. This has proved extremely useful in the synthesis of a wide range of nanostructured inorganic materials, including high surface area transition metals, alloys, carbides, oxides, and sulfides, as well as colloids of nanometer cluster.

Ultrasound is known to enhance the reaction rate, thus minimizing the duration of a reaction. A large number of published examples, which highlight this observation, are shown in Appendix 2.1. Apart from this, it is known to induce specific reactivity, known as "sonochemical switching." Ando et al. (1984) reported that benzyl bromide, on treatment with alumina impregnated with potassium cyanide, yielded benzyl cyanide on sonication, while, without sonication, on heating the reaction mixture yielded diphenylmethanes (see Fig. 2.2). This work was the first experimental evidence that ultrasonic irradiation induces a particular

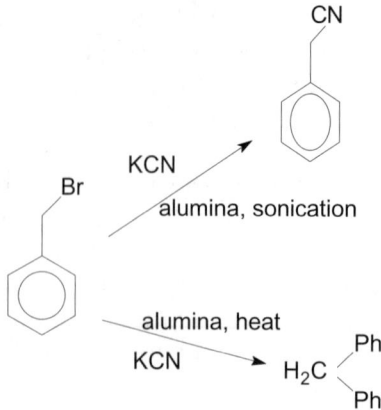

Figure 2.2. Sonochemical switching.

reactivity. Further studies on sonochemical induction indicated that

- Reactions activated by sonication are those that proceed via a *radical or radical ion* intermediate (electron transfer).
- *Ionic reactions* (polar) mostly remain unaffected.

Use of Microwaves for Synthesis

In synthetic chemistry, 1986 was an important year for the use of microwave devices. Since that year, countless syntheses initiated by microwaves have been carried out on a laboratory scale. The result is often a drastic reduction in the reaction time with comparable product yields, if microwaves are used instead of classical methods of energy input. Unwanted side reactions can often be suppressed and solvents dispensed with.

Numerous reactions, such as esterifications, Diels Alder reactions, hydrolyses, or the production of inorganic pigments, have been investigated in recent years. Reactions listed in Appendix 2.2 illustrate nicely the advantages of this nonclassical means of energy input. Apart from the obvious advantages of the use of microwaves in chemical syntheses, microwave technologies are being tested as energy- and cost-saving alternatives. Hopes are high, for example, in the field of green extraction of pollutants from contaminated soil, or for the improvement of the breakdown of biomass waste by fermentation as part of green biorefinery.

Electro-Organic Methods

Over the past 25 to 30 years, the use of electrochemistry as a synthetic tool in organic chemistry has increased remarkably. According to Pletcher and Walsh (1993), more than 100 electro-organic synthetic processes have been piloted at levels ranging from a few tons up to 10^5 tons. Such examples include reductive dimerization of acrylonitrile, hydrogenation of heterocycles, pinacolization, reduction of nitro aromatics, the Kolbe reaction, Simons fluorination, methoxylation, epoxidation of olefins, oxidation of aromatic hydrocarbons, etc. Many excellent reviews and publications highlight the synthetic utility of electro-organic methods (Lund and Baizer, 1991). These cover a broad spectrum of applications of electrochemical methods in organic synthesis, including their use in the pharmaceutical industry. Mild reaction conditions, ease of control of solvent and counter-ions, high yields, high selectivities, as well as the use of readily available equipment, simply designed cells, and regular organic glassware make the electrochemical syntheses very competitive to the conventional methods in organic synthesis. The use of sacrificial anodes is an effective way for the preparation of metallo-organic compounds by cathodic generation of organic anions and anodic generation of metal cations. This approach was very successful for synthesis of the organosilicon compounds (Fry and Touster, 1989). A large variety of fluorinated organosilicon compounds can be synthesized using a sacrificial Al anode and a stainless steel cathode under very mild conditions and in good yields (Bordeau et al., 1997).

Discoveries of new types of electro-organic reactions based on coupling and substitution reactions, cyclization and elimination reactions, electrochemically promoted rearrangements, recent advances in selective electrochemical fluorination, electrochemical versions of the classical synthetic reactions, and successful use of these reactions in multistep targeted synthesis allow the synthetic chemist to consider electrochemical methods as one of the powerful tools of organic synthesis.

Elegant and Cost-Effective Synthetic Design

The *heart* of synthesis is in the design of the synthetic scheme for the given target molecule. All the technological advances (discussed above) can only supplement the synthetic scheme. Innovation, elegance, and brevity in the synthetic design are essential

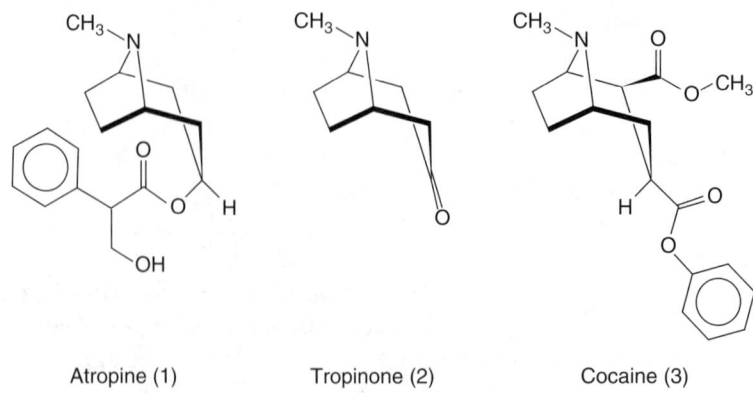

Atropine (1)   Tropinone (2)   Cocaine (3)

Figure 2.3. Structures of atropine, tropinone, and cocaine.

primary requisites to *green* chemical processes. A classic example to highlight the importance of the synthetic scheme can be understood by analyzing the two divergent synthetic schemes developed for the synthesis of tropinone, a precursor of an alkaloid atropine, a close analogue of the well-known local anesthetic "cocaine" (see Fig. 2.3).

Richard Willstätter achieved the first synthesis of tropinone in 1901 (see Fig. 2.4). At that time, structure determination was not always equivocal, and final proof could only be established by unambiguous synthesis of the compound with the suspected structure followed by comparison with an authentic sample of the natural product. Thus, synthesis was often a matter of utilitarian necessity rather than the creative, elegant art form illustrated by the work of many of the great synthetic chemists such as Woodward and Corey.

Willstätter's preparation of tropinone is a competent but long synthesis that demonstrates one of the fundamental difficulties involved in the preparation of complex organic molecules. Although the individual steps in the synthesis generally give good to excellent yields, there are many steps, which means that the overall yield becomes diminishingly small, of the order of 1%. As a result, the early steps in the synthesis have to be carried out on inconveniently large quantities of material and, despite this, usually have to be repeated several times in order to obtain sufficient material to carry out the later stages on an acceptable scale. In 1917, Robinson approached the synthesis in a totally radical way. Tropinone was obtained by condensation of succinaldehyde with acetone and methylamine in aqueous solution (see Fig. 2.5). An improvement

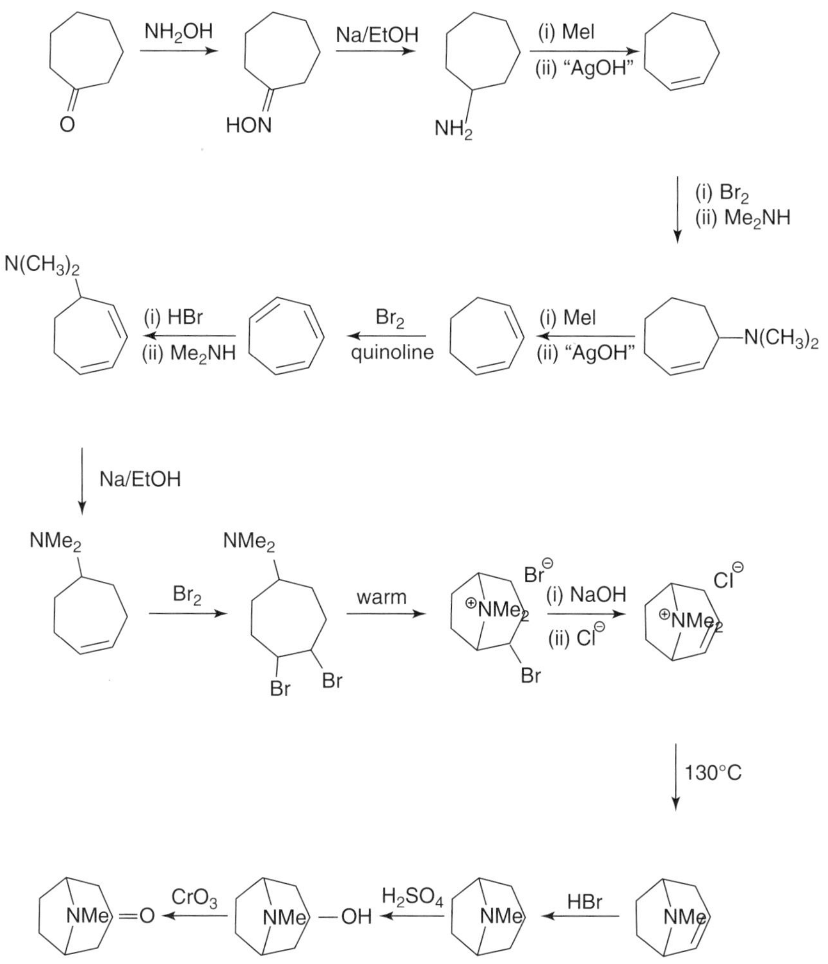

Figure 2.4. Willstätter's synthesis of tropinone.

followed with the replacement of acetone by a salt (calcium) of acetone dicarboxylic acid. The initial product was a salt of tropinone dicarboxylic acid, and this loses two molecules of carbon dioxide with the formation of tropinone when the solution was acidified and heated.

In fact, we can view this synthesis of tropinone as one of the earliest examples of *multicomponent reactions* (MCR). MCRs are convergent reactions in which three or more starting materials react to form a product, where basically all or most of the atoms contribute to the newly formed product. In an MCR, a product is assembled according to a cascade of elementary chemical

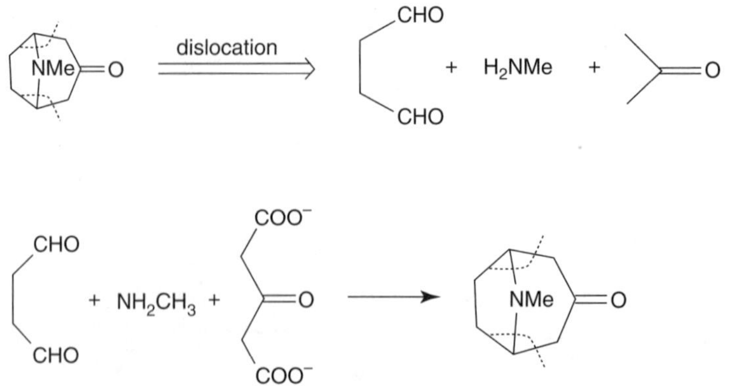

Figure 2.5. Robinson's synthesis of tropinone.

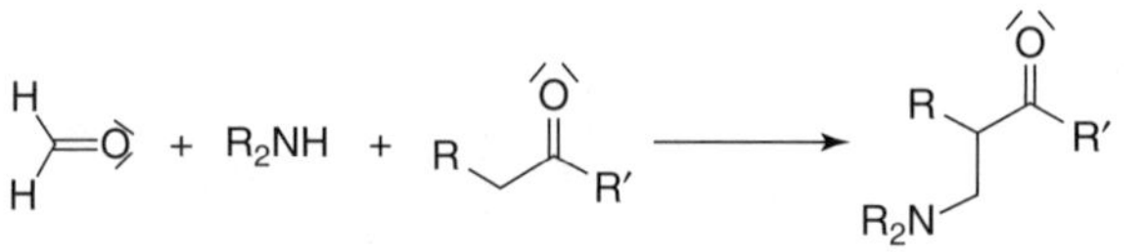

Figure 2.6. Multicomponent reaction.

reactions. Thus, there is a network of reaction equilibria, which all finally flow into an irreversible step, yielding the product. Carbonyl compounds played a crucial role in the early discovery of multicomponent reactions. Some of the first multicomponent reactions to be reported function through derivation of carbonyl compounds into more reactive intermediates, which can react further with a nucleophile. One example is the Mannich reaction (see Fig. 2.6). Some of the well-known (or "name") MCRs are listed in Table 2.2.

In the more specific applications to the drug discovery process, MCRs offer many advantages over traditional approaches. Thus, the chemistry development time, which can typically take up to 6 months for a linear six-step synthesis, is considerably shortened. With only a limited number of chemists and technicians, more scaffold synthesis programs can be achieved within a shorter time. With one-pot reactions, each synthesis procedure (weighing of reagents, addition of reagents, reaction/time control) and work-up procedure (quenching, extraction, distillation, chromatography,

TABLE 2.2
Common Multicomponent Reactions

Name of the Reaction	Reactants/ Components	Predominant Product
Mannich	Carbonyl compounds + amines	Ketoamines
Biginelli	Carbonyl compounds + esters + amines	Diazine derivatives
Bucherer–Bergs	Carbonyl compounds + cyanides + ammonium salts	Imidazolium derivatives
Gewald	Carbonyl compounds + cyanides + sulphur	Thiophene derivatives
Hantzsch–dihydropyridine synthesis	Carbonyl compounds + active methylenes + amines	Dihydropyridine derivatives
Kabachnik–Fields	Carbonyl compounds + amines + phosphates	Aminophosphates
Strecker	Carbonyl compounds + HCN + mineral acid	Amino acids
Kindler thioamide synthesis	Carbonyl compounds + sulphur + amines	Thioamide
Passerini	Carbonyl compounds + carboxylic acid + isocyanide	Amino esters
Ugi	Carbonyl compounds + amines + isocyanides	Keto amines

weighing, analysis) needs to be performed only once, in contrast to multistep syntheses.

Conclusions

The various reaction types most commonly used in synthesis can have different degrees of impact on human health and the environment. Addition reactions, for example, completely incorporate the starting materials into the final product and, therefore, do not produce waste that needs to be treated, disposed of, or otherwise dealt with. Substitution reactions, on the other hand, necessarily generate stoichiometric quantities of substances as byproducts and waste. Elimination reactions do not require input of materials during the course of the reaction other than the initial input of a starting material, but they do generate stoichiometric quantities of substances that are not part of the final target molecule. As such, elimination reactions are among the least atom-economical transformations. For any synthetic transformation, it is important to evaluate the hazardous properties of all substances necessarily being generated from the transformation, just as it is important to evaluate the hazardous properties of all starting materials and reagents that are added in a synthetic transformation.

The atom-economy of various reaction types is shown in Fig. 2.7.

The most atom-economy–suited reactions are condensations, multicomponent reactions, and rearrangements. Hence, where possible, these reaction types should be adopted, in order to ensure

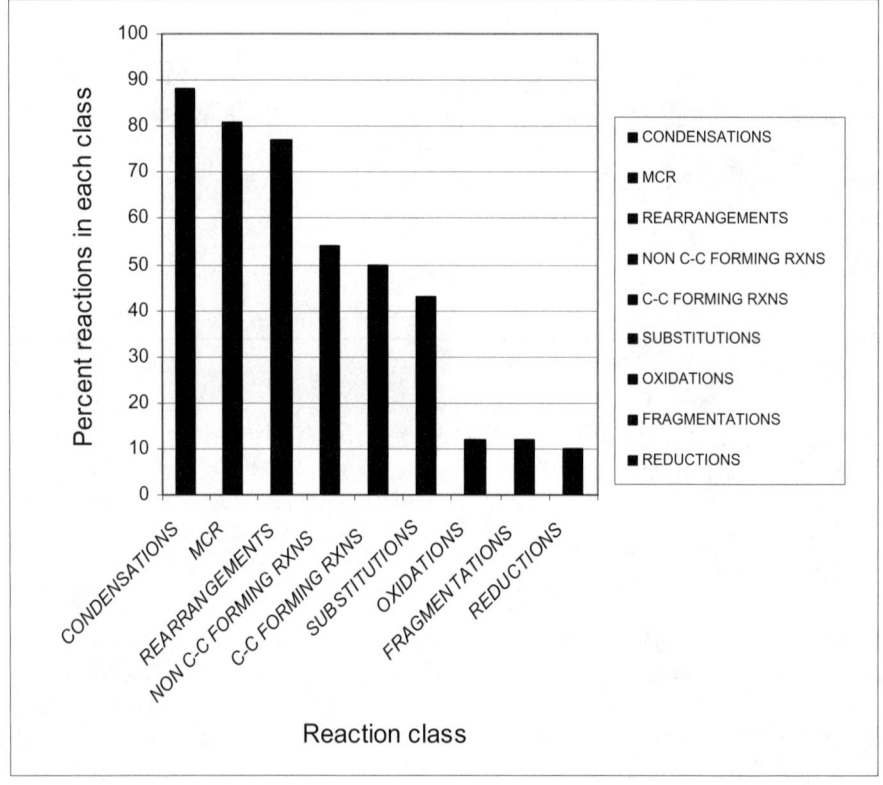

Figure 2.7. Atom-economy of various reaction types.

efficient synthesis. The challenges for designing a synthetic route can therefore be listed as

- Minimize overall number of steps.
- Maximize yield per step.
- Maximize atom-economy per step.
- Use stoichiometric conditions.
- In multistep syntheses, perform the following:
  Maximize frequency of condensations, MCRs, rearrangements, C-C and non-C-C bond-forming reactions.
  Minimize frequency of substitutions (protecting group strategies) and redox reactions.
- If forced to use oxidations, opt for hydrogen peroxide as oxidant.
- If forced to use reductions, opt for hydrogen as reductant.
- Devise electrochemical transformations.
- Devise catalytic methods where catalysts are recycled and reused.
- Devise regio-/stereoselective synthetic strategies.
- Opt for solventless reactions, recycle solvents, or use benign solvents (ionic liquids).
- Minimize energy demands: heating, cooling, reactions under pressure.

Thus, a judicious use of suitable synthetic transformations coupled with an energy-efficient kinetic activation is the way to eco-friendly and cost-effective chemical synthesis.

References

Ando T., Shinjiro S., Takehiko K., Junko I. (née Yamawaki), and Terukiyo H., *J. Chem. Soc., Chem. Commun.*, **10**: 439–440, 1984.

Bordeau, M., Biran, C., Serein-Spirau, F., Leger-Lambert, M.-P., Deffieux, D., and Dunogues, J., *The Electrochemical Society Meeting Abstracts*, **97**(1): 1167, 1997.

Fry, A. J. and Touster, J., *J. Org. Chem.* **54**: 4829, 1989.

Lund, H. and Baizer, M. M., *Organic Electrochemistry*, Marcel Dekker, New York, 1991.

Margulis, M. A. and Mal'tsev, A. N., *Russ. J. Phy. Chem.* **42**: 751–757, 1968.

Pletcher, D. and Walsh, F. C., *Industrial Electrochemistry*, 3rd ed., Blackie Academic & Professional, London, 1993.

Appendix 2.1

Comparison of Thermal Vs. Ultrasound Assisted Reactions (shown above and below the arrow)

1. $C_{15}H_{31}\text{-}COCHN_2 \xrightarrow[\text{90\%}]{C_6H_5COOAg,\ NEt_3, CH_3OH,\ )))),\ 1min,\ r.t.} C_{15}H_{31}\text{-}CH_2COOCH_3$

silent, > 1h, 180°C, 81%

2.

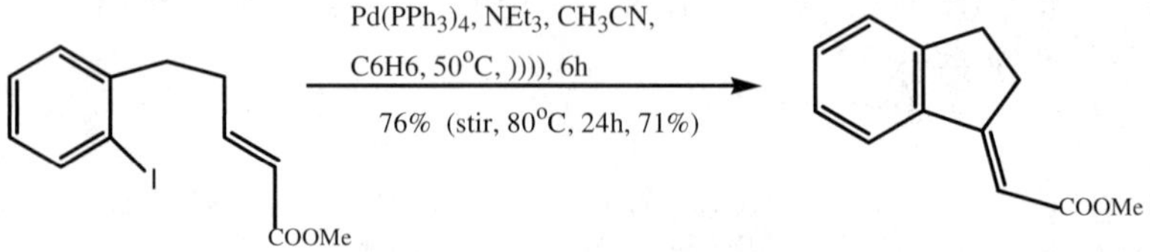

I
COOMe
Pd(PPh3)4, NEt3, CH3CN,
C6H6, 50°C,)))), 6h
76% (stir, 80°C, 24h, 71%)
COOMe

3.

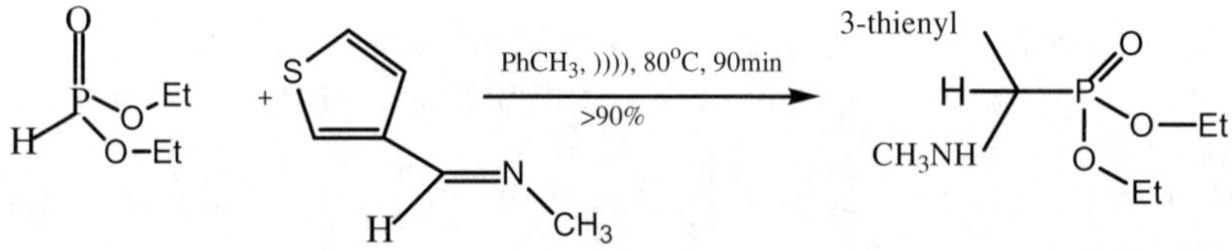

O
H
P
O
Et
O-Et
+
S
N
H
CH3
PhCH3,)))), 80°C, 90min
>90%
3-thienyl
H
CH3NH
P
O
O—Et
O
Et

Thermal reaction exhibits an induction period of *ca.* 1 h, then proceeds to give 50% of the adduct after 2 h.

4.

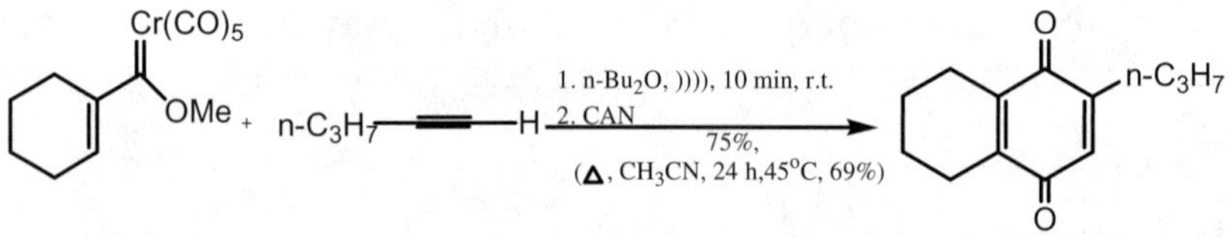

Cr(CO)5
OMe
+
n-C3H7
H
1. n-Bu2O,)))), 10 min, r.t.
2. CAN
75%,
(Δ, CH3CN, 24 h,45°C, 69%)
O
n-C3H7
O

5.

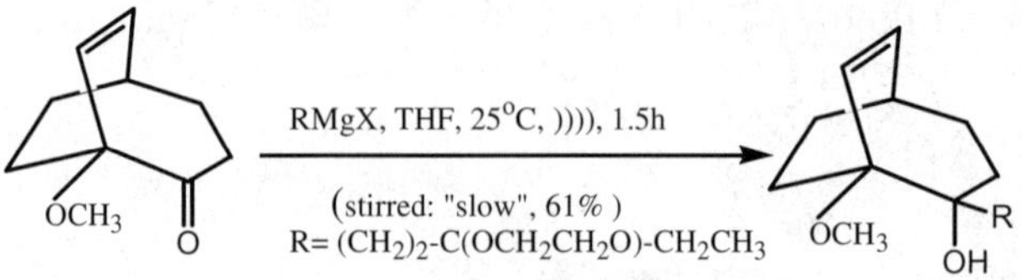

OCH3
O
RMgX, THF, 25°C,)))), 1.5h
(stirred: "slow", 61%)
R= (CH2)2-C(OCH2CH2O)-CH2CH3
OCH3
R
OH

6.

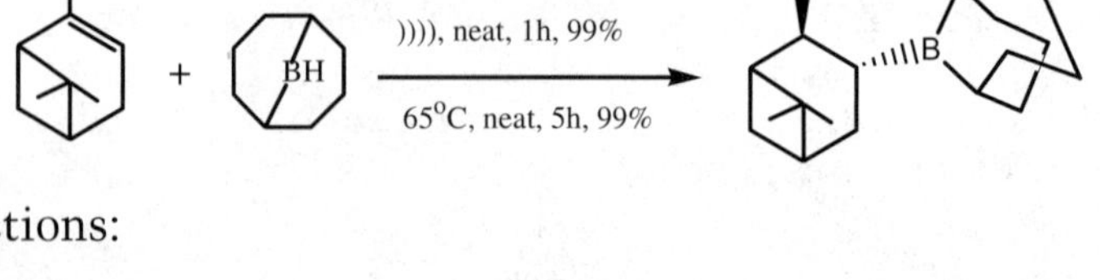

+
BH
)))), neat, 1h, 99%
65°C, neat, 5h, 99%
B

Reactions:

7.

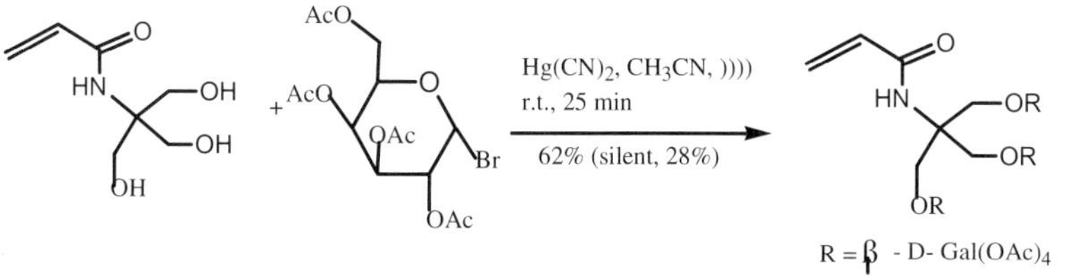

$$\text{Hg(CN)}_2, \text{CH}_3\text{CN},)))) \quad \text{r.t., 25 min}$$
$$62\% \text{ (silent, 28\%)}$$

R = β - D- Gal(OAc)$_4$

8. 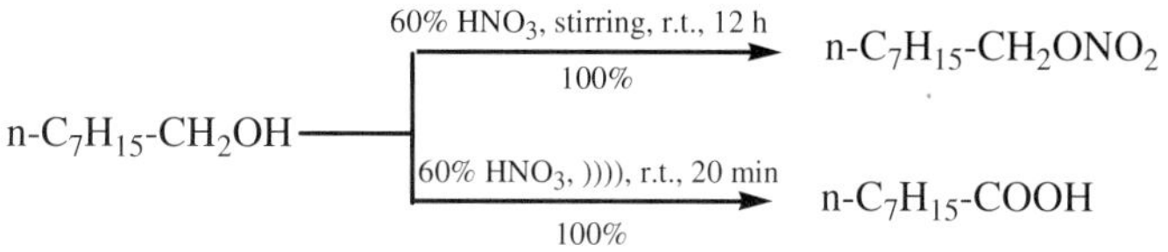 n-C$_7$H$_{15}$-CH$_2$OH

$$\xrightarrow[100\%]{60\% \text{ HNO}_3, \text{ stirring, r.t., 12 h}} \text{n-C}_7\text{H}_{15}\text{-CH}_2\text{ONO}_2$$

$$\xrightarrow[100\%]{60\% \text{ HNO}_3,)))), \text{ r.t., 20 min}} \text{n-C}_7\text{H}_{15}\text{-COOH}$$

9.

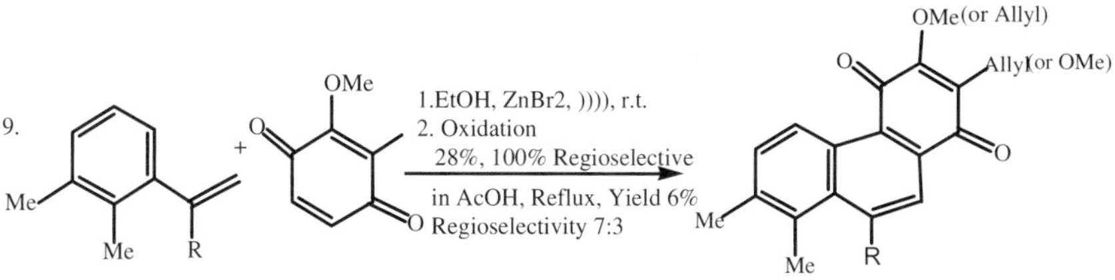

1. EtOH, ZnBr2,)))), r.t.
2. Oxidation
28%, 100% Regioselective
in AcOH, Reflux, Yield 6%
Regioselectivity 7:3

10.

$$\xrightarrow{\text{AcOH, 50}^\circ\text{C}} \text{Pb(OAc)}_4$$

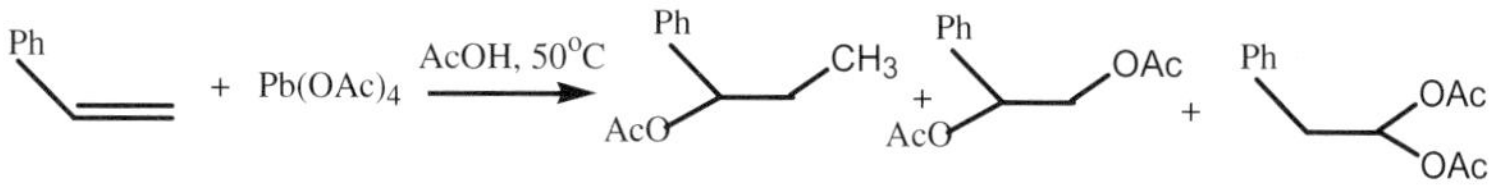

Conditions			
Stirring, 50°C, 1h	0	0	5
)))), 50°C, 1h	38	12	3
)))), 50°C, 1h, radical scavanger	38	12	3

11. 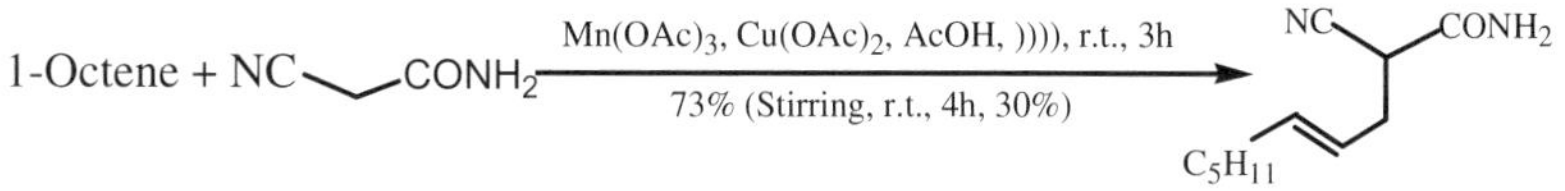 1-Octene + NC⌒CONH$_2$ $\xrightarrow[73\% \text{ (Stirring, r.t., 4h, 30\%)}]{\text{Mn(OAc)}_3, \text{Cu(OAc)}_2, \text{AcOH},)))), \text{r.t., 3h}}$

12.

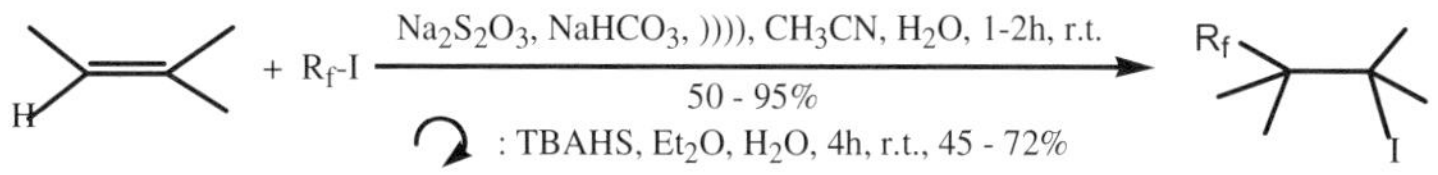

$$+ \text{R}_f\text{-I} \xrightarrow[50 - 95\%]{\text{Na}_2\text{S}_2\text{O}_3, \text{NaHCO}_3,)))), \text{CH}_3\text{CN, H}_2\text{O, 1-2h, r.t.}}$$

↻ : TBAHS, Et$_2$O, H$_2$O, 4h, r.t., 45 - 72%

13.

$$\xrightarrow{\text{A or B}}$$

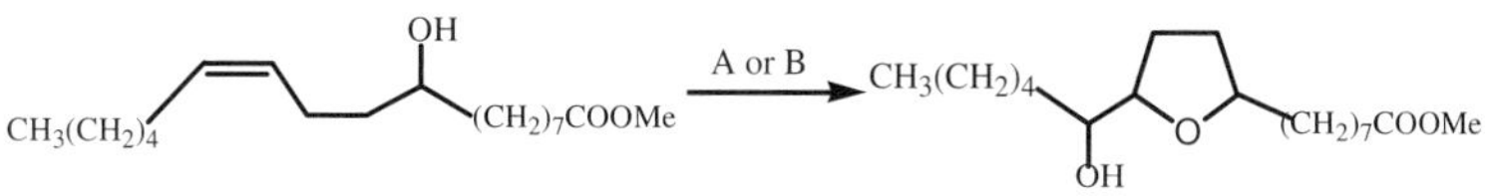

References

1. Winum, J. Y., Kamal, M., Leydet, A., Roque, J. P., and Montero, J. L. *Tetrahedron Lett.*, **37**: 1781–1782, 1996; Rehorek, D. and Janzen, E. G., *J. Prakt. Chem.* **326**: 935–940, 1984.
2. O'Connor, B., Zhang, Y., Negishi, E., Luo, F. T., and Cheng, J. W., *Tetrahedron Lett.*, **29**: 3903–3906, 1984; Cheng, J. and Luo, F., *Bull. Inst. Chem. Acad. Sin.*, **36**: 9–15, 1989.
3. Hubert, C., Oussaid, B., Etemad, G., Koenig, M., and Garrigues, B., *Synthesis*, 51–55, 1994; Hubert, C., Munopz, A., Garrigues, B., and Luvhe, J. L., *J. Org. Chem.*, **60**: 1488–1490, 1995.
4. Harrity, J. P. A., Kerr, W. J., and Middlemiss, D., *Tetrahedron*, **49**: 5565–5576, 1993.
5. Uyehara, T., Yamada, J., Furuta, T., and Kato, T., *Chem. Lett.*, 609–612, 1986.
6. Smith, K. and Pelter, A., in *Comprehensive Organic Synthesis*, Trost, B. M., ed.; Pergamon Press, Oxford, Vol. 8 (Fleming, I., vol. ed.), pp. 703–731, 1991. Brown, H. C. and Racherla, U. S., *Tetrahedron Lett.*, **26**: 2187–2190, 1985.
7. Polidori, A., Pucci, B., Maurizis, J. C., and Pavia, A. A., *New J. Chem.*, **18**: 839–848, 1994.
8. Einhorn, C., Einhorn, J., Dickens, M. J., and Luche, J. L., *Tetrahedron Lett.*, **31**: 4129–4130, 1990.
9. Zhang, Z., Flachsmann, F., Moghaddam, F. M., and Ruedi, P., *Tetrahedron Lett.*, **35**: 2153–2156, 1994.
10. Ando, T., Bauchat, P., Foucaud, A., Fujita, M., Kimura, T., and Sohmiya, H., *Tetrahedron Lett.*, **32**: 6379–6382, 1991; Ando, T., Fujita, M., Bauchat, P., Foucaud, A., Sohmiya, H., and Kimura, T., *Ultrasonics Sonochemistry*, **1**: S33–S35, 1994.
11. Bosman, C., D'Annibale, A., Resta, S., and Trogolo, C., *Tetrahedron*, **50**: 13847–13856, 1994.
12. Rong, G. and Keese, R., *Tetrahedron Lett.*, **31**: 5615–5616, 1990.
13. Lie Ken Jie, M. S. F. and Lam, C. K., *Ultrasonics Sonochemistry*, **2**: S11–S14, 1995.
14. Fuentes, A., Marinas, J. M., and Sinisterra, J. V., *Tetrahedron Lett.*, **28**: 4541–4544, 1987.
15. Silveira, C. C., Perin, G., and Braga, A. L., *J. Chem. Res.*, 492–493, 1994.
16. Moon, S., Duchin, L., and Cooney, J. V., *Tetrahedron Lett.*, 3917–3920, 1979.
17. Grigon-Dubois, M., Diaba, F., and Grellier-Marly, M. C., *Synthesis*, 800–804, 1994.
18. Silveira, C. C., Braga, A. L., and Fiorin, G. L., *Synth. Commun.*, **24**: 2075–2080, 1994.
19. Ezquerra, J. and Alvarez-Builla, J., *J. Heterocyclic Chem.*, **25**: 917–925, 1988; Alvarez-Builla, J., Galvez, E., Cuadro, A. M., Florencio, F., and Garcia Blanco, S., *J. Heterocyclic Chem.*, **24**: 917–926, 1987.

20. Davidson, R. S., Patel, A. M., Safdar, A., and Thornthwaite, D., *Tetrahedron Lett.*, **24**: 5907–5910, 1983.
21. Mason, T. J., Lorimer, J. P., Paniwnik, L., Harris, A. R., Wright, P. W., Bram, G., Loupy, A., Ferradou, G., and Sansoulet, J., *Synth. Commun.*, **20**: 3411–3420, 1990; Mason, T. J., Lorimer, J. P., Turner, A. T., and Harris, A. R., *J. Chem. Res.*, 80–81, 1988; Davidson, R. S., Safdar, A., Spencer, J. D., and Robinson, B., *Ultrasonics*, **25**: 35–39, 1987.
22. Preston Reeves, W. and McClusky, J. V., *Tetrahedron Lett.*, **24**: 1585–1588, 1983.
23. Ando, T., Kawate, T., Ishihara, J., and Hanafusa, T., *Chem. Lett.*, 725–728, 1984.
24. Singh, A. K., *Synth. Commun.*, **20**: 3547–3551, 1990.
25. Casiraghi, G., Cornia, M., Rassu, G., Zetta, L., Fava, G. G., and Belicchi, M. F., *Carbohydr. Res.*, **191**: 243–251, 1989.
26. Han, B. H. and Boudjouk, P., *Tetrahedron Lett.*, **23**: 1643–1646, 1982.
27. Kumar, D., Singh, O. V., Prakash, O., and Singh, S. P., *Synth. Commun.*, **24**: 2637–2644, 1994.
28. Yamawaki, J., Sumi, S., Ando, T., and Hanafusa, T., *Chem. Lett.*, 379–380, 1983.
29. Adams, L. L. and Luzzio, F. A., *J. Org. Chem.*, **54**: 5387–5390, 1989.
30. Morey, J. and Saa, J. M., *Tetrahedron*, **49**: 105–112, 1993.
31. Farooq, O., Farnia, S. M. F., Stephenson, M., and Olah, G. A., *J. Org. Chem.*, **53**: 2840–2843, 1988.
32. Harrowven, D. C. and Dainty, R. F., *Tetrahedron Lett.*, **37**: 7659–7660, 1996.
33. Lickiss, P. D. and Lucas, R., *Polyhedron*, **15**: 1975–1979, 1996.
34. Petrier, C. and Luche, J. L., *Tetrahedron Lett.*, **28**: 2347–2352, 1987.
35. Lillwitz, L. D. and Karachewski, A. M., U.S. Patent 5198594, *Chem. Abstr.*, **119**: 72339r, 1993.

Appendix 2.2

Comparison of Thermal Vs. Microwave-Assisted Reactions

1.

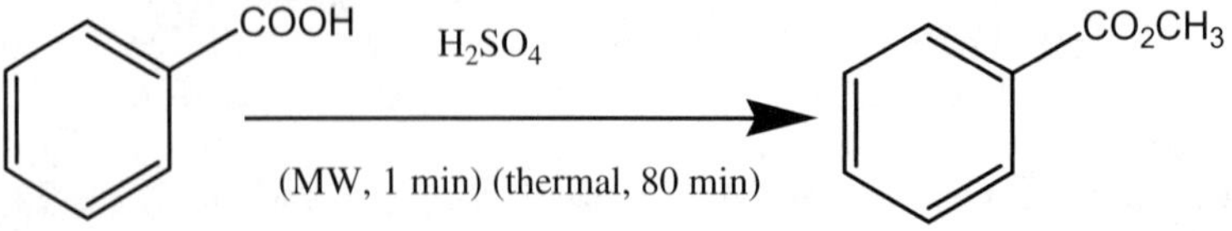

benzene-COOH $\xrightarrow[\text{(MW, 1 min) (thermal, 80 min)}]{H_2SO_4}$ benzene-CO$_2$CH$_3$

2.

naphthalene-OMe $\xrightarrow[\text{(thermal, 72 hours)}]{HBr \text{ (MW, 5 min)}}$ naphthalene-OH

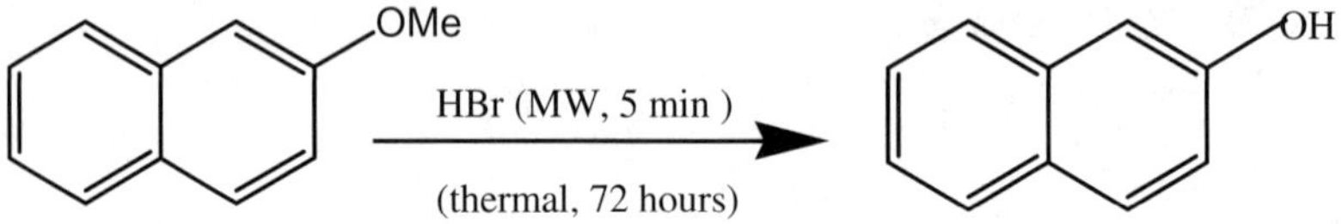

3.   CoCO$_3$ + Al$_2$O$_3$ $\xrightarrow[\text{(MW, 4 min) (thermal, more)}]{[KCl], 1100^\circ C}$ CoAl$_2$O$_4$ + CO$_2$

4.

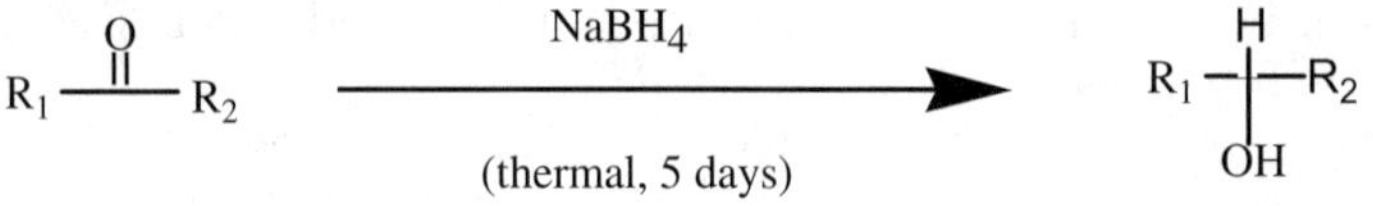

$$R_1 \overset{O}{\underset{}{\overset{\|}{-C-}}} R_2 \xrightarrow[\text{(thermal, 5 days)}]{NaBH_4} R_1-\overset{H}{\underset{OH}{-C-}}-R_2$$

a: $R_1 = R_2 = $ Ph
b. $R_1 = $ trans-PhCH=CH ; $R_2 = $ Ph
c. $R_1 = $ 2 naphthyl ; $R_2 = $ Me
d. $R_1 = $ PhCH$_2$; $R_2 = $ Ph
e. $R_1 = $ PhCH(OH) ; $R_2 = $ Ph

5.

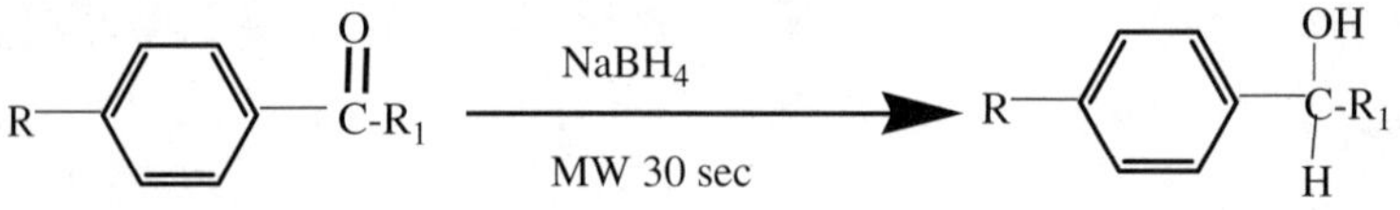

$$R-\text{C}_6H_4-\overset{O}{\overset{\|}{C}}-R_1 \xrightarrow[\text{MW 30 sec}]{NaBH_4} R-\text{C}_6H_4-\overset{OH}{\underset{H}{C}}-R_1$$

a. R = Me ; R_1 =H
b. R = Cl ; R_1 = H
c. R = NO$_2$; R_1 = H
d. R = R_1 = Me

6.

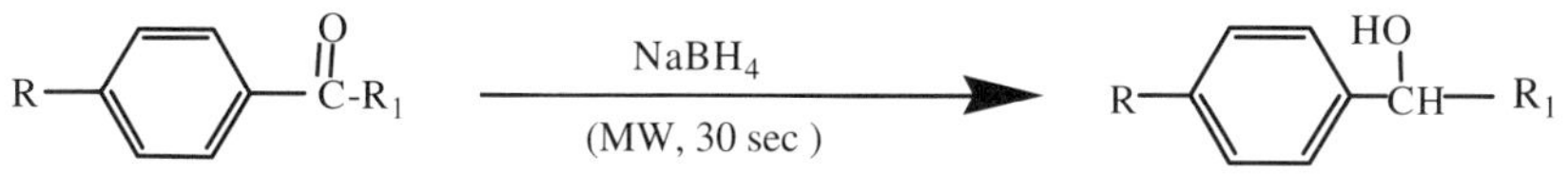

a: R=Me , R_1=H
b: R=Cl , R_1 =H
c: R =NO_2 , R_1=H
d: R =R_1=Me

e: R =H , R_1=Me
f: R =H , R_1=Ph
g: R =OCH_3 , R_1=CH(OH)$C_6H_4OCH_3$P
h: R = H , R_1 =CH(OH)Ph

7.

$$R\text{-CHO} \xrightarrow[\text{NaOH, (MW, 25sec)}]{\text{HCHO}} R\text{-}CH_2OH$$

a: R = Ph
b: R= 4-C_6H_4
c: R= 4- MeOC_6H_4
d: R= 4- $Me_2NC_6H_4$
e: R= 4- MeC_6H_4
f: R= 4- $O_2NC_6H_6$

g: R= 3-$O_2NC_6H_4$
h: R= 2-O2NC6H4
i: R= PhCH=CH
j: R= 2-furyl
k: R= 2-thienyl

$$R\text{-CHO} \;+\; (CH_2O)_n \xrightarrow[\text{MW}]{\text{Ba(OH)}_2.8H_2O} R\text{-}CH_2OH \;+\; R\text{-COOH}$$

$$RCHO \;+\; (CH_2O)_n \xrightarrow[\text{(MW, ~1min)}]{\text{Ba(OH)}_2.8\,H_2O} RCH_2OH \;+\; RCOOH$$

a: R=Ph
b: R=4-ClC_6H_4
c: R=4-BrC_6H_4
d: R=4-FC_6H_4

e: R=2-FC_6H_4
f: R=2-HOC_6H_4
g: R=4-MeC_6H_4
h: R=PhCH=CH

8.

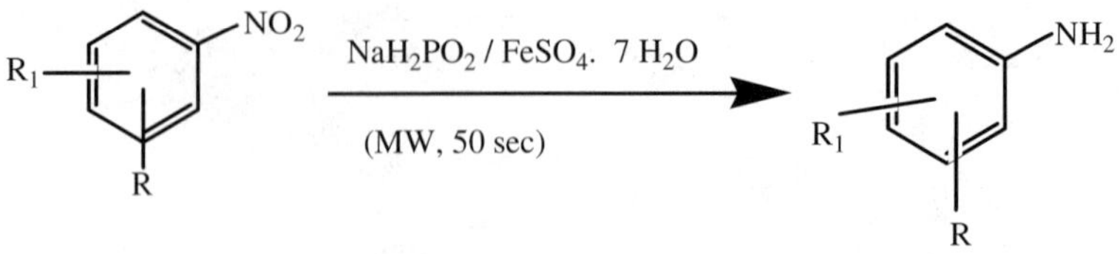

$$R_1 \underset{R}{\overbrace{\qquad}} NO_2 \quad \xrightarrow[\text{(MW, 50 sec)}]{\text{NaH}_2\text{PO}_2 / \text{FeSO}_4 \cdot 7\,\text{H}_2\text{O}} \quad R_1 \underset{R}{\overbrace{\qquad}} NH_2$$

R,R_1= H, Me, OH, $CONH_2$, Ph, COOH, CN, NH_2

9.

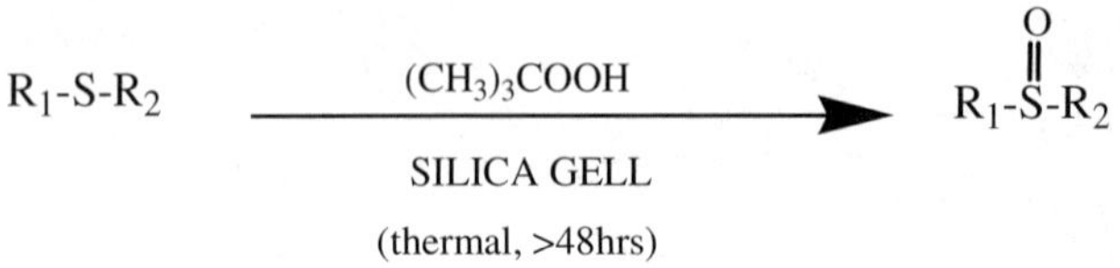

$$R_1\text{-}S\text{-}R_2 \quad \xrightarrow[\substack{\text{SILICA GELL} \\ \text{(thermal, >48hrs)}}]{(CH_3)_3COOH} \quad R_1\text{-}\overset{\overset{\displaystyle O}{\|}}{S}\text{-}R_2$$

a: R_1=R_2= n-Bu
b: R_1=Ph , R_2=CH_2Ph
c: R_1=R_2=Ph

10. $$R_1\text{-}S\text{-}R_2 \quad \xrightarrow[\text{(MW, stipulated time)}]{20\% \text{ NaIO}_4\text{- silica}} \quad R_1\text{-}\overset{\overset{\displaystyle O}{\|}}{\underset{\|}{S}}\text{-}R_2$$

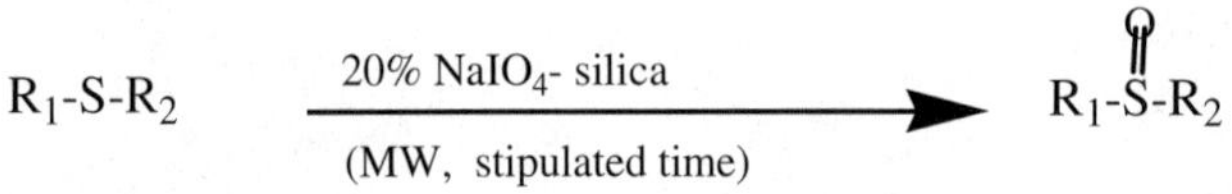

a: R_1=$PhCH_2$, R_2=Ph
b: R_1=R_2=$PhCH_2$
c: R_1=R_2=Ph
d: R_1=Ph , R_2=Me
e: R_1=R_2=Bu

11.   PhCHO + $\overset{\displaystyle CN}{\underset{\displaystyle R}{\diagup}}$ $\xrightarrow[\substack{\Delta, 100°C \\ \text{(specified time)}}]{ZnCl_2}$ $\underset{H}{\overset{Ph}{\diagdown}}C=C\underset{R}{\overset{CN}{\diagup}}$

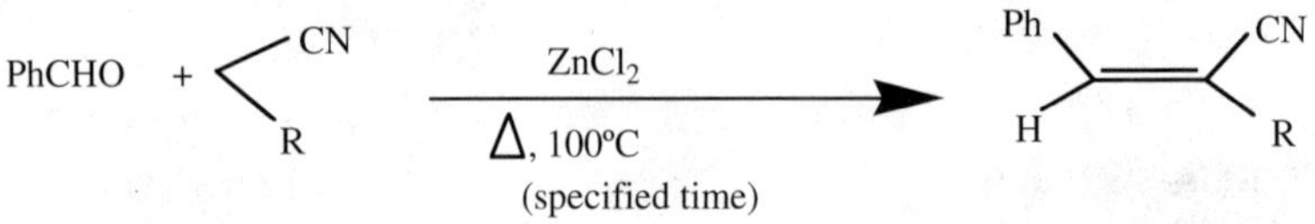

R=CN, $CONH_2$, COOEt

12.   PhCHO   +   [CN / R]   →(LiCl, MW, given time)→   Ph–CH=C(CN)(R)

R=CN, COOEt

13.   ArCHO + [pyrrole]   →(60° alumina, thermal, 4 hrs)→   [tetraaryl porphyrin, Ar]

14.   PhCHO
       +   [pyrrole]   →(MW, 10 min)→   [tetraphenyl porphyrin, Ph]

15   Ph–C(=O)–CH(Ph)–OH   →(MnO₂ or BaMnO₄, thermal, 2 hrs)→   Ph–C(=O)–C(=O)–Ph

16

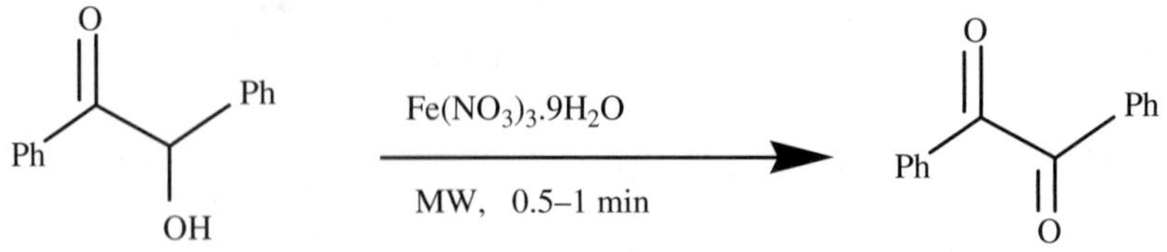

17

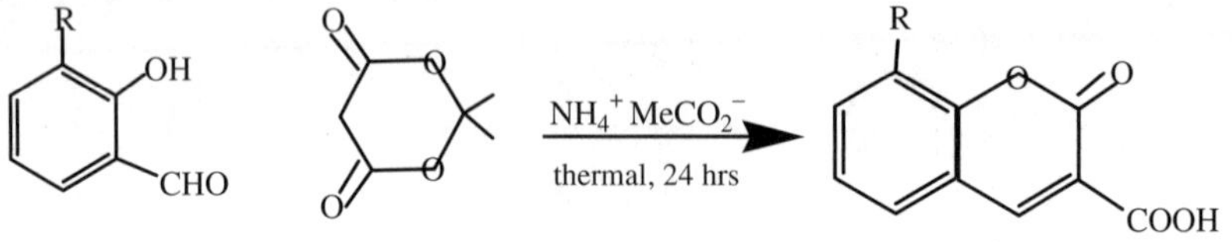

18

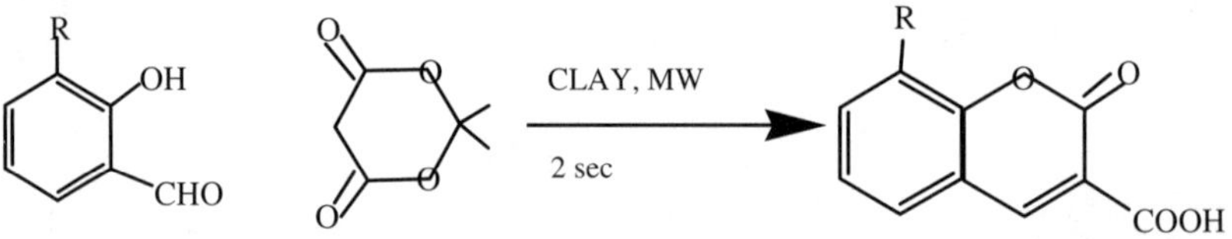

19.

$$R_1CHO \quad + \quad R_2CH_2P^+Ph_3 \xrightarrow[\text{thermal}]{\text{NaOH}} R_1CH=CH\,R_2$$

$R_1=Fc$ $\qquad$ $R_2= C_6H_5, Fc$ $\qquad$ $X= Cl, I, Br$

20.

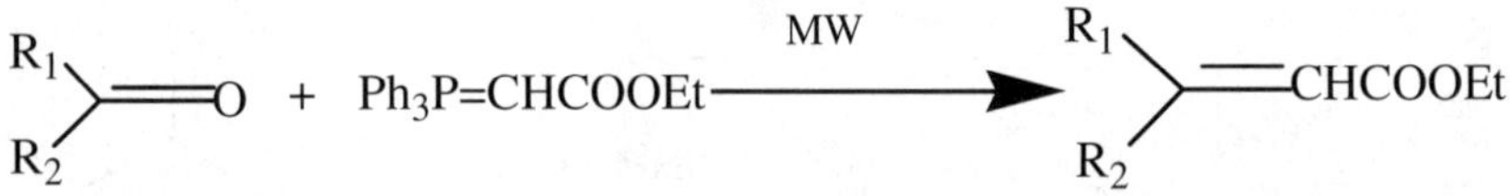

$R_1=R_2=Ph$
$R_1=Ph, R_2=Me$

21.

$$R=Ph \qquad Y=PhNH$$

Reagents: H$_2$NY, Si Gel / NaOH

22

$$R=4\text{-}FC_6H_4, \; 4\text{-}ClC_6H_4 \qquad Y=OH$$

Reagents: H$_2$NY, MW / (15 sec)

23.

$$R_1=R_2= -(CH_2)_5\text{-}, \; -(CH_2)_6$$

Reagents: HONH$_2$, HCl; CaO / thermal, few minutes

24.

$$R_1= Ph, \quad R_2= H, Me$$

Reagents: silica gel, MW, 2min / 98% pure

25.

$$R_1=Ph, \; R_2=H, \; R_3=-(CH_2)_3\text{-}$$

Reagents: R$_3$SH, CdI$_2$ / MW, 75sec

26.

$$R_3SH, LiOSO_2$$

thermal, 90–110°C

$R_1=C_6H_5$, $R_2=H$, $R_3=-(CH_2)_3-$

References

1. Emelingmeier, A., *Römpp Lexikon Chemie, Mikrowelle*, Version 1.3, Georg Thieme Verlag, Stuttgart/New York, 1997.
2. Mingos, M. P. and Baghurst, D. R., *Chem. Soc. Rev.*, **20**: 1, 1991.
3. Brockhaus, F. A., *ABC Physik, Auflage*, Brockhaus-Verlag, Leipzig, 586, 1989.
4. Kingston, H. M. and Haswell, S. J. (Hrsg.), *Microwave-Enhanced Chemistry*, American Chemical Society, Washington, DC, 1997.
5. Toda, F., Kiyoshige, K., and Yagi, M., *Angew. Chem. Int. Ed. Engl.*, **28**: 320, 1989.
6. Varma, R. S. and Naicker, K. P., *Org. Lett.*, **1**: 189, 1999.
7. Varma, R. S. and Saini, R. K., *Tetrahedron Lett.*, **38**: 4337, 1997.
8. Thakuria, J. A., Baruah, M., and Sandhu, J. S., *Chem. Lett.*, 995, 1999.
9. Varma, R. S., Naicker, K. P., and Liesen, P. J., *Tetrahedron Lett.*, **39**: 8437, 1998.
10. Meshram, H. M., Ganesh, Y. S. S., Sekhar, K. C., and Yadav, J. S., *Syn. Lett.*, 993, 2000.
11. Kropp, P. J., Breton, G. W., Fields, J. D., Tung, J. C., and Loomir, B. R., *J. Am. Chem. Soc.*, **122**: 4280, 2000.
12. Varma, R. S., Meshram, H. M., and Saini, R. K., *Tetrahedron Lett.*, **38**: 6525, 1997.
13. Rao, P. S. and Venkatratnam, R. V., *Tetrahedron Lett.*, **32**: 5821, 1999.
14. Sabitha, G., Reddy, B. V. S., Satheesh, R. S., and Yadav, J. S., *Chem. Lett.*, 773, 1998.
15. Gross, Z., Galili, N., Simkhovich, L., Saltsman, I., Botoshansky, M., Blaser, D., Boese, R., and Goldberg, I., *Org. Lett.*, **1**: 599, 1999.
16. Firouzabani, H., Karimi, B., and Abbasi, M., *J. Chem. Soc.*, 236, 1999.
17. Warner, M. G., Succaw, G. L., and Hutchison, J. E., *Green Chem.*, **3**: 267, 1999.
18. Scott, J. L. and Ratoson, C. L., *Green Chem.*, **2**: 245, 2000.
19. Bandgar, B. P., Uppalla, L. S., and Kurule, D. S., *Green Chem.*, **1**: 243, 1999.

20. Liu, W., Xu, Q., Ma, Y., Liang, Y., Dong, N., and Guan, D., *J. Organomet. Chem.*, **625**: 128, 2001.
21. Spinella, A., Fortunati, T., and Soriente, A., *Synlett.*, 93, 1997.
22. Hajipour, A. R., Mohammadpoor-Baltork, I., and Bigdeli, M., *J. Chem. Res.*, 570, 1999.
23. Bandger, B. P., Sadavarte, V. S., Uppalla, L. S., and Govande, R., *Monatsh. Chem.*, **132**: 143, 2001.
24. Shaghi, H. and Sarvani, M. N., *J. Chem. Res.*, 24, 2000.
25. Hajipour, A. R., Mallakpour, S. E., and Imanzadeh, G., *J. Chem. Res.*, 228, 1999.
26. Firouzabani, H., Karimi, B., and Eslami, S., *Tetrahedron Lett.*, **40**: 4055, 1999.
27. Laskar, D. D., Prajapati, D., and Sandhu, J. S., *J. Chem. Res.*, 331, 2001.

CHAPTER 3

Catalysis and Green Chemistry

Since the 1990s, the scientific community has progressively changed its approach toward dealing with regulations for environmental protection. This evolution in the chemical industries and R & D labs has led to the development of green chemistry. In the last decade, green chemistry has been widely recognized and accepted as a new means for *sustainable development*. As the name suggests, green chemistry is a scientific approach that is inherently safer than the old-fashioned cleanup chemistry, which, prior to the 1990s, was used to meet environmental protection laws.

Industries are often forced to pay heavy prices to meet with the standards set by the pollution regulatory boards while using the traditional methods of treating or recycling waste. Also, with growing environmental problems, law-making boards are now looking more critically at the possible hazardous effects of a larger number of chemical substances. Thus, the combined effect of existing environmental safety regulations, the evolving scientific understanding of previously little-known toxic chemicals and their long-term effect on the biota, and the industries' monetary interests have turned attention from end-of-pipe cleanup to environmentally safer production processes through the green chemistry approach. A simple definition of green chemistry as given by the U.S. EPA is, "Use of chemistry for pollution prevention and design of chemical products and processes that are more environmentally benign" www.epa.gov/greenchemistry/. The growing importance now given to green chemistry can be attributed to the ability of this approach to bridge ecoefficiency and economic growth.

Catalysis and Green Chemistry

In general, catalysis plays a major role in making industrial processes more efficient and economically profitable. This can be quite obviously attributed to three general characteristics of catalysts:

1. Catalytic reagents reduce the energy of the transition state, thereby reducing the energy input required for a process.
2. Catalysts are required in small quantities. In the case of bio-catalysts, the number of catalysts (generally enzymes) needed compared to the quantity of reactants is very low.
3. The regeneration and reversibility of catalysts are good for green processes.

As much as it is a key in achieving economic objectives, catalysis is also a powerful tool in realizing the goals of green chemistry. Innovation in the field of catalysis is driven by both profit motives and efforts to make more eco-efficient processes. Most often profits are markedly improved with the development of green processes. An important concept of green chemistry that can be addressed by the use of catalysis is atom-efficiency, also known as atom-economy.

The growing environmental consciousness has resulted in the paradigm shift in viewing process efficiency. The focus is now shifting from the traditional chemical yields to the *environmental quoefficient* (EQ) factors and the *atom-economy* (or efficiency) (AE) factor. The AE factor can be simply defined as the quantity of the waste generated per kg of product, considering waste as everything other than the desired product produced. AE calculation is a quick means of composing two alternative roots to a specific product. It is calculated by dividing the molecular weight of the desired product by the sum total of the molecular weight of all substances produced in the stoichiometric equation for the reactions involved. Some authors also describe it as the number of atoms of all the reactants that are converted into atoms of the desired product in a reaction. Processes that involve stoichiometric reagents are less atom-efficient compared to the catalytic alternative. For instance, when replaced with cleaner catalyzed oxidation, traditional oxidations using oxidants such as permanganate or chromium reagent (as shown in Fig. 3.1) improve atom-economies.

$$3 \, PhCH(OH)CH_3 + 2 \, CrO_3 + 3 \, H_2SO_4 \longrightarrow 3PhCOCH_3 + Cr_2(SO_4)_3 + 6H_2O$$

FIGURE 3.1. Jones oxidation of secondary alcohol.

$$PhCH(OH)CH_3 + 1/2 \, O_2 \xrightarrow{\text{catalyst}} PhCOCH_3 + H_2O$$

FIGURE 3.2. Atom-economical oxidation of secondary alcohol.

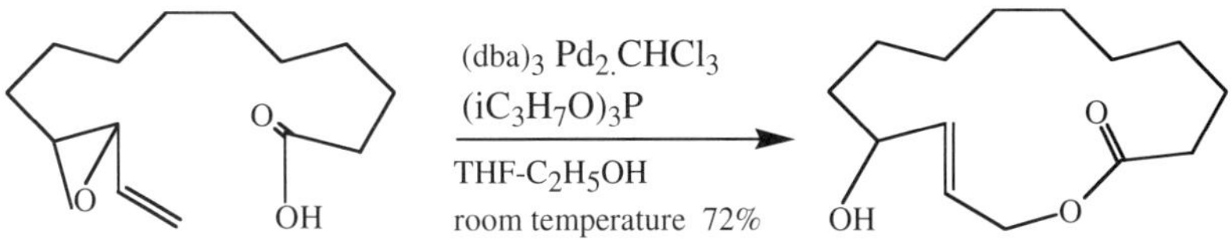

FIGURE 3.3. Macrolactone synthesis with high atom-efficiency (AE).

The oxidation contained in Fig. 3.1 of a secondary alcohol to a ketone using stoichiometric reagents has an atom-efficiency of 42%. The same product obtained through catalytic oxidation improves the atom-efficiency by more than double, to 87% (see Fig. 3.2).

Trost and co-workers (Trost, 1998) used a variety of palladium catalysts to effect allylic alkylation reaction. Synthesis of macro-lactones from corresponding carboxylic acid catalyzed by palladium complex is a 100% atom-efficient reaction (Fig. 3.3). The reaction, as it occurs at room temperature, is also an example of catalysis reducing energy usage.

Ibuprofen, an analgesic (marketed under the brand names Advil™ and Matrix™), was traditionally synthesized in six stoichiometric steps involving an atom-efficiency of less than 40%. A new catalytic process designed for the synthesis of Ibuprofen by BHC (BHC, 1997) involves three steps, with an atom-efficiency of 80%. Though the usage of HF, a toxic substance, is a drawback of the process, the recovery of HF is effected with 99.9% efficiency. This is a fine example of an eco-efficient process being commercialized.

The process shown in Fig. 3.4, also known as the Hoeschst–Celanere process, is a typical example of a catalyzed carboxylation reaction. Carboxylation reactions are generally 100% atom-

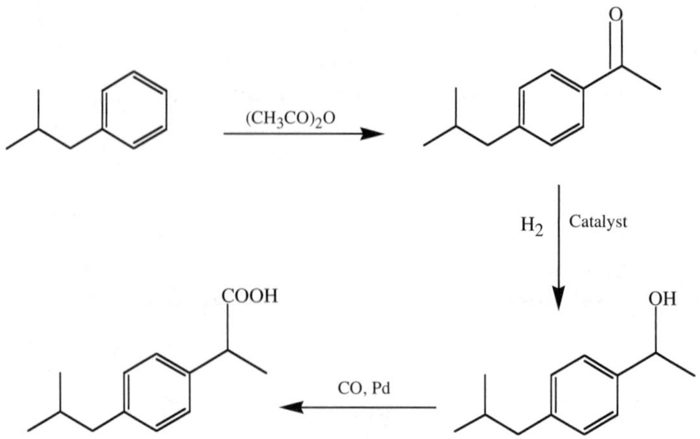

FIGURE 3.4. Hoeschst–Celanere process.

$$R_1CHO \ + \ R_2\text{-NH-}\underset{\displaystyle O}{\overset{\displaystyle O}{C}}\text{-}R_3 \ + \ CO \ \xrightarrow[\text{LiBr , H+}]{[Pd]} \ HOOC\diagup \underset{R_2}{\overset{R_1}{N}}\diagdown R_3$$

$$[Pd] = PdBr_2 \ \ (or) \ \ (Ph_3P)_2PdBr_2$$

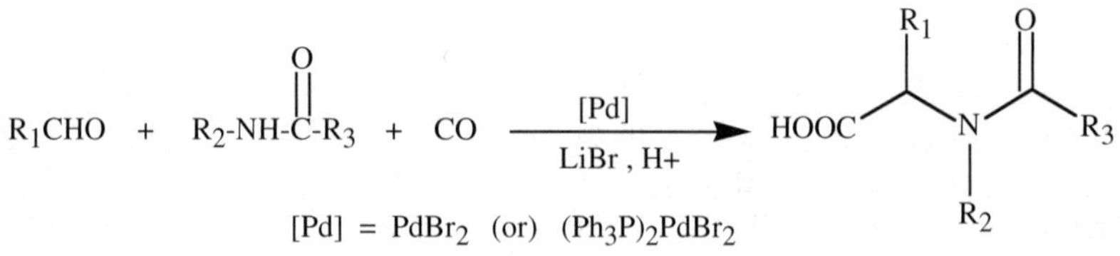

FIGURE 3.5. Amidocaboxylation.

efficient, like the elegant one-step conversion of an aldehyde, CO, and an amide to an acylamino acid. Amidocaboxylation (see Fig. 3.5) is also a palladium-catalyzed reaction (Beller, 1997).

The use of the catalyst/solvent HF in the Hoeschst–Celanere process does not follow green chemistry principles. This leaves a need for truly catalytic procedures: for instance, using a solid acid or an ionic liquid. Use of zeolites in an acid-catalyzed rearrangement of epoxides to carbonyl compounds (Elings et al., 1997; Holderich et al., 1997; Kurkeler et al., 1998) is a good example of the use of recyclable solid acid catalysts. Traditionally, Lewis acids such as $ZnCl_2$ were used in stoichiometric amounts for the type of reaction displayed in Fig. 3.5. The following examples are two commercially relevant processes. The products are precursors of chemicals used for their fragrance (see Fig. 3.6).

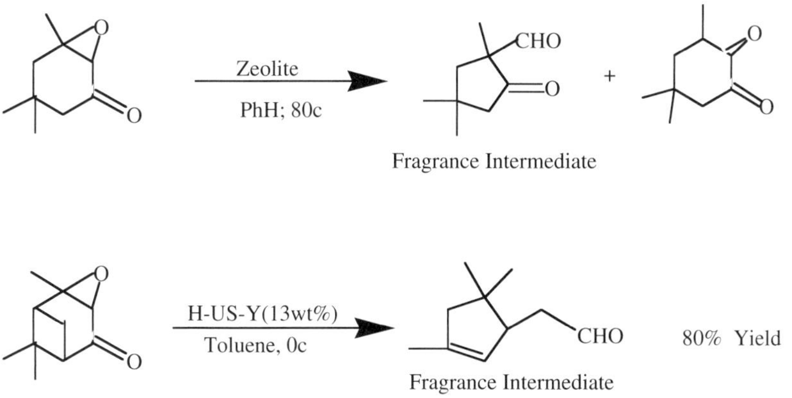

Zeolite
PhH; 80c
CHO
O
O
+
O
O
O
Fragrance Intermediate
O
O
H-US-Y(13wt%)
Toluene, 0c
CHO
80% Yield
Fragrance Intermediate

FIGURE 3.6. Zeolites and clay-catalyzed, high-AE reactions.

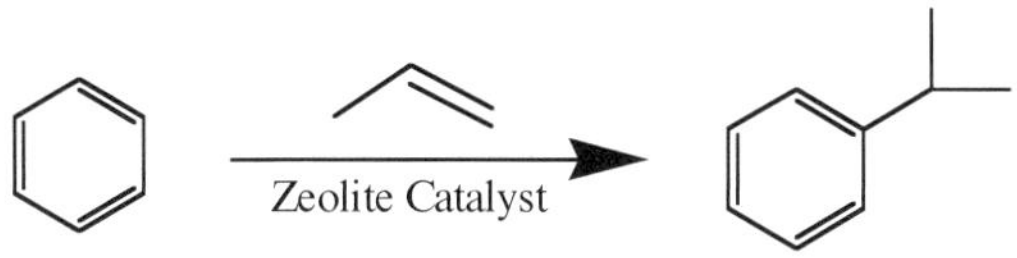

Zeolite Catalyst

FIGURE 3.7. Mobil/Badger cumene process.

The use of zeolites in the manufacture of cumene is of immense importance. About 7 million metric tons of cumene are produced annually worldwide. The earlier-used process involved alkylation of benzene over a solid phosphoric acid or an aluminum chloride catalyst. Cumene Production, U.S. Patent 4008290. Both catalysts are toxic in nature. The Mobil/Badger cumene process (Mobil Technology Co., 1997) uses the less toxic carozine zeolite catalyst (see Fig. 3.7). In addition, it also generates less waste and requires less energy than the earlier catalysts, thus simultaneously satisfying various conditions of green chemistry.

The use of zeolites in making industrial processes eco-compatible is growing with the widespread research on using these as catalysts. One such example of zeolite being used to better the existing process is that of the Meerwin–Ponndorf–Verly (MPV) reduction. The MPV reduction process is an extensively used technology for reducing aldehydes and ketones to their corresponding alcohols. In practice, the reduction involves a reaction of the substrate with a hydrogen donor (usually isopropanol), in the presence

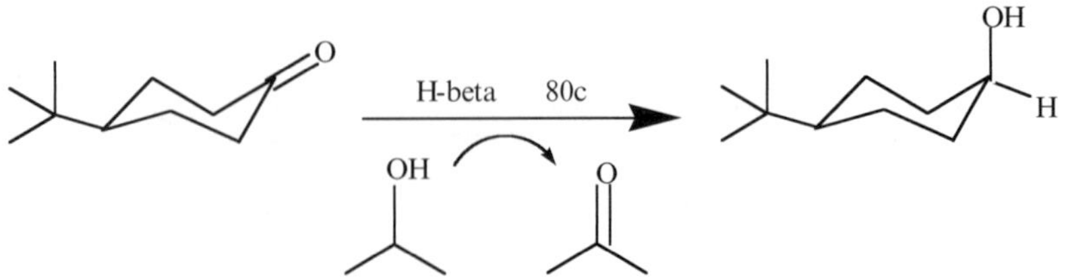

FIGURE 3.8. MPV reduction using zeolite.

of an aluminum alkoxide. The stoichiometric requirement of aluminum alkoxide, due to the slow exchange of the alkoxy group, was an inherent drawback in the method. Creghton et al. (1997) have shown that zeolite beta is able to catalyze the MPV reduction (see Fig. 3.8). Unlike the earlier one, this process is truly catalytic, with the possibility of regaining the catalyst with simple filtration.

The example in Fig. 3.8, the reduction of 4-tert-butyl cyclohexanone, brings the next important aspect of catalysis: selectivity of catalyzed reactions. In this reaction, the *trans*-alcohol was the preferred product in the traditional MPV reduction. The zeolite-catalyzed reaction forms the thermally less stable *cis*-isomer, which is an important fragrance chemical intermediate. Catalysis offers an edge over stoichiometric reactions in achieving selectivity in production, when mono substitution is preferred over disubstitution, when one stereo-isomer is preferred over another or one regioisomer over another. Hence, by driving the reaction to a preferred product, catalyzed reactions decrease the amount of waste generated while reducing the energy requirements, as mentioned earlier.

The contribution of Spiney and Gogate (Spivey and Gogate, 1998) in developing heterogeneous catalysts for the condensation of acetone to methyl isobutyl ketone (MIBK) is commendable. The reaction typically requires stoichiometric amounts of base and could also result in considerably overcondensed products. The catalysts tested for the process are nickel/alumina (see Fig. 3.9), palladium, zirconia, nickel, niobium, and ZSM-5 with palladium, which have exhibited various levels of selectivity and a degree of conversion.

In the production of biologically active molecules (pharmaceuticals and pesticides), there is often a need to produce chiral molecules as the pure enantiomer. This is due to the fact that one

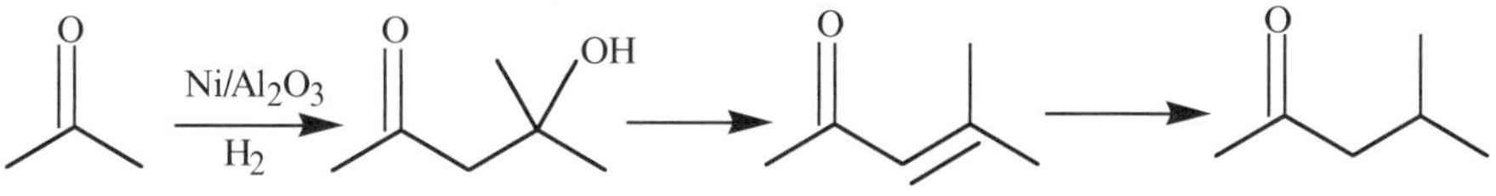

FIGURE 3.9. Ni/Al$_2$O$_3$ catalyzed reaction with high AE.

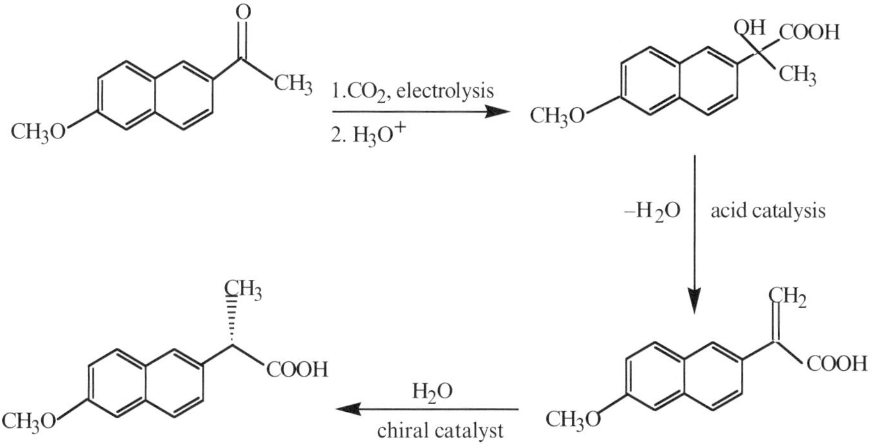

FIGURE 3.10. Synthesis of the anti-inflammatory drug naproxen.

stereoisomer of a bioactive molecule may not have the same activity as the other stereoisomer and, in some cases, could even lead to certain side effects. This need has directed the focus onto asymmetric catalysis using chiral metal complexes and enzymes. The use of a catalyst containing BINAP [2,2^1–bis(diaryl phospheno)-1, 1^1-binaphthyl] in the synthesis of the anti-inflammatory drug naproxen (Simmons, 1996) is an example of an industrially used chiral metal catalyst (see Fig. 3.10). The ligand BINAP has a restricted rotation due to steric hindrance, and the drug is obtained in 97% yield under high pressure in this method.

The Novartis process for the synthesis of the optically active herbicide(s)-metachlor (Blaser and Spindler, 1998) involves a chiral metal complex as a catalyst (see Fig. 3.11). An iridium(I) complex of a chiral ferrocenyldiphosphine catalyzes the asymmetric hydrogenation of a prochiral imine, a key step in the process. The substrate/catalyst ratio for this step is 750,000, with high turnover, thus making the process industrially viable.

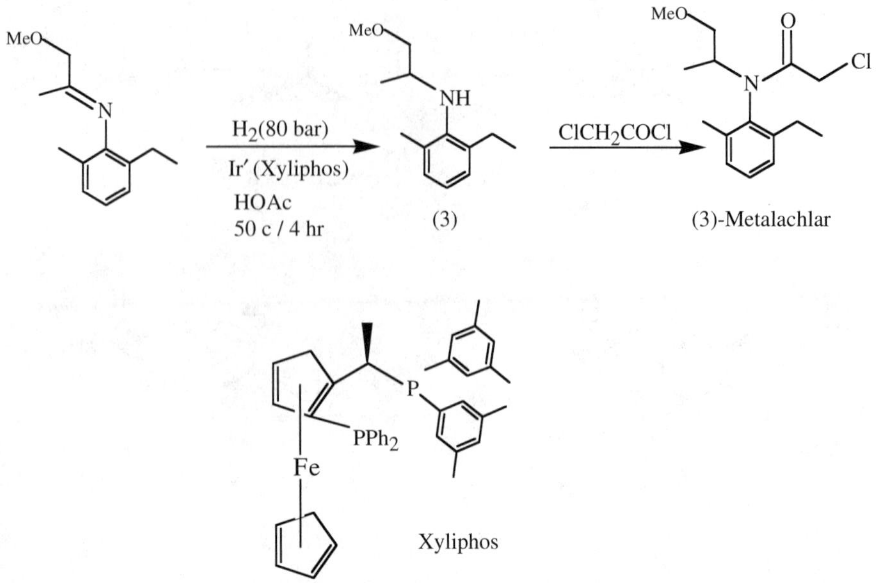

FIGURE 3.11. Novartis process for the synthesis of the optically active herbicide.

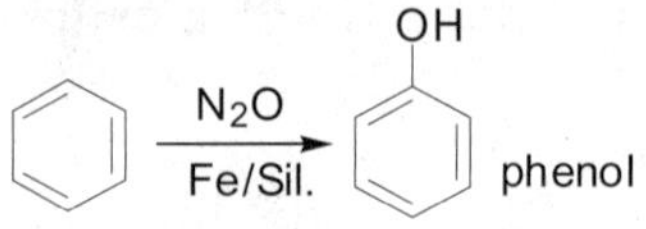

FIGURE 3.12. Production of phenol from benzene.

Solutia (USA), in joint work with the Boreskov Institute of Catalysis, Russia, developed a one-step process to manufacture phenol from benzene using nitrous oxide as the oxidant (see Fig. 3.12). Nitrous oxide (a greenhouse gas) is a waste product from Solutia's adipic acid process. The preferred catalysts are acidified ZSM-5 and ZSM-11 zeolites containing iron or a silica/alumina ratio of 100:1 containing 0.45 wt% iron(III) oxide. The catalyst's half-life is 3 to 4 days, and it can be restored by passing air through the bed at high temperatures.

The production of cumene from benzene using β-zeolite has been developed by Enichem (see Fig. 3.13). This catalytic process

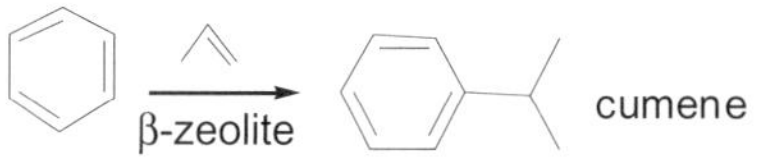

FIGURE 3.13. Production of cumene from benzene.

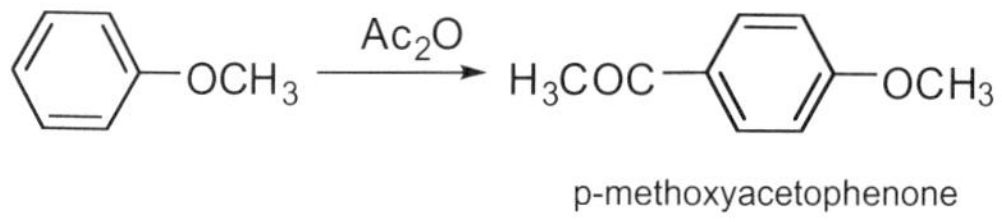

p-methoxyacetophenone

FIGURE 3.14. Production of p-methoxyacetophenone from methoxy-benzene.

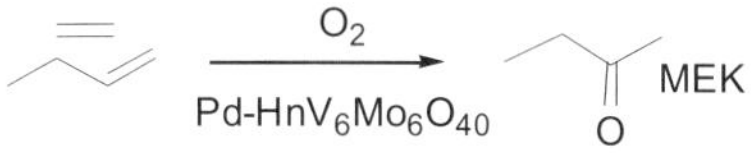

FIGURE 3.15. Manufacture of methylethyl ketone (MEK) from ethylene and butylenes.

reduces polyalkylate waste generated by the traditional chemical route.

The Rhodia process for the production of p-hydroxyacetophenone from methoxybenzene using clay as the catalyst eliminates the use of toxic chemicals such as $AlCl_3$ and BF_3 and also eliminates toxic waste (see Fig. 3.14).

The Catalytic process for the manufacture of methylethyl ketone (MEK) from ethylene and butylenes uses a mixture of palladium, vanadium, and molybdenum oxides as catalyst (see Fig. 3.15). The original process used chlorinated chemicals, which led to a large amount of chlorinated waste that posed several problems during disposal.

The Enichem process for the preparation of propene oxide from propylene involves using H_2O_2 as the oxidizing agent using titanium silicate catalyst (see Fig. 3.16).

The Avetis process for preparing halo benzaldehyde is to oxidize corresponding halo toluene using air and a mixture of iron, vanadium, and molybdenum oxide catalyst (see Fig. 3.17). The catalytic process eliminates the formation of chlorinated byproducts.

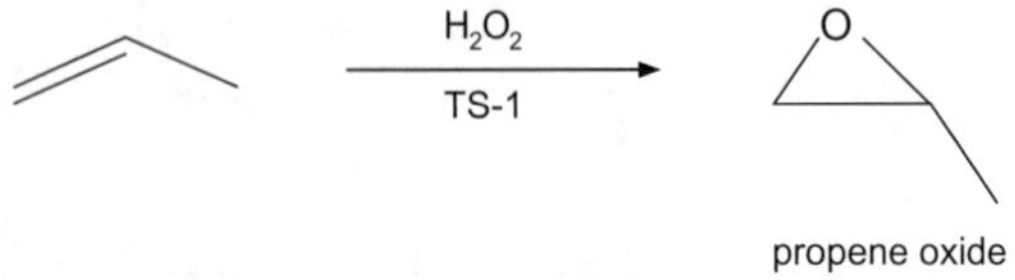

FIGURE 3.16. Preparation of propene oxide from propylene.

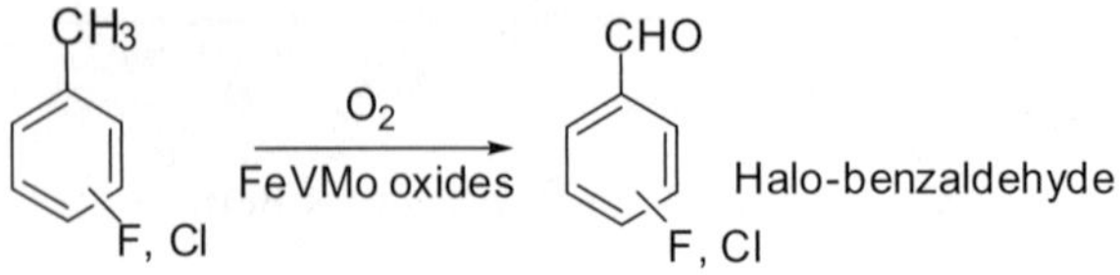

FIGURE 3.17. Preparation of propene oxide from propylene.

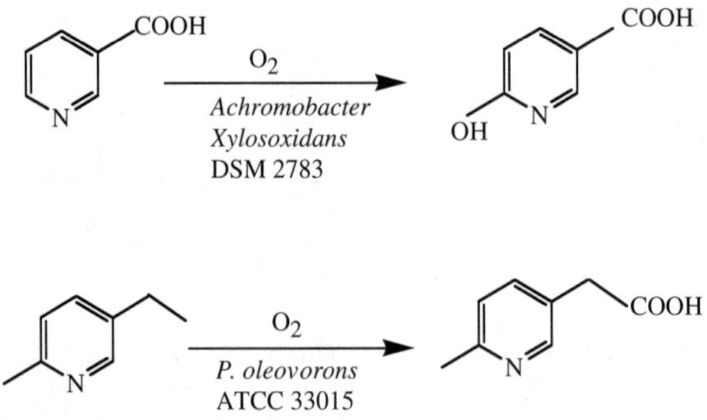

FIGURE 3.18. Microbial mediated aromatic ring hydroxylation.

Biocatalysis is the other option when selectivity (sterio or regio) is a priority in a reaction. The various aspects of biocatalysis are discussed elsewhere in the book; the following are some examples of biocatalysts that have been used in important synthesis. Kirner (1995) conducted microbial ring hydroxylation and side chain oxidation of hetero-aromatics (see Fig. 3.18). Such selectivity is difficult to achieve in one step in traditional chemical synthesis.

As the example in Fig. 3.18 shows, apart from achieving selectivity, biocatalysis also renders steps such as protection, deprotection, or activation redundant, thereby increasing atom-efficiency, reducing waste, and minimizing energy requirements.

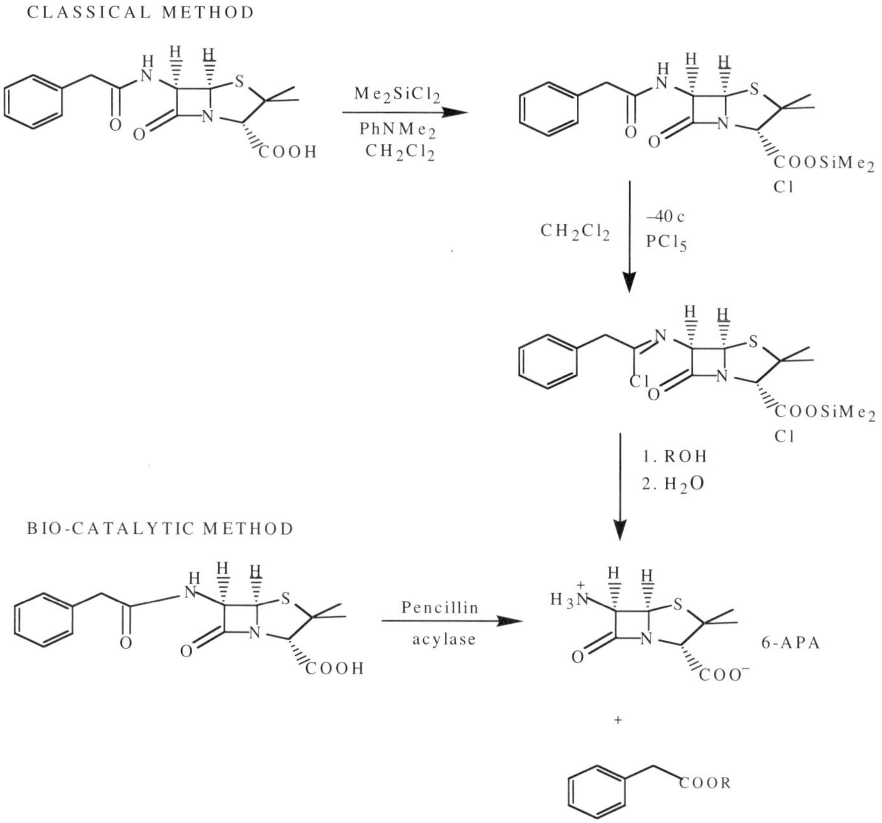

CLASSICAL METHOD

BIO-CATALYTIC METHOD

FIGURE 3.19. Synthesis of Penicillin-G.

The synthesis of 6-aminopenicillonic acid (6-APA) (Penicillin-G) is an example that can be cited in this context (see Fig. 3.19). The classical method calls for the protection of the carboxy group of Penicillin-G, making it a four-step process. Enzymatically, this conversion can be achieved in a single step (Sheldon, 1994).

Genetic engineering also comes in handy when dealing with chemical reactants that are not biological substrates. Stewart (1998) used genetically engineered Baker's yeast to perform a Bayer–Villiger reaction (see Fig. 3.20). It involves the conversion of a ketone into a lactone commonly using the reagent m-chloro-peroxybenzoic acid (m-CPBA). This reagent is both sensitive to shocks and explosive. This is a classic example of biocatalysis making a reaction eco-compatible. The yeast, a safe and nonpatho-genic organism, performs the oxidation of ketone using atmo-

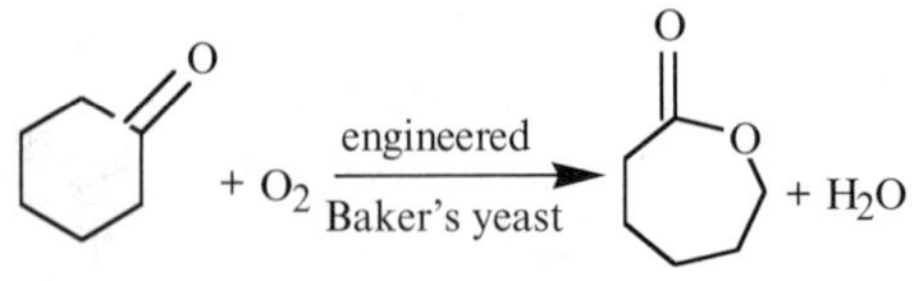

FIGURE 3.20. Yeast-catalyzed Bayer–Villiger reaction.

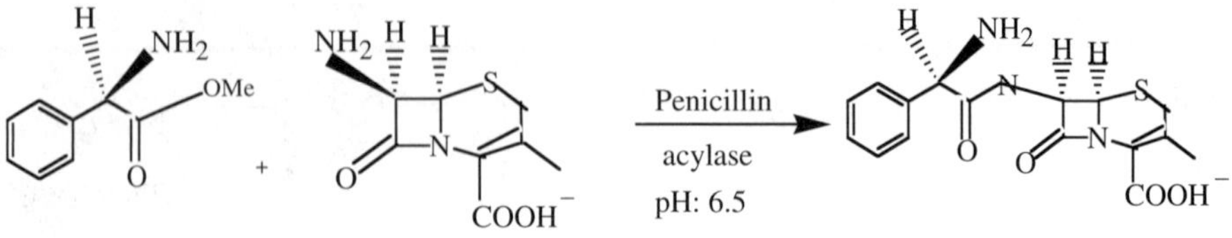

FIGURE 3.21. Synthesis of cephalexin through the use of CLECs.

spheric oxygen, with water being the only byproduct formed. The reaction is also run in an aqueous medium. The industrial scope of the reaction is under study.

Enzymes do have their disadvantages. Their solvent incompatibility and instability restrict their industrial use. However, researchers strive to find means to overcome such drawbacks to make use of the catalytic efficiency of enzymes. Altus Biologics (1997) has developed *cross-linked enzyme crystals* (CLECs) to increase the versatility of enzymes in organic reactions. CLECs exhibit a high level of stability in extreme conditions of temperature and pH and in exposure to both aqueous and organic solvents. The synthesis of the antibiotic cephalexin was carried out using CLECs (see Fig. 3.21). The N-protection step of methyl phenyl glycinate in the classical synthesis was eliminated.

Genetically engineered microbes have been used by Draths and Frost (1998a, b) to synthesize common but important chemicals such as adipic acid and catechol (see Fig. 3.23). The noteworthy aspect of this work is that the starting materials were renewable feedstock. The principles of green chemistry state that "a raw material of feedstock should be renewable rather than depleting wherever technically and economically practicable" (Anastas and Warner, 1998). This reaction addresses this principle and more, as it can be seen. Classical catechol synthesis beginning with benzene (obtained from petroleum, a nonrenewable feedstock) involves a multistep process (see Fig. 3.22).

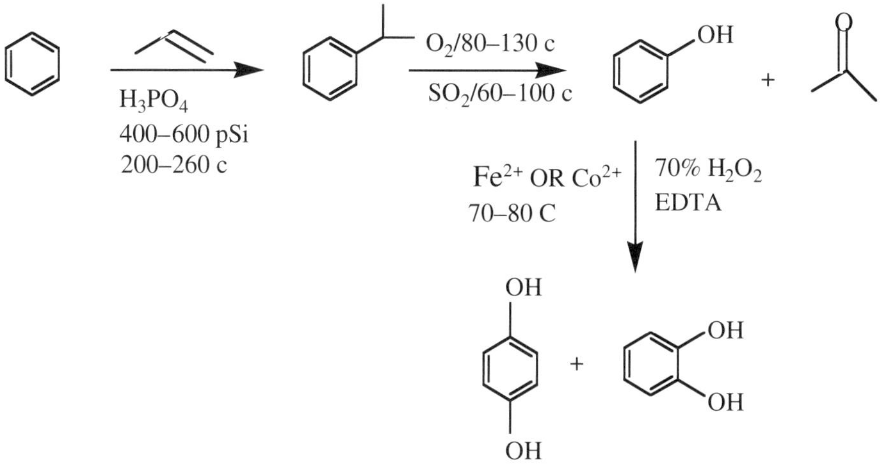

FIGURE 3.22. Classical synthesis of catechol.

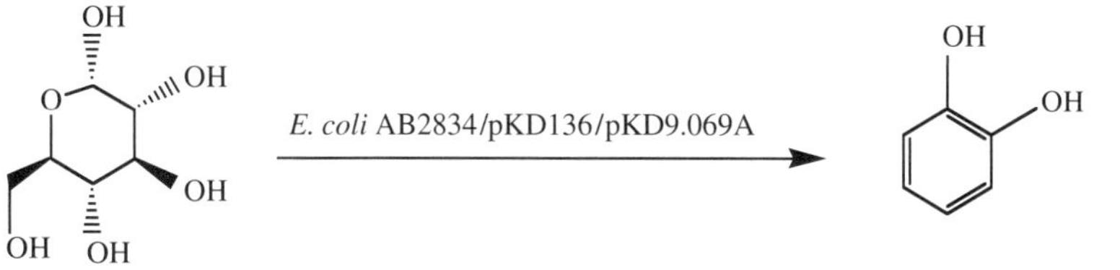

FIGURE 3.23. Biocatalysis for the synthesis of catechol from a renewable source.

The biocatalyzed reaction is a far better process than the classical one, as it replaces the hazardous starting chemical, benzene, with D-glucose and tremendously decreases the energy demands apart from replacing a nonrenewable feedstock with a renewable one.

In a similar effort, Ho and colleagues (1998) have succeeded in creating recombinant *Saccharomyces* yeast that can ferment glucose and xylose simultaneously to ethanol (see Fig. 3.24). Cellulose biomass (made of materials such as grasses, woody plants, etc.) was used as the feedstock.

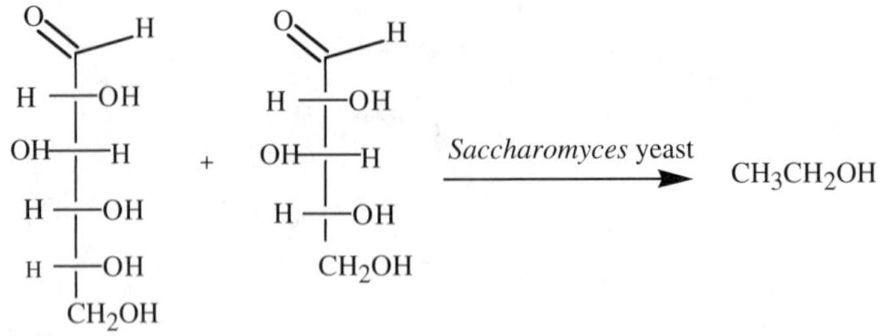

FIGURE 3.24. Recombinant yeast for fermentation of both glucose and xylose.

Conclusions

Catalysts play a significant role in green chemistry by decreasing energy requirements, increasing selectivity, and permitting the use of less hazardous reaction conditions. The central role these catalysts play in directing the course of a reaction, thereby minimizing or eliminating the formation of side products, cannot be disputed. Hence, catalysis—or rather, designed catalysis—is the mainstay of green chemical practices.

References

Altus Biologics, Inc., The Presidential Green Chemistry Challenge Awards program, summary of 1997 award entries and recipients, EPA 744-S-97-001, U.S. Environmental Protection Agency, Office of Pollution Prevention and Toxics, Washington, DC, p. 13, 1997.

Anastas, P. T. and Warner, J. C., *Green Chemistry: Theory & Practice*, Oxford University Press, New York, 1998.

Beller, M., Eckert, M., Vollimiiller, F., Bogdanovic, S., and Geissler, H., *Angew. Chem. Intl. Ed. Engl.*, **36**: 1494, 1997.

BHC Company, The Presidential Green Chemistry Challenge Awards program, summary of 1997 award entries and recipients, EPA 744-S-97-001, U.S Environmental Protection Agency, Office of Pollution Prevention and Toxics, Washington, DC, p. 2, 1997.

Blaser, H. U. and Spindler, F., *Topics in Catalysis*, **5**: 275, 1998.

Creghton, E. J., Ganeshie, S. N., Downing, R. S., and Van Bekkum, H., *J. Mol. Catal. A: Chemical*, **115**: 457, 1997.

Draths, K. M. and Frost, J. W., Chapter 9 in Anastas, P. T. and Williamson, T. C., eds., *Green Chemistry: Frontiers in Benign Chemical Synthesis and Processes*, Oxford University Press, New York, 1998a.

Draths, K. M. and Frost, J. W., The Presidential Green Chemistry Challenge Awards program, summary of 1998 award entries and recipients, EPA 744-S-98-001, U.S. Environmental Protection Agency, Office of Pollution Prevention and Toxics, Washington, DC, p. 3, 1998b.

Elings, J. A., Lempers, H. E. B., and Sheldon, R. A., *Stud. Surf. Sci. Catal.*, **105**: 1165, 1997.

Ho, N. W. Y., The Presidential Green Chemistry Challenge Awards program, summary of 1998 award entries and recipients, EPA 744-S-98-001, U.S. Environmental Protection Agency, Office of Pollution Prevention and Toxics, Washington, DC, p. 21, 1998.

Holderich, W. F., Roseler, J., Heitmann, G., and Liebens, A. T., *Catal. Today*, **37**: 353, 1997.

Kirner, A., *Chem. Tech.*, pp. 31–35, 1995.

Kurkeler, P. J., Vanderwal, J. C., Bremmer, J., Zuardirg, B. J., Downing, R. S., and Van Bekkum, H., *Catal. Lett.*, **53**: 135, 1998.

Mobil Technology Company, The Presidential Green Chemistry Challenge Awards program, summary of 1997 award entries and recipients, EPA 744-S-97-001, U.S. Environmental Protection Agency, Office of Pollution Prevention and Toxics, Washington, DC, p. 33, 1997.

Sheldon, R. A., *Chem. Tech.*, p. 38, 1994.

Simmons, M. S., Chapter 10 in Anastas, P. C. and Williamson, T. C., eds., *Green Chemistry; Designing Chemistry for the Environment*, American Chemical Society, Washington, DC, 1996.

Spivey, J. J. and Gogate, M. R., *Proc. 2nd Annual Green Chemistry & Engineering Conf.*, Washington, DC, 1998.

Stewart, J. D., The Presidential Green Chemistry Challenge Awards program, summary of 1998 award entries and recipients, EPA 744-S-98-001, U.S. Environmental Protection Agency, Office of Pollution Prevention and Toxics, Washington, DC, p. 12, 1998.

Trost, B. M., Chapter 6 in Anastas, P. C. and Williamson, T. C., eds., *Green Chemistry; Frontiers in Benign Chemical Synthesis and Processes.* Oxford University Press, New York, 1998.

CHAPTER 4

Biocatalysis: Green Chemistry

Introduction

The use of enzymes and microorganisms, which nature has developed, is undoubtedly an ideal choice toward "greening" chemical reactions. *Biotransformations* have been known since the early stages of human civilization and have been used since then to make fermented foods and beverages. However, biotransformations planted firm roots in industrial processes only when an industrial interest in steroid modifications emerged in the 1950s. The industrial use of enzymes and microorganisms in food processing and in the production of fine chemicals, drugs, and detergents has been undergoing a drastic development in the last decades (Stinson, 1999; Thayer, 2001). Stereospecific synthesis—the main stay of biotransformations—is of immense importance in the drug industry, with an estimated market of about US$100 billion worldwide (Stinson, 1999; Krishna, 2002). Moreover, the homochiral drug market is expected to grow even bigger, increasing from having 25% to accounting for 70% of the total fine chemical market during the 21st century (Schulze and Wubbolts, 1999). These numbers suggest the important role of biocatalysis to the chemical industry. In addition to stereospecific synthesis, biocatalysis is unique in that a drastic difference between the molecular weights of the catalyst and reaction products could become a basis for large-scale, cost-efficient separation of those compounds. Chemical reactions performed by microorganisms or catalyzed by enzymes are essentially the same as those carried out in

TABLE 4.1
General Characteristics of Enzymatic and Chemical Reactions

S. No.	Reaction Conditions and Characteristics	Enzymatic Reaction	Chemical Reaction
1	Reaction conditions		
	Temperature	Physiological	High
	Pressure	Physiological	High
2	Source of the reaction energy	Making and breaking of van der Waals bonds, hydrogen bonds, electrostatic interaction	Predominantly thermal
3	Solvent	Water	Predominantly organic solvents
4	Specificity		
	Substrate specificity	High	Low
	Stereospecificity	High	Medium to low
	Regiospecificity	High	Medium
5	Concentration of substrate and/or product	Low	High
6	Reaction under drastic conditions	Inert or deactivated	Good activity

conventional inorganic and organic chemistry. The most striking differences between enzymes and chemical catalysts are summarized in Table 4.1.

Advantages Within Industrial Applications

Microorganisms are undoubtedly the most superior enzyme sources among living organisms. They show high adaptability to new environments and high growth rates. These characteristics are especially useful for easy handling and large-scale cultivation without a high cost. At present, several new techniques such as extractive biocatalysis, immobilization, biocatalysis in organic solvents, and recombinant DNA technology for enzyme engineering are rapidly being developed in order to make biocatalysis industrially viable. Furthermore, protein engineering and cell technology, such as cell fusion, will become useful techniques for microbial transfor-

TABLE 4.2
Applications of Biocatalysis in Industry

Compounds Transformed/ Synthesized	Types of Transformations/Reactions	Important Reference
Steroids and sterols	Hydroxylation, dehydrogenation, side chain degradation, hydrolysis, peroxidation, reduction, isomerization, and conjugation	Maxon (1985)
Semisynthetic antibiotics	Synthesis of penicillins and cephalosporins (acylases), modification of carbapenem side chain	Kubo et al. (1984)
Organic acids	Hydration of fumaric acid, oxidation of alkenes, oxidation of isobutyric acid, reduction of 3-chloroacetic acid ester, aminolysis of histidine	Hasegawa (1982)
Sugars	Isomerization of glucose, hydrolysis of starch, hydrolysis of cellulose, hydrolysis of sucrose	Rehm and Reed (1984)
Peptides and proteins	Synthesis of plastin, semisynthesis of human insulin, synthesis of aspartame	Ichishima (1983)
Commodity chemicals	Synthesis of alkene oxides, ketones, aldehydes, pyrogallol, and amides	Harrison et al. (1980)
Nucleic acid-related compounds	*trans*-N-ribosylation, phosphorylation of nucleosides, synthesis of nucleotides, synthesis of co-enzymes, nucleosidylation of L-methionine and L-homocysteine	Kawaguchi et al. (1980)

mations in the near future. A large number of biologically and chemically useful compounds are prepared through microbial transformations. They are briefly summarized in Table 4.2.

Challenges to Make Biocatalysis Industrially Viable

Many of the unique features of the enzymatic reactions prove to be limitations for their commercial use. For example, high selectivity of the enzyme catalysis, which is one of the most important characteristics that make enzymes promising as catalysts in

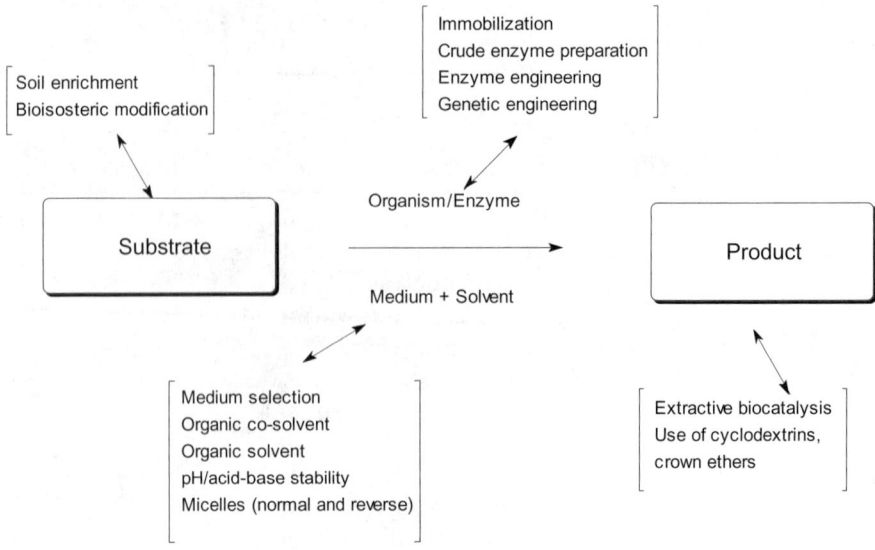

FIGURE 4.1. Approaches to overcome the limitations of biocatalysis.

synthetic chemistry, may also be a disadvantage, because the selectivity implies limitation. Many of these problems have been addressed by a large variety of approaches, all of which can be summarized, as shown both in Fig. 4.1 and text below.

- To address the limitation of substrate specificity leading to narrow range applicability—soil enrichment and bioisosteric modifications have been developed to broaden the range of substrates which could be catalyzed by an enzyme.
- To overcome the problem of product inhibition—extractive biocatalysis, use of supramolecular systems for forming inclusion complexes with the product, such as cyclodextrins and crown ethers, have been attempted.
- In order to maintain the enzyme activity in extreme conditions (conditions other than physiological conditions), also to recycle (cost-reduce)—immobilization and the use of crude enzyme preparation have been attempted. Enzyme engineering helps in designing the enzyme for a given transformation.
- To address the limitation of poor solubility of organic substrates in aqueous systems—medium selection (increase in sucrose

concentration), use of organic co-solvent or micelles and carrying out the catalysis in organic solvent have been attempted with considerable success.

Although the classification and arrangement of the various approaches to improvise and tune the biocatalytic processes for industrial use (as shown in Fig. 4.1) are not very precise, they do give a comprehensive account. The approaches adopted could overlap; for example, the soil enrichment technique could be used to enhance the range of substrates that the enzyme could act on, while the same technique could be adopted for enzyme engineering to develop newer modified enzymes (that could function under nonaqueous conditions). Numerous reviews on conventional approaches, such as immobilization techniques, genetic engineering, and extractive biocatalysis, have appeared in the literature at regular intervals. Here we shall briefly discuss nonconventional and relatively recent approaches.

Soil Enrichment

Enrichment culture techniques have long been used in the field of microbiology and are responsible for the development of enzymes used in laundry detergents, new antibiotics, bacteria that can withstand extreme heat or extreme cold, and many other advancements. For example, enrichment culture techniques were used to isolate microorganisms capable of growing at temperatures as high as 113°C. The subsequent isolation of DNA polymerases that can function at high temperatures has revolutionized the biotechnology industry.

Bioremediation is often the most cost-effective means of cleaning up contaminated soil and water. However, bioremediation may not always be viewed as an appropriate treatment option due to the chemical nature of the contaminant or a mixture of contaminants present at a site. Yet microbial biodiversity is so immense that it is usually possible to either isolate from nature, or evolve in the laboratory, a microbial culture capable of treating almost any type or mixture of environmental contaminants.

Applications for Enrichment Cultures

One example of the use of enrichment culture techniques was the isolation of a microbial culture that could degrade a chemical warfare agent. The chemical agent mustard gas—or HD

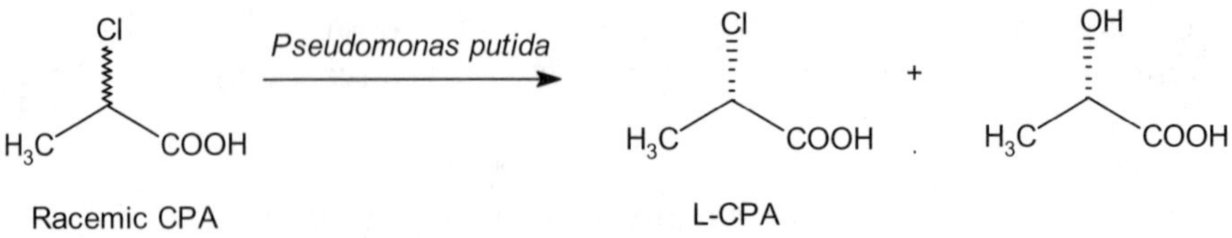

FIGURE 4.2. Resolution of CPA by soil-enriched *Pseudomonas putida*.

(2,2'-dichlorodiethyl sulfide)—is one of the 10 deadliest and most toxic chemicals known. Still, researchers were able to use enrichment culture techniques to isolate a bacterial culture that cannot only survive exposure to this deadly compound and its derivatives but can also use the chemical as a food source for growth. Besides their role in degrading unwanted chemicals and pollutants, enrichment culture techniques can also be used to isolate microbial cultures that possess biochemical pathways that are useful for making chemicals by biocatalysis. For example, consider the conversion of racemic 2-chloropropanoic acid (CPA) to L-CPA, by the dehalogenase from *Pseudomonas putida*—the necessary strain (AJ1) was isolated from the environment with high-chlorine-containing compounds—the road tanker off-loading point (see Fig. 4.2).

Enrichment culture techniques can also be used for bioremediation to detoxify xenobiotic pollutants such as polycyclic aromatic hydrocarbons (PAHs), heterocyclic polyaromatics, and halogenated aromatics in soils and sediments through microbial degradation. An effective way to do this is by isolating microbes through enrichment cultures with the substrate one wants to detoxify as a limiting compound. Once this is proven in the laboratory, it can be taken full-scale to the field.

Other applications include

- the search for life in extreme environments,
- isolation of bacterial cultures with novel biochemical abilities,
- isolation of microorganisms that produce novel antibiotics,
- isolation of cultures that are representative of previously uncultivated phylogenetic groups.

Approach

Enrichment culture techniques rely on creating a condition in which the survival and growth of bacterial cultures, with whatever traits are desired, are favored. The nutritional composition of the microbial growth media can be adjusted so that an environmental contaminant serves as the only available source of food and energy or the growth conditions favor the growth of only those bacteria that can grow at a certain temperature or in the presence of other chemicals. In these ways, the conditions can be controlled in the laboratory to allow for the selection of those bacteria that can provide solutions to various problems.

In addition to selecting naturally occurring microbial cultures that possess a desired metabolic trait, it is also possible to use enrichment culture techniques to develop microbial cultures with unique biochemical traits. *The substrate range of enzymes catalyzing a certain reaction can be expanded through the use of enrichment culture techniques.*

This process of evolving new biochemical traits in the laboratory can also be accelerated by the use of *directed evolution*. In directed evolution, the genes that encode a biochemical trait of interest are subjected to specific mutagenesis, and then enrichment culture techniques are used to isolate derivatives that contain the desired improvements.

Bioisosteric Modifications

In general, even though it represents a subtle structural change, the *isosteric* replacement results in a modified profile, and some properties of the parent molecule will remain unaltered, while others will be changed. The similar shape and polarity within a series of substrates of different reactivity (*bioisosteres*) eliminate effects due to differences between enzyme-substrate binding (ES), which is *hence a good method of extending the range of substrates that can be chosen for the transformation.*

A number of instances can be cited from the literature wherein the *isosteres* had similar transformations. Bacterial dioxygenase-catalyzed *cis*-dihydroxylation of the tetracyclic arene benzo[c]phenanthrene was found to occur exclusively at fjord region (cavity region) bonds. The isosteric compounds benzo[b]naphthol [1,2-d]furan and benzo[b]naphthol[1,2-d]thiophene were also similarly *cis*-dihydroxylated at the fjord region bonds by bacterial dioxygenases (Boyd et al., 2001) (see Fig. 4.3). The isosteres 1,2-dihydronaphthalene, 2,3-dihydrobenzothiophene, and 2,3-dihydro-benzofuran gave similar corresponding diol products on incubation with *Pseudomonas putida* UV4. Microbes that possess the metabolic pathways to metabolize benzene, when substituted by

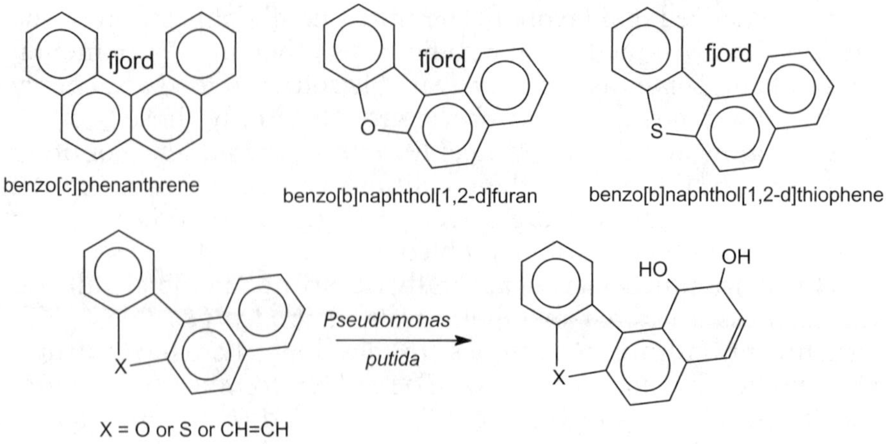
fjord
benzo[c]phenanthrene
fjord
O
benzo[b]naphthol[1,2-d]furan
fjord
S
benzo[b]naphthol[1,2-d]thiophene
X
X = O or S or CH=CH
Pseudomonas
putida
HO
OH
X

FIGURE 4.3. Hydroxylation of isosteres.

benzenes and phenols, were found to metabolize fluorinated benzenes (isosteres) in a similar manner.

Homogenous Biocatalysis in Water–Organic Solvent Mixtures and in Organic Solvents

Biocatalysis in organic solvents has unique advantages compared to traditional aqueous enzymology/fermentation. Often times in nonaqueous media enzymes exhibit properties drastically different from those displayed in aqueous buffers. These novel properties are given in Table 4.3. In addition to those mentioned in Table 4.3, the solubility of hydrophobic substrates and/or products increases in organic solvents, which diminishes diffusional barriers for bioconversions, and thus speeds up the reactions and improves the potential for direct applications in industrial chemical processes. Once organic solvent becomes a reaction medium, there cannot be contamination, which thus precludes release of proteolytic enzymes by microbes and favors the direct application of the process in an industrial setting. Most proteins (enzymes) inherently function in an aqueous environment, and hence their behavior in nonaqueous solvents is completely different due to the loss in the three-dimensional structure. Thus, only polar solvents

TABLE 4.3
Effects of Organic Co-Solvent in Enzyme-Catalyzed Synthesis

S. No.	Effects
1	Enhances the reaction rates. In many cases maximal rate of the reaction in water–organic mixture is higher than the rate of the same reaction in aqueous buffers (Khmelnitsky et al., 1991).
2	Changes the reaction pathway by promoting change in substrate cleavage and product synthesis (Pal and Gertler, 1983; Blankeney and Stone, 1985).
3	There are cases when stability of enzymes drastically improved in water–organic solvent mixtures as compared to aqueous media (Guagliardi et al., 1989).
4	Shift in the direction of the biocatalyzed reaction (Deschrevel et al., 2002; Zhang et al., 2003).
5	When water is one of the products of reaction (esterification), water–organic solvents enhance the rate and yield of the product (ester) (Plou et al., 2002).
6	Modification of enantioselectivity of the biocatalyzed conversion (Watanabe and Ueji, 2002).

capable of H-bond formation with the protein, such as dimethyl formamide, dimethyl sulphoxide, acetonitrile, pyridine, glycerol, and methanol, were found suitable.

Homogenous biocatalysis in organic solvents requires the solubility of enzymes in nonaqueous media. Since proteins inherently function in aqueous environments, initial efforts were to study biocatalysis in water–organic mixtures. Biocatalysis in nonaqueous systems using water–miscible organic solvents was studied in detail and has been reviewed previously (Butler, 1979; Blinkovsky et al., 1992). In general, enzyme activity in a homogenous mixture of water–organic solvent is extremely sensitive to the nature and amount of organic solvent (Budde and Khmelnitsky, 1999). It is interesting to note that in many cases the maximal rate of the reaction in a water–organic mixture is higher than the rate of the same reaction in aqueous buffers (Khmelnitsky et al., 1991). The most obvious reason for shifting to water–organic mixtures as a reaction medium is to enable bioconversion of substrates poorly soluble in water. Versari et al. (2002) exploited this approach by using an engineered papain to convert/hydrolyze organonitriles, which are a very toxic family of compounds of industrial interest (e.g., nitrile herbicides, acrylonitrile) that are most poorly soluble in water and aqueous buffers as well as resistant to chemical hydrolysis. Application of water–organic mixtures often enables a shift in the direction of a biocatalyzed reaction due to a decrease in the content of water, a reaction substrate. For example, synthesis of dipeptides using chymotrypsin and procine pancreatic lipase present good examples of reverse reaction becoming predominant while moving from aqueous media to water–organic mixtures (Deschrevel et al., 2002; Zhang et al., 2003). The authors reported an increase in the dipeptide concentration in reaction medium concurrent with the decrease in the water content. It is worth mentioning that bioconversions in water–organic solvent mixtures are not limited to monomeric enzymes. Aspartate transcarbamylase (ATCase) from *E. coli* and malate NADP oxireductase from the extremophile *Sulfolobus solfataricus* are excellent examples of allosteric enzymes active in water–organic solvent mixtures (Dreyfus et al., 1984; Guagliardi et al., 1989).

In recent years, the development of ionic liquids, which are composed of organic molecules derived from 1-alkyl-3-methylimidazolium cation, such as [BMIm][BF$_4$], [MMIm][MeSO$_4$], and [EtPy][CF$_3$COO], has attracted much attention to nonconventional biocatalysis (Husum et al., 2001; Kaftzik et al., 2002; Zhao and Malahortra, 2002) (see Fig. 4.4). Studies using miscible water–ionic liquid mixtures gave controversial results, thus suggesting significant dependence of the solvent effect on the properties of the enzymes. Ionic liquids in mixtures with water display a potential to modify properties of biocatalyst. For instance, enzymatic resolution of N-acetyl amino acids using subtilisin Carlsberg in 15% N-ethyl pyridinium trifluroacetate in water was higher compared to acetonitrile under the same experimental conditions (Zhao and

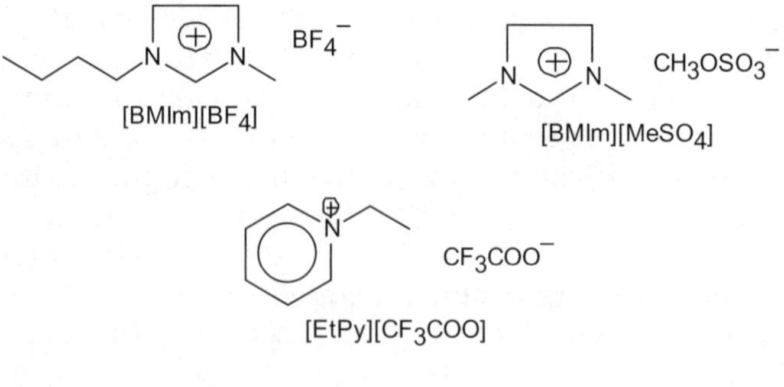

BF_4^-

[BMIm][BF$_4$]

$CH_3OSO_3^-$

[BMIm][MeSO$_4$]

CF_3COO^-

[EtPy][CF$_3$COO]

FIGURE 4.4. Common ionic liquids.

Malhorta, 2002). The development of new, more "enzyme-friendly," ionic liquids should open new possibilities by tailoring the systems to optimize the performance of biocatalysts in non-aqueous environments.

The chemical modification of enzymes to solubilize them in completely nonaqueous, nonpolar media has been attempted with some success. All methods of chemical modification of enzymes share the same principle: the attachment of amphipatic reagents to the polar groups on the exterior of the native protein to stabilize the enzyme against denaturation by hydrophobic organic solvents. Depending on the nature of the interaction between the enzyme and the solubilizing agent, chemical modifications of enzymes can be classified into covalent and noncovalent ones. Various methods of enzyme modification adopted are summarized in Table 4.4.

Conclusions

Biocatalysis in nonaqueous systems has proven itself as a powerful tool. HIP allows significant improvement in enzyme activity in nonaqueous systems. A combination of directed evolution and rational enzyme design is likely to result in many more exciting developments in the near future.

Use of Cyclodextrins

The use of enzymes as valuable catalysts in organic solvents has been well documented. However, some of their features limit their application in organic synthesis, especially the frequently lower-enzyme activity under nonaqueous conditions, which constitutes a major drawback in the application of enzymes in organic solvents. In addition, many enzymatic reactions are subject to substrate or product inhibition, leading to a decrease in the reaction rate and enantioselectivity. To overcome these drawbacks and to make enzymes more appealing to synthesis, cyclodextrins are used. The effects of the cyclodextrins range from increasing the availability of insoluble substrates to reducing substrate inhibition to limiting product inhibition. In each case, the effects of the cyclodextrins are interpreted in terms of the formation of inclusion complexes. It is thus demonstrated that cyclodextrins can be used rationally to increase the utility of enzymes in organic synthesis. In an interesting study, cyclodextrins were used as regulators for the *Pseudomonas cepacia* lipase (PSL) and macrocyclic additives to enhance the reaction rate and enantioselectivity E in

TABLE 4.4
Chemical Modification of Enzymes

Modification Technique	Modifying Reagents	Examples	Remarks/Comments	References
Noncovalent modification	Amphipathic polymers	Polyethylene glycols (PEGs)	PEG-coated enzymes generally exhibit much less catalytic activity than unmodified ones do.	Secundo et al. (1999)
	Polyelectrolytes	Polybrene	This maintains activity, but is not suited for pure organic solvents.	Izumrudov et al., 1984; Bruke et al. (1993)
	Synthetic lipids	Didodecyl N-D-gluconyl-D-glutamate	Very helpful in resolution of racemic compounds with high enantioselectivity.	Okahata and Mori (1997)
	Ionic surfactants (hydrophobic ion pairing, HIP)	Sodium dodecyl sulfate, Sodium bis(2-ethylhexyl) sulfosuccinate	HIP results in highly active and stable preparations.	Pardkar and Dordick (1994)
Covalent modification	Amphipatic reagents	Polyethylene glycols (pegylation), polysaccharides	Covalent modification enables easy separation and recycling of biocatalysts.	De Santis and Jones (1999); Kobayashi and Takatsu (1994)
Reverse micelles	Surfactants	Sodium bis(2-ethylhexyl) sulfosuccinate, cetyltrimethyl ammonium bromide	Activity depends on water content and organic solvent. Activity range from zero to several-fold higher is observed.	Martinek et al. (1981), Luisi et al. (1988)
Directed evolution	Mutations in enzymes	Molecular biology techniques	Involves techniques of molecular biology, and success in this area depends on screening methods.	Arnold (2001), Jaeger et al. (2001)

lipase-catalyzed enantioselective transesterification of 1-(2-furyl) ethanol in organic solvents (Ghanem, 2003). Both the reaction rate and the enantioselectivity were significantly enhanced by several orders of magnitude when a co-lyophilized lipase was used in the presence of cyclodextrins. The observed enhancements were tentatively interpreted in terms of their ability to give certain flexibility to the enzyme and to form a host–guest complex, thus avoiding product inhibition and leading to enhancement of both reaction rate and enantioselectivity. An innovative method of forming an inclusion complex with the product was reported by Easton et al. (1995), in which the rate of conversion of (S)-phenylalanine to *trans*-cinnamate catalyzed by PAL is enhanced by the addition of β-cyclodextrin.

Use of Crown Ethers

Today it is well established that enzymes can be catalytically active in organic solvents. Compared to aqueous solutions, the use of an organic reaction medium can have some interesting advantages, such as the enhanced thermal stability of the enzyme, the easy separation of the suspended enzyme from the reaction medium, the increased solubility of substrates, the favorable equilibrium shift to synthesis over hydrolysis, the suppression of water-dependent side reactions, and possibly the new (stereo-) selectivity properties of the enzyme. An important drawback of the use of organic solvents for enzyme reactions is that the activity of the enzyme is generally several orders of magnitude lower than in aqueous solution. Prior lyophilization of the enzyme from an aqueous solution, which is buffered at the pH of optimal aqueous enzyme activity, if necessary in the presence of an inhibitor, improves the activity in organic solvent. It is proposed that both the optimal pH and the inhibitor contribute to the fixation of the enzyme in a catalytically active conformation during lyophilization (Zaks and Klibanov, 1988). As discussed earlier, several strategies have been developed to enhance the enzyme activity in organic solvents, like protein engineering, chemical modification of the enzymes, immobilization, and addition of "water-mimicking" compounds like formamide, glycol, or DMF.

In recent years the effects of crown ethers on enzyme reactions in organic solvents have been investigated. Depending on their ring size and structure, crown ethers can form complexes with metal ions, ammonium groups, guanidinium groups, and water, species that are all common in enzymatic reactions.

Moreover, the strongest interactions of crown ethers with these guest species occur in organic solvent. Therefore, these results show that pretreatment of enzymes by lyophilization with crown ethers or by simply adding 18-C-6 to the organic solution can enhance the enzyme activity to a level where they are suitable for practical applications. Moreover, it was reported that for relatively reactive substrates the enantioselectivity of proteases in organic solvent is very sensitive for small changes in solvent composition. This offers the possibility to tune the enantioselectivity and to apply these enzymes as catalysts for conversions of both the L- and the D-enantiomers (Johan et al., 1996). It was also reported that co-lyophilization or co-drying of subtilisin Carlsberg with the crown ethers 18-crown-6, 15-crown-5, and 12-crown-4 substantially improved enzyme activity in THF, acetonitrile, and 1,4-dioxane in the transesterification reactions of N-acetyl-L-phenylalanine ethylester and 1-propanol and that of (+/−)-1-phenylethanol and vinylbutyrate (Santos et al., 2001). The acceleration of the initial rate, $V(0)$, ranged from less than 10-fold to more than 100-fold. All crown ethers activated subtilisin substantially, which excludes a specific macrocyclic effect from being responsible. This suggests that molecular imprinting is likely the primary cause of subtilisin activation by crown ethers, as recently suggested (Santos et al., 2001). Similarly, lipases from *Candida rugosa* (CRL) and *Pseudomonas cepacia* (PCL) were co-lyophilized with cyclic oligoethers, including four crown ethers and nine cyclodextrins (CyDs), and their transesterification activity and enantioselectivity in organic solvents were evaluated. The PCL co-lyophilized with each additive showed simultaneous enhanced enzyme activity and enantioselectivity when compared to the native lipase lyophilized from buffer alone; in contrast, such enhancement was not observed for the co-lyophilized CRL. The initial rate determined for the transesterification between racemic 2,2-dimethyl-1,3-dioxane-4-methanol and vinyl butyrate in diisopropyl ether at 30°C increased by up to 17-fold and the enantioselectivity represented by E could be doubled. Hence, this method seems to be of practical use for the large-scale production of optically active compounds (Mine et al., 2003).

Process Design

Now that the various methodologies to overcome the limitations of biocatalysis have been discussed, a brief account of process design will give the necessary information to make these processes industrially viable. Three categories of questions need to be answered for any process design:

1. How much material do you need to supply to customer—100 g, 1 kg, 10 kg? How much product can you obtain per unit of reaction volume (substrate/product solubility)?
2. How long does one batch take to run in process per unit of reaction volume? How long do you have to make your delivery?
3. How large is the reaction vessel? How long can the vessel be practically and safely operated?

The answers to these questions determine the nature of the process and the configuration of the final design. The basic strategy should be as follows, for both purified enzymes and whole cells:

- Select a catalyst.
- Establish a baseline process.
- Optimize the process.
- Determine how to run the process to meet a target delivery.

Details of these four steps are given in Table 4.5.

Future Trends

Organic synthesis—chemistry as it is done in the laboratory or manufacturing plant—traditionally uses a step-by-step approach. In a typical sequence, a starting material A is converted into a final product D, and intermediate products B and C have to be isolated and purified in each conversion step.

Such multistep organic syntheses, which are still quite common in today's fine chemical industry, suffer from several disadvantages. They are often carried out noncatalytically using relatively large amounts of reagents that produce many kilograms of waste per kg of final product. The separation and purification steps needed after each conversion step produce waste heat as energy is consumed. They require extra energy to overcome the thermodynamic hurdles to produce and isolate intermediates B and C if they lie in high-energy states.

On the other hand, biosynthesis—chemistry as nature performs it in the cells of organisms—goes through a multistep cascade to convert starting material A to final product D without separation of intermediates B and C.

TABLE 4.5
Steps to Determine Process Design

Enzyme Process	Whole-Cell Process
1. Select a catalyst (screening):	
Choose 10 to 20 enzymes, based on commercial availability.	Choose 100 to 200 strains from pre-existing library, more from environmental library.
Set up identical reactions and keep enzyme loading (mass) constant.	Add same amount of substrate and then normalize to cell mass.
Choose enzyme with good conversion and high selectivity.	Choose cell with good specific activity, high selectivity, and (if possible) low byproducts.
2. Establish *baseline* process:	
Optimize pH, temperature, enzyme modifications necessary, use of cyclodextrins, crown ethers, etc.	Optimize *growth* phase (pH, temperature, medium) for target activity.
	Optimize *conversion* phase (pH, temperature) and use of cyclodextrins, crown ethers, etc.
Determine amount of enzyme needed to achieve target conversion in target time. Take initial read on kinetics.	Determine amount of time needed to achieve target conversion with fixed catalyst (max cell density). Take initial read on kinetics.

3. Optimize process:
Address challenges (e.g., low solubility, substrate/product inhibition, yield limits).
Determine operational limits (max achievable substrate charge and product titer).

4. Design for delivery:
Address questions of "how much," "how long," and "how large."
Determine reactor parameters, such as batch reactor time to achieve target conversion of fixed substrate amount (delivered as one charge or fed) for given vessel size.
Determine total processing time required for delivery.

Address challenges (e.g., low solubility, substrate/product inhibition, yield limits).
Determine operational limits (max achievable substrate charge and product titer).

Address questions of "how much," "how long," and "how large."
Determine reactor parameters, such as batch reactor time to achieve target conversion of fixed substrate amount (delivered as one charge or fed) for given vessel size.
Determine total processing time required for delivery.

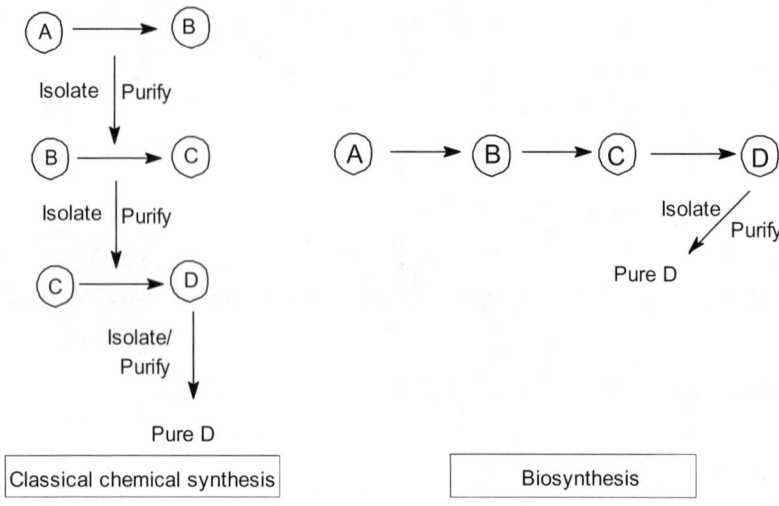

FIGURE 4.5. Comparison of classical chemical synthesis versus biosynthesis.

   Such multistep, combined syntheses are common in everyday life. They are carried out in a fully catalytic way by using enzymes with relatively limited amounts of reagents (cofactors) and thus produce much less waste. The mutual compatibility and high selectivity of the enzymatic conversions make it possible to proceed without intermediate recovery steps. They save energy by avoiding the separation and isolation of intermediates *B* and *C*. Figure 4.5 compares classical chemical synthesis with biosynthesis.
   For catalysis of the next generation of organic synthesis, the challenges are to

- combine the power of chemical, enzymatic, and microbial conversions,
- overcome problems associated with the disparities in the sizes of catalytic species,
- search for multistep conversions that do not require recovery steps—one-pot multicatalytic procedures—such as those nature uses,
- fine-tune reaction conditions and catalytic systems to permit the desired concerted reactions without intermediate isolation or purification steps.

The result of these efforts should drastically diminish the costs of energy consumption and waste treatment characteristic of current multistep processes in the fine chemical industry.

A Multienzyme One-Pot Example

An elegant, multistep, one-pot approach recently developed (Schoevaart et al., 1999) is the four-enzyme–catalyzed, four-step, one-pot conversion of glycerol into a D-heptose sugar, in which a pH switch method is applied to temporarily turn off the phytase enzyme during the second and third steps of the concerted synthesis (see Fig. 4.6).

The four consecutive enzymatic conversion steps in one reactor without any separation of intermediates consist of

1. Phosphorylation. Glycerol is phosphorylated with pyrophosphate by phytase at pH 4.0 at 37°C. Racemic glycerol 3-phosphate is obtained in 100% yield (based on pyrophosphate) in 95% glycerol after 24 h.
2. Oxidation. By raising the pH to 7.5, phytase activity is "switched off" and hydrolysis is prevented. Oxidation of 1-glycerol

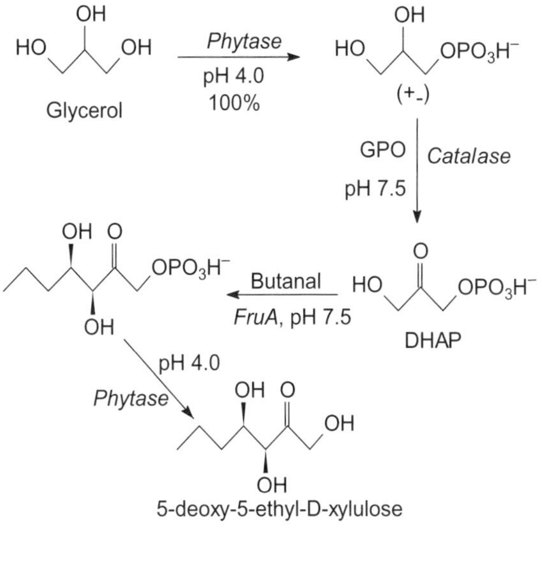

FIGURE 4.6. Four-step, one-pot synthesis of carbohydrates from glycerol.

3-phosphate to dihydroxyacetone phosphate (DHAP) by glyc-erol-L-phosphate oxidase (GPO) at 55 vol% glycerol is quantita-tive. Catalase is added to suppress the buildup of hydrogen peroxide. The D-isomer is converted back to glycerol and phos-phate in the last step.
3. Aldol reaction. More than 20 aldehydes are known to be sub-strates for the aldolases from *Staphylococcus carnosus* and *S. aureus*. Stereo-selectivity of the aldolases must be looked at for each acceptor substrate, because isomers are formed in dif-ferent proportions. The oxidation and aldol reaction can be carried out simultaneously.
4. Dephosphorylation. Lowering the pH back to 4 "switches on" phytase's activity, and hydrolysis of the aldol adduct is initiated.

Combined with the broad substrate specificity of DHAP aldo-lases, this sequence constitutes a simple procedure for the synthe-sis of a wide variety of carbohydrates from readily available glycerol and pyrophosphate.

A Synthesis Renaissance

In conclusion, we are at the threshold of a renaissance in synthesis methods by integration of bio- and organic syntheses for fine chemicals by one-pot, multistep catalytic procedures and by meta-bolically engineering the microbes that perform the catalysis (Heijnen et al., 2000). In this respect, future clean synthesis methods should be inspired by the achievements in the field of modern detergent formulations that have up to six different enzymes (McCoy, 2001): an advanced multicatalytic one-pot con-version of "dirty laundry" to "clean laundry" plus "dirt," with the washing machine as the in-house catalytic reactor that simul-taneously separates the product (clean laundry) from waste (dirt). Economic and environmental goals will cause chemical and bio-technological conversion methods to merge into integrated biology–chemistry routes based on optimum feedstock. Conserva-tion of matter and energy from starting material to end product is required to achieve sustainable conversion processes. Neither chemistry nor biotechnology, neither fossil fuels nor renewable feedstock, will be the ultimate winner—only the combination of these disciplines and resources.

References

Arnold, F. H., Combinatorial and computational challenges for biocatalyst design, *Nature*, **409**: 253–257, 2001.

Blankeney, A. B. and Stone, B. A., Activity and action pattern of *Bacillus licheiformis α-amylase* in aqueous ethanol, *FEBS*, **186**: 229–232, 1985.

Blinkovsky, A. M., Martin, B. D., and Dordick, J. S., Enzymology in monophasic organic media, *Curr. Opin. Biotech.*, **3**: 124–129, 1992.

Boyd, D. R., Sharma, N. D., Harrison, J. S., Kennedy, M. A., Allen, C. C. R., and Gibson, D. T., Regio- and stereo-selective dioxygenase-catalysed cis-dihydroxylation of fjord-region polyclic arenas, *J. Chem. Soc., Perkin Trans. I*, 1264–1269, 2001.

Bruke, C. J., Volkin, D. B., Mach, H., and Middaugh, C. R., Effect of polyanions on the unfolding of acidic fibroblast growth factor, *Biochemistry*, **32**: 6419–6426, 1993.

Budde, C. L. and Khmelnitsky, Y. L., Aldolase stability in the presence of organic solvents, *Biotechnol. Lett.*, **21**: 77–80, 1999.

Butler, L. G., Enzymes in non-aqueous solvents, *Enzyme Microb. Tech.*, **1**: 253–259, 1979.

De Santis, G. and Jones, J. B., Chemical modification of enzymes for enhanced functionality, *Curr. Opin. Biotech.*, **10**: 324–330, 1999.

Deschrevel, B., Vincent, J. C., Ripoll, C., and Thellier, M., Thermodynamic parameters monitoring the equilibrium shift of enzyme-catalyzed hydrolysis/synthesis reactions in favor of synthesis in mixtures of water and organic solvent, *Bioeng. Biotech.*, **81**: 167–177, 2002.

Dreyfus, M., Fires, J., Tauc, P., and Herve, G., Solvent effects on allosteric equilibria: Stabilization of T and R conformations of *Escherichia coli* aspartate transcarbamylase by organic solvents, *Biochem.*, **23**: 4852–4859, 1984.

DSM Magazine, **147**: 18–21, 1998.

Easton, C. J., Harper, J. B., and Lincoln, S. F., *J. Chem. Soc., Perkin. Trans.*, **1**: 2525, 1995.

Ghanem, A., The utility of cyclodextrins in lipase-catalyzed transesterification in organic solvents: Enhanced reaction rate and enantioselectivity, *Org. Biomol. Chem.*, **1**(8): 1282–1291, 2003.

Glow, F. J., Cruces, M. A., Ferrer, M., Fuentes, G., Pastor, E., Bernabe, M., Christensen, M., Comelles, F., Parra, J. L., and Ballesteros, A., *J. Biotechnol.*, **96**: 55–66, 2002.

Guagliardi, A., Manco, G., Rossi, M., and Bartolucci, S., Stability and activity of a thermostable malic enzyme in denaturants and water-miscible organic solvents, *Eur. J. Biochem.*, **183**: 25–30, 1989.

Hasegawa, J., Ogura, M., Kanema, H., Noda, N., Kawahrada, H., and Watanabe, K., *J. Ferment. Tech.*, **60**: 501, 1982.

Heijnen, J. J., Haasnoot, C. A. G., Bruggink, A., Bovenberg, R. A. L., Meijer, E. W., Feringa, B. L., Driessen, A., Integration of biosynthesis and organic synthesis, a new future for synthesis; Programme proposal of the National Scientific Foundation, 2000.

Ichishima, E. (ed.), *Proteases*, Gakkai Shuppan Center, Tokyo, p. 237, 1983.

Izumrudov, V. A., Zezin, A. V., and Kabanov, V. A., Monomolecular exchange in solutions of complexes of globular proteins with artificial polyelectrolytes, *Russian Proc. Natl. Acad. Sci.*, **275**: 1120–1123, 1984.

Jaeger, K. E., Eggert, T., Eipper, A., and Reetz, M. T., Directed evolution and creation of enantioselective biocatalysts, *Appl. Microbiol. Biotech.*, **55**: 519–530, 2001.

Johan, F. J., Engbersen, J. B., Verboom, W., and Rcinhoudt, D. N., Effects of crown ethers and small amounts of cosolvent on the activity and enantioselectivity of a-chymotrypsin in organic solvents, *Pure & Appl. Chem.*, **68**: 2171–2178, 1996.

Kaftzik, N., Wassercheid, P., and Kragl, U., Use of ionic liquids to increase the yield and enzyme stability in the beta-galactosidase catalyzed synthesis of N-acetylactosamine, *Org. Proc. Res. Dev.*, **6**: 553–557, 2002.

Kawaguchi, K., Kawai, H., and Tochikura, T., *Methods Carbohydr. Chem.*, **8**: 261, 1980.

Khmelnitsky, Y. L., Mozhaev, V. V., Belova, A. B., Sergeeva, M. V., and Martinek, K., Denaturation capacity: A new quantitative criterion for the selection of organic solvents as reaction media in biocatalysis, *Eur. J. Biochem.*, **198**: 31–41, 1991.

Kobayashi, M. and Takatsu, K., Cross-linked stabilization of trypsin with dextran-dialdehyde, *Biosci. Biotech. Biochem.*, **58**: 275–278, 1994.

Krishna, S. H., Developments and trends in enzyme catalysis in nonconventional media, *Biotech. Adv.*, **20**: 239–267, 2002.

Luisi, P. L., Grommi, M., Pilen, M. P., and Robinson, B. H., Reverse micelles as hosts for proteins and small molecules, *Biochem. Biophys. Acta.*, **947**: 209–246, 1988.

Martinek, K., Levashov, A. V., Klyachko, N. L., Pantin, V. I., and Berezin, I. V., The principles of enzyme stabilization, *Biochim. Biophys. Acta.*, **657**: 277–294, 1981.

Maxon, W. D., *Ann. Rep. Ferment. Processes*, **8**: 171, 1985.

Mine, Y., Fukunaga, K., Itoh, K., Yoshimoto, M., Nakao, K., and Sugimura, Y., Enhanced enzyme activity and enantioselectivity of lipases in organic solvents by crown ethers and cyclodextrins, *J. Biosci. Bioeng.*, **95**(5): 441–447, 2003.

Okahata, Y. and Mori, T., Lipid-coated enzymes as efficient catalysts in organic media, *TIBECH*, **15**: 50–54, 1997.

Pal, P. K. and Gertler, M. M., The catalytic activity and physical properties of bovine thrombin in presence of dimethyl sulfoxide, *Thromb. Res.*, **29**: 175–185, 1983.

Pardkar, V. M. and Dordick, J. S., Activity of α-chymotrypsin dissolved in nearly anhydrous organic solvents, *J. Am. Chem. Soc.*, **116**: 5009–5010, 1994.

Plou, F. J., Cruces, M. A., Ferrer, M., Fuentes, B., Pastor, E., Bernabé, M., Christensen, M., Comelles, F., Parra, J. L., and Ballesteros, A., *J. Biotechnol.*, **96**: 55–66, 2002.

Santos, A. M., Vidal, M., Pacheco, Y., Frontera, J., Baez, C., Ornellas, O., Barletta, G., and Griebenow, K., Effect of crown ethers on structure, stability, activity, and enantioselectivity of subtilisin Carlsberg in organic solvents, *Biotech. Bioeng.*, **74**(4): 295–308, 2001.

Schoevaart, R., van Rantwijk, F., and Sheldon, R. A., *Tetrahedron: Asymmetry*, **10**: 705–711, 1999.

Schulze, B. and Wubbolts, M. A., Biocatalysis for industrial production of fine chemicals, *Curr. Opin. Biotech.*, **10**: 609–615, 1999.

Secundo, F., Carrea, G., Vecchio, G., and Ziambiainachi, F., Spectroscopic investigation of lipase from *Pseudomonas cepacia* solubilized in 1,4-dioxane by non-covalent complexation with metoxypoly(ethylene glycol), *Biotech. Bioeng.*, **64**: 624–629, 1999.

Stinson, S. C., Chiral drugs interactions, *Chem. Eng. News*, **77**: 101–120, 1999.

Thayer, A. M., Biocatalysis, *Chem. Eng. News*, **79**: 27–34, 2001.

Versari, A., Menard, R., and Lortie, R., Enzymatic hydrolysis of nitrides by an engineered nitrile hydratase (papain Gln19Glu) in aqueous-organic media, *Biotech. Bioeng.*, **79**: 9–14, 2002.

Watanabe, K. and Ueji, S., Dimethyl sulfoxide induced high enantioselectivity of subtilisin Carlsberg for hydrolysis of ethyl 2-(4-substituted phenoxy) propionates, *Biotech. Lett.*, **22**: 599–603, 2002.

Zaks, A. and Klibanov, A. M., *J. Biol. Chem.*, **263**: 3194, 1988.

Zhang, L. Q., Xu, L., Yang, X. C., Wu, X. X., and Zhang, X. Z., Lipase-catalyzed synthesis of precursor dipeptides of RGD in aqueous water-miscible organic solvents, *Prep. Biochem. Biotech.*, **33**: 1–12, 2003.

Zhao, H. and Malahorta, S. V., Enzymatic resolution of amino acid esters using ionic liquid N-ethyl pyridinium trifluoroacetate, *Biotech. Lett.*, **24**: 1257–1260, 2002.

CHAPTER 5

Alternate Solvents

Chemical reactions are diverse and are known to occur at a wide range of temperature and pressure conditions. In most reactions, the reaction vessel provides the following three components (see Fig. 5.1):

- solvent,
- reagent/catalysts,
- energy input.

Since chemical systems are irreversible, nonequilibrium systems, they can be labeled "chaotic." Hence, they may be viewed as macroscopic analogues to the quantum uncertainties. Just as in a chaotic system, every component—minor or major—affects the outcome of the change or transformation, every component affects the outcome of the reaction.

Developing more benign synthetic procedures in chemical synthesis is important in moving toward sustainable technologies, as part of the rapidly emerging field of green chemistry. In reducing the amount of waste, the energy usage, and the use of volatile, toxic, and flammable solvents, several approaches are available, including avoiding the use of organic solvents for the reaction media. At the heart of green chemistry are alternative reaction media. They are the basis of many of the cleaner chemical technologies that have reached commercial development. Most well-known among these alternate reaction media being

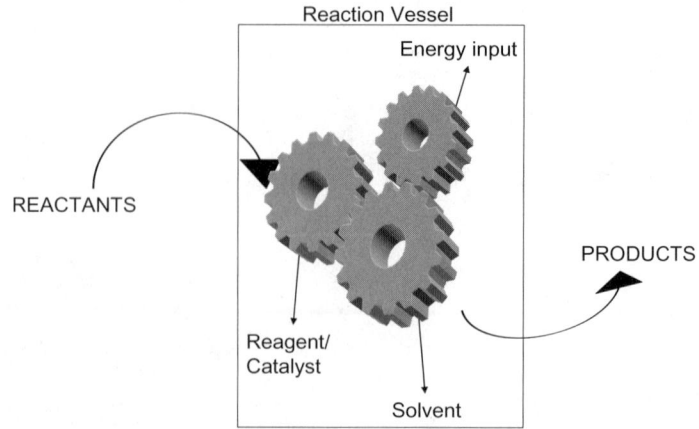

FIGURE 5.1. Basic components of a reaction.

- use of safer solvents,
- use of water as solvent,
- reactions under solventless/solvent-free conditions,
- supercritical carbon dioxide (31.1°C, 73 atm),
- supercritical water (374°C, 218 atm),
- room-temperature ionic liquids.

Safer Solvents

A major concern with regard to sustainability is the release of hazardous substances into the environment. Green chemistry can have a significant impact in this area. Solvents, for example, are ubiquitous in academic, industrial, and government laboratories. In the chemical process, industrial solvents are used in various process operation steps, such as

- separation,
- reaction (reaction medium),
- degreasing,
- washing,
- reactants,
- carrier,
- crystallization.

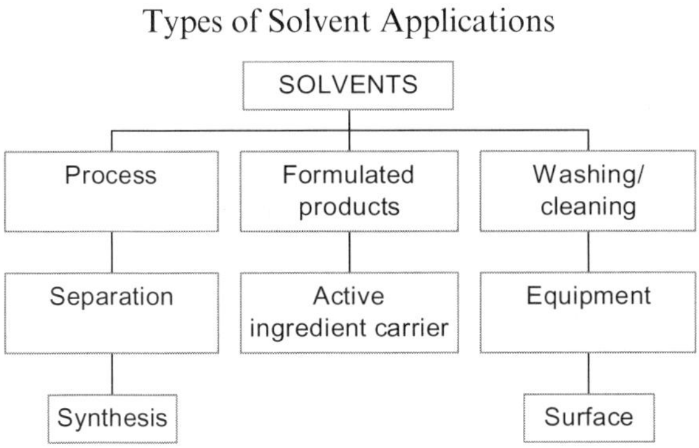

Types of Solvent Applications

FIGURE 5.2. Applications of solvent in chemical processes.

Various types of solvent applications in all types of industries and academic laboratories are illustrated in Fig. 5.2.

Many of the commonly used solvents are volatile organic compounds (VOCs), hazardous air pollutants (HAPs), flammable, and/or toxic. They also pose serious environmental, health, and safety (EHS) concerns, including human and eco-toxicity issues, process safety hazards, and waste management issues.

One of the 12 principles of green chemistry states that "the use of auxiliary substances (e.g., solvents, separation agents, etc.) should be made unnecessary wherever possible and innocuous when used" (Anastas and Warner, 1998). The auxiliary substances, although necessary as reaction media for reactions, are not incorporated in the final product. Many of the commonly used solvents (benzene, chlorinated organic solvents, etc.) are known carcinogens, and many others pose hazardous threats to the environment. Hence, the use of solvents should be contained; when its use is inevitable, the solvent should be nontoxic, nonflammable, and eco-compatible, like water. Alternatives to organic solvents are needed to decrease the negative environmental impact of these substances. One consequence is the need to avoid the use of organic solvents where possible. When elimination is not feasible, efforts should be made to optimize, minimize, and recycle solvents. Solvents that are stable, inexpensive, and readily available, with an acceptable environment impact, are the most suitable. The selection of the solvent should consider the following:

- What type of application is it for?
- Can the solvent dissolve the solute(s), and can it be recycled with minimum solvent loss?
- Is the solvent stable, low-cost, and readily available, and is its environmental impact acceptable?
- What are the physical and chemical properties of pure solvent(s)?
- What are the interactions among different solvents used in multistep processes?

Solvents having an average to high degree of safety are listed in Table 5.1.

As water is immiscible with most organic substrates, most reactions involving water are done with liquid–liquid biphasic systems. The use of biphasic organometallic catalysts to catalyze aqueous-phase reactions is a novel method to address this issue. The catalyst in such reactions is a water-soluble transition metal complex with substrates that are partially water-soluble. The Ruhrchemie–Rhône–Poulenc process, which involves hydroformylation of propylene to *n*-butanol, is an example of biphasic organometallic catalysts being used on an industrial scale (Cornils and Kuntz, 1995). The catalyst employed is a water-soluble Rhodium (I) complex of trisulfonated triphenylphosphine (tppts) (see Fig. 5.3).

TABLE 5.1
Safe Solvents

S. No.	Solvent
1	Isoamyl alcohol
2	2-Ethylhexanol
3	2-Butanol
4	Ethylene glycol
5	1-Butanol
6	Diethylene glycol butyl ether
7	*t*-Butyl acetate
8	Butyl acetate
9	*n*-Propyl acetate
10	Isopropyl acetate
11	Dimethylpropylene urea
12	Propionic acid
13	Ethyl acetate
14	Methyl isobutyl ketone

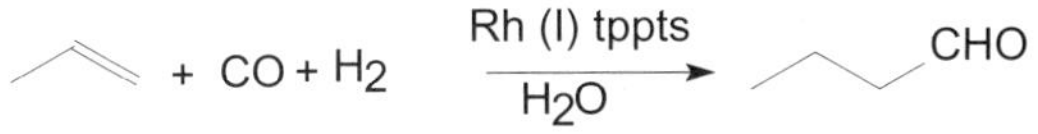

$$\text{\textbackslash} + CO + H_2 \xrightarrow[\text{H}_2\text{O}]{\text{Rh (I) tppts}} \text{\textbackslash} CHO$$

FIGURE 5.3. Ruhrchemie–Rhône–Poulenc process.

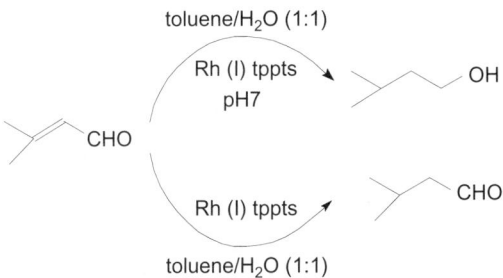

FIGURE 5.4. Chemo-selective hydrogenation in biphasic system.

The same concept was used in the chemo-selective hydrogenation of unsaturated aldehydes (see Fig. 5.4).

Green Solvents

Green solvents are environmentally friendly solvents, or biosolvents, which are derived from the processing of agricultural crops. The use of petrochemical solvents is the key to the majority of chemical processes but not without severe implications on the environment. Green solvents were developed as a more environmentally friendly alternative to petrochemical solvents. Ethyl lactate, for example, is a green solvent derived from processing corn. Ethyl lactate is the ester of lactic acid. Lactate ester solvents are commonly used solvents in the paints and coatings industry and have numerous attractive advantages including being 100% biodegradable, easy to recycle, noncorrosive, noncarcinogenic, and nonozone-depleting. Ethyl lactate is a particularly attractive solvent for the coatings industry as a result of its high solvency power, high boiling point, low vapor pressure, and low surface tension. It is a desirable coating for wood, polystyrene, and metals and also acts as a very effective paint stripper and graffiti remover. Ethyl lactate has replaced solvents such as toluene, acetone, and xylene, resulting in a much safer workplace. Other applications of ethyl lactate include being an excellent cleaner for the polyurethane industry. Ethyl lactate has a high solvency power, which means it is able to dissolve a wide range of polyurethane resins. The excellent cleaning power of ethyl lactate also means it can be used to clean a variety of metal surfaces, efficiently removing greases, oils, adhesives, and solid fuels. The use of ethyl lactate is highly valuable, as it has eliminated the use of chlorinated solvents.

Water as Solvent

The use of water as solvent for organic reactions is one of the finest solutions to the problem of solvent toxicity and disposal. The chemistry in natural systems (biochemical reactions) is based on water. The use of water as solvent for synthetic chemistry holds great promise for the future in terms of the cheaper and less hazardous production of chemicals. Researchers in this area are discovering that reactions in water may be predisposed to favor transition states that optimize hydrophobic interactions, thereby achieving unusual, unique selectivity in organic reactions (Sijbren and Engberts, 2003). It was discovered in pioneering studies on Dies–Alder reactions in water by Breslow's and Biscoe's research groups that such reactions often proceed with much higher rates and higher endoselectivity and exoselectivity than in organic solvents. Water as a solvent favored a more compact endo-transition state in Dies–Alder reactions. The accelerating effect of water has been ascribed to a number of factors, including the hydrophobic effect as well as hydrogen bonding between water molecules and reactants (Breslow, 2004). Breslow and co-workers showed that hydrophobic interactions might determine the ratio of O versus C alkylations of phenoxide ions in water (Breslow et al., 2002). They also used hydrophobic borohydrides to control the regio-selective reduction of the sulfated, naturally occurring steroid. Reduction proceeded with significant selectivity (87:13) for the intrinsically more reactive 17-keto group when the reaction was performed in water using $LiBH_4$. A remarkable reversal of selectivity was observed when the reduction was performed instead with $LiC_6F_5BH_3$ in 4M $LiCl/D_2O$, due to efficient hydrophobic packing at the 6-keto group (Biscoe et al., 2004). In another remarkable study, Biscoe and Breslow showed that epoxidations of olefins could be selectively accelerated by the use of hydrophobic oxidizing agent (2005). Use of hydrophobic oxaziridinium salts for epoxidation of cinnamic and crotonic acid derivatives showed a very large increase in selectivity for the hydrophobic cinnamic acid derivatives.

Solvent-Free Conditions

Several advantages are associated with the use of a solvent-free system over the use of organic solvent. These include

1. There is no reaction media to collect, dispose of, or purify and recycle.
2. On a laboratory's preparative scale, there is often no need for specialized equipment.
3. Extensive and expensive purification procedures such as chromatography can often be avoided due to the formation of sufficiently pure compounds.
4. Greater selectivity is often observed.
5. Reaction times can be rapid, often with increased yields and lower energy usage.
6. Economic considerations are more advantageous, since cost savings can be associated with the lack of solvents requiring disposal or recycling.

Not surprisingly, solvent-free synthesis has recently drawn attention from the wider synthetic community. Reactions epitomizing the simplicity, versatility, high-yielding, and selective nature of solvent-free systems include aldol condensations, sequential aldol and Michael additions, Stobbe condensations, O-silylation of alcohols with silyl chlorides, and clay-catalyzed syntheses of transchalcones. A potential criticism of the solvent free approach, which may be inhibiting its widespread introduction into routine synthesis, is the possibility of producing "hotspots," which can lead to runaway reactions and consequently the increased likelihood of unwanted side reactions. Thus, measurement of heat of reaction in solvent-free systems is important, as is effective heat dissipation. If highly exothermic reactions are identified, which are otherwise suited to solventless conditions, the problem could be addressed through advanced reactor design.

Ionic Liquids

An ionic liquid generally consists of a large nitrogen-containing organic cation and a smaller, inorganic anion. The asymmetry reduces the lattice energy of the crystalline structure and results in a low-melting-point salt. These simple liquid salts (single anion and cation) can be mixed with other salts (including inorganic salts) to form multicomponent ionic liquids. There are estimated to be hundreds of thousands of simple ion combinations to make ionic liquids and a near endless (10^{18}) number of potential ionic liquid mixtures. This implies that it should be possible to design an ionic liquid with the desired properties to suit a particular application by selecting anions, cations, and mixture concentrations. Ionic liquids can be adjusted or tuned to provide a specific melting point, viscosity, density, hydrophobicity, miscibility, etc. for specific chemical systems. The first room-temperature ionic liquid ($EtNH_3$) (NO_3) was discovered in 1914, but its discovery did not initiate a huge amount of interest in ionic liquids until the development of binary ionic liquids from mixtures of aluminum (III) chloride and N-alkylpyridinium or 1,3-dialkylimidazolium chloride.

The components of ionic liquids (ions) are constrained by high coulombic forces and thus exert practically no vapor pressure above the liquid surface. Importantly, the near-zero vapor pressure (nonvolatile) property of ionic liquids means they do not emit the potentially hazardous volatile organic compounds (VOCs) associated with many industrial solvents during their transportation, handling, and use. (It should be noted, however, that the decomposition products of ionic liquids from excessive temperatures can have measurable vapor pressures.) In addition, they are nonexplosive and nonoxidizing (nonflammable). These characterizations could contribute to the development of new reactions and processes that provide significant environmental, safety, and health benefits compared to existing chemical systems.

Ionic liquids have been found useful for a wide range of chemical reactions and processes, including hydrogenation reactions, biocatalysis reactions such as transesterification and hydrolysis, and electrochemical applications such as battery electrolytes. Another use for ionic liquids is as a medium for separation of biologically produced feedstock from a fermentation broth, such as acetone, ethanol, or butanol. In the alkylation reactions, ionic liquids gave better results than sulfuric acid or aluminum trichloride, with the added benefit that the ionic liquid can be recovered and reused. Reactions in ionic liquids also occur at significantly lower temperatures and pressures than conventional reactions, resulting in lower energy costs and capital equipment costs. Ionic liquids can act as both catalyst and solvent. In many systems, the reaction products can be separated by simple liquid–liquid extraction, avoiding energy-intensive and costly distillation.

However, each ionic liquid is likely to be classified as a "new chemical" and may require significant environment, safety, and health (ES&H) studies prior to widespread use. Structural similarities among certain ionic liquids, herbicides, and plant growth regulators have been noted. These similarities raise significant ES&H concerns and give ionic liquids a potentially large unknown risk factor. New health and safety concerns could also result from ionic liquid residuals in polymers, particularly those used for packaging food and personal care products. Broad commercialization of ionic liquids will require a sound, science-based understanding of their environmental, safety, and health impacts. The development of exposure and handling guidelines for ionic liquid production, transportation, storage, use, and disposal are required. Generally accepted environmental, safety, and health protocols for ionic liquids will also have a significant impact in attracting employees to support accelerated scientific research, increasing manufacturer willingness to produce the materials, increasing chemical manufacturer willingness to incorporate them in their production processes, and increasing consumer willingness to buy products containing trace quantities of these materials.

Supercritical Carbon Dioxide: $scCO_2$

During the late 1980s it was found that reactions usually carried out in chlorofluorocarbon solvents can be done in liquid or supercritical CO_2. It was also observed that the polymerization of tetrafluoroethylene in CO_2 using fluorinated initiators gave good yields. CO_2-based processes can also be used for dry cleaning, metal cleaning, and textile processing. Liquid CO_2 is also used in the microelectronics industry to spin-coat photoresists instead of using traditional organic solvents. Another example is the use of CO_2 to clean integrated circuits and flat-panel displays during manufacturing rather than using large amounts of water and organic solvents.

Carbon dioxide, as a final product of deep oxidations, behaves fully inert under oxidizing conditions, making this reaction medium especially attractive for aerobic oxidations. In addition, its high specific-heat capacity ensures efficient heat transfer in these mostly highly exothermic oxidation reactions. Biphasic catalytic oxidation of alcohols using PEG-stabilized palladium nanoparticles in $scCO_2$ was studied. This catalytic system shows high activity, selectivity, and stability in the conversion of structurally diverse primary and secondary alcohols to their corresponding aldehydes and ketones.

The oxidation of sulfides to sulfoxides is a key routine organic transformation. Oakes et al. have developed conditions to enable this reaction to be carried out in supercritical carbon dioxide, with clean sulfoxide formation and no overoxidation to the sulfone (1999). Using TBHP as the oxidant and Amberlyst 15 as a heterogeneous acid catalyst, a range of alkyl, aryl, and benzyl sulfides have been selectively oxidized to the sulfoxide in quantitative yield.

Supercritical Water

Water has obvious attractions as a solvent for clean chemistry. Both near-critical and supercritical water (scH_2O) have increased acidity, reduced density, and lower polarity, greatly extending the possible range of chemistry that could be carried out in water. scH_2O has already been studied extensively as a medium for the complete destruction of hazardous and toxic wastes (Savage et al., 1995).

In an interesting report for the synthesis of terephthalic acid, scH_2O was used as a solvent. The traditional synthesis of terephthalic acid uses *p*-xylene and air with acetic acid as the solvent and a manganese/cobalt catalyst system at 190°C and 20 atm of pressure. The process is highly selective, but it can be energy-intensive. In addition, terephthalic acid is insoluble in acetic acid, and some 10% of the acetic acid is oxidized during the reaction. By contrast, *p*-xylene, oxygen, and terephthalic acid are all soluble in supercritical water. Thus, in the new method, oxygen from hydrogen peroxide was used as the oxidant with 1000 ppm of manganese dibromide as the catalyst. Isolated yields of terephthalic acid of 70% and above with selectivity better than 90% were reported (Dunn et al., 2002).

A key advantage of chemistry with supercritical water (SCW) is the possibility of varying the properties of the reaction medium over a wide range solely by changing the pressure and temperature and of optimizing the reaction in this way without changing solvent. This can be seen particularly clearly in the variation of the relative static dielectric constant and the pKW value as a function of temperature and pressure, two physical properties that have a decisive effect on polarity and acid/base catalytic properties. Furthermore, the reaction kinetics can be strongly affected in the supercritical region by varying the pressure (kinetic pressure effect). In addition, many nonpolar organic substances (e.g., cyclohexane) and gases (e.g., oxygen) are highly soluble in SCW so that mass transfer restrictions due to phase boundaries do not apply. In summary, SCW has great potential with regard to the optimization of chemical syntheses. There are, however, drawbacks from working at high pressures (high investment costs), the problem of corrosion (expensive materials), and the lack of kinetic and thermodynamic data. For these reasons, applied research in the field of chemistry in SCW must not remain confined to synthetic chemistry, rather the issues concerning materials and thermodynamics must also be addressed in the research at the same time.

Conclusions

To achieve near-total "greenness" of chemical processes, we need to focus on every aspect of the chemical reaction. Until now, most companies' responses to environmental regulation and protection have been of the circumstantial variety, such as lessening human exposure to a dangerous substance. But an approach that is instead intrinsic (for instance, employing molecular design) can be not only more reliable but also less costly. Some of the challenges facing chemists in this regard could be summarized as the following:

- employ solvents that also serve as catalysts,
- employ solvents that can be easily separated downstream,
- employ reagents that will be effective in catalytic concentrations,
- employ reagents that are recyclable,
- employ solvents and reagents that are biodegradable.

The aim of all research must be to study a piece of reality so as to *completely* realize and experience all aspects of it. In the normal secular mode, we do not experience much at all. We read about a subject, we study it, and at the most we discover a few underlying and hidden laws and principles, but we have not experienced the joy of it. We have not experienced the *totality of this reality* and realized that this complexity of subjects and fields of study relates to only that one reality. But once we make the proper kind of study, seeing them all as perspectives, it will take us to this principle of delight and give us the total experience of reality. And once we have that, life will change all its colors, and research will enrich our sensibility. It will have made us know life as it ought to be known and studied. We see that what we call, in the cold, dead language of our educational system, "academic research," is, in truth, in the living language of spirit, perspectives in which you see the reality revealed. In other words, many new insights will be generated. Research itself will be raised to a new height and a new freshness, and moreover, it will no longer be "dead learning." We'll have experienced the joy ourselves, and once we do that we have really fulfilled ourselves and ignited a number of others. Hence, every step in this direction will go a long way in environmental protection.

References

Anastas, P. T. and Warner, J. C., Green Chemistry: Theory & Practice, Oxford University Press, 1998.

Biscoe, M. and Breslow, R., *J. Am. Chem. Soc.*, **125**: 12718, 2003.

Biscoe, M. and Breslow, R., *J. Am. Chem. Soc.*, **127**: 10812, 2005.

Breslow, R., *Acc. Chem. Res.*, **37**: 471, 2004.

Breslow, R., Groves, K., and Mayer, M. U., *J. Am. Chem. Soc.*, **124**: 3622, 2002.

Cornils, B. and Kuntz, E. G., *J. Org. Met. Chem.*, **502**: 177–186, 1995.

Dunn, J. B., Urquhart, D. I., and Savage, P. E., *Advanced Synthesis and Catalysis*, **344**: 385–392, 2002.

Oakes, S. R., Clifford, A. A., Bartle, K. D., Pett, M. T., and Rayner, C. M., *Chem. Comm.*, 247, 1999.

Savage, P. E., Gopaian, S., Mizan, T. I., Martino, C. J., and Brock, E. E., *A. Inst. Chem. Eng. J.*, **41**: 1725, 1995.

Sijbren, O. and Engberts, J. B. F. N., *Org. Biomol. Chem.*, **1**: 2809, 2003.

CHAPTER 6

Process and Operations

When applied, the design philosophy known as *process intensification* (PI) can lead to savings in energy, to a reduction in capital expenditure, land usage, and associated chemicals, and to environmental and safety benefits. All these benefits arise due to the reduction in plant size, which is generally of the order of three or four, and reduction in mass and heat transfer resistances and reaction time, which is of the order of several hundreds (van den Berg, 2003; Akay, 2005). The concepts of PI were originally pioneered in the 1970s by Colin Ramshaw and his co-workers at ICI, and it was defined as a "reduction in plant size by at least several orders of magnitude" (*Green Chemistry*, Feb. 1999, G15–G17). The "HiG concept" was invented about 25 years ago by Prof. Colin Ramshaw of ICI, while at the same time the compact heat exchanger in the form of the printed circuit/diffusion bonded unit was developed by Tony Johnstone. The latter approach is now the accepted design basis of microreactors, where the fine channels promote both rapid heat and mass transfer and give the unit a powerful multifunctional capability (see Fig. 6.1). Originally PI was developed for the bulk chemical industry, but it has been extended to value-added chemicals and pharmaceutical active ingredient manufacture (Wegeng et al., 1996; Wess et al., 2001).

The energy efficiency of a process is determined by the ability to transfer heat in a cost-effective manner. This energy savings could be achieved through the design of processes using compact heat exchangers (using PI) or using innovative heat-exchanger designs. Also, heat transfer is related to the flow diameter; hence,

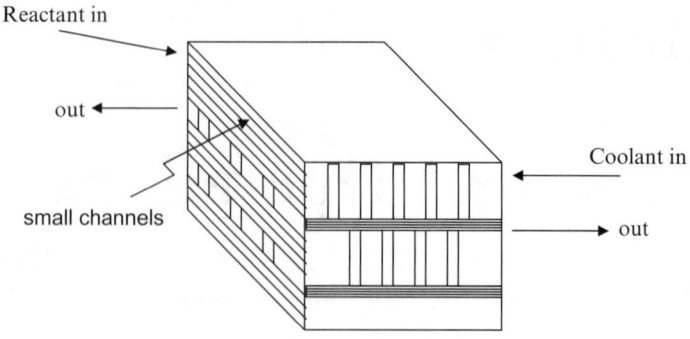

FIGURE 6.1. Micro reactor and heat exchanger (heat transfer coefficient: ~20 kW/m²/k).

a small-diameter tube will lead to higher heat transfer. So, compact heat exchangers will perform better and are less expensive than conventional shell and tube heat exchangers. These heat exchangers are made of ceramics or polymers or use new designs such as printed circuit heat exchangers or multistream heat exchangers.

Process intensification has several advantages (Stankiewicz and Moulijn, 2000; Simons, 2004), including

- reduced process plant size by orders of 10-fold to 1000-fold,
- mobile plants,
- reduced capital and operating costs,
- reduced inventory and risk of hazards,
- improved safety (Hendershot, 2000),
- environmentally friendly concept (Benson and Ponton, 1993),
- elimination of side reactions,
- design of novel processes.

Process intensification is achieved through the application of one or more of the following *intensification principles*:

- high to ultrahigh body forces,
- high to ultrahigh pressures,
- electric field,
- ultrasonic,
- carrier mediation (surfactant-based separations),
- reduction of diffusion and conduction pathway,
- interactions of flow fields and fluid microstructures,
- size-dependent phenomena.

Processes can be intensified either (1) through the application of a high or ultrahigh processing field (physical intensification) or (2) through the enhancement of selectivity (phenomenon-based intensification). The processing volume is reduced in both cases, while in the latter case, improved selectivity is often achieved through intensified fields as well as with drastically reduced processing volume.

PI not only leads to reductions in capital expenditure (Green, 1999) but also produces several other benefits:

1. better product quality,
2. just-in-time manufacture,
3. distributed manufacture,
4. lower waste amounts,
5. reduced downstream purification costs,
6. valuable for "green" manufacture,
7. smaller inventories, leading to intrinsic safety,
8. lower energy use,
9. better process control.

Figure 6.2 shows the relationships among the principal features of PI, which are listed here:

1. Move from batch to continuous mode of processing, which may be very relevant to the pharmaceutical and fine chemical industries. Fine chemicals accounted for U.S.$70 billion in sales in the year 2003, and 60% of those sales came from the pharmaceutical sector (Mullin, 2004).
2. Use of intensive reactor technologies with high mixing and high heat and mass transfer rates in place of conventional stirred tanks, which have poor heat and mass transfer coefficients, inefficient mixing, and large radial thermal gradients.
3. Multidisciplinary approach, which considers opportunities to improve the process technology and underlying chemistry simultaneously.
4. "Plug and play" process technology to provide flexibility in a multiproduct environment (Ehrfeld et al., 2000).
5. High selectivity and rate and reduced batch cycle time.

Industry Perception

Industry leaders believe that PI would typically provide the following benefits:

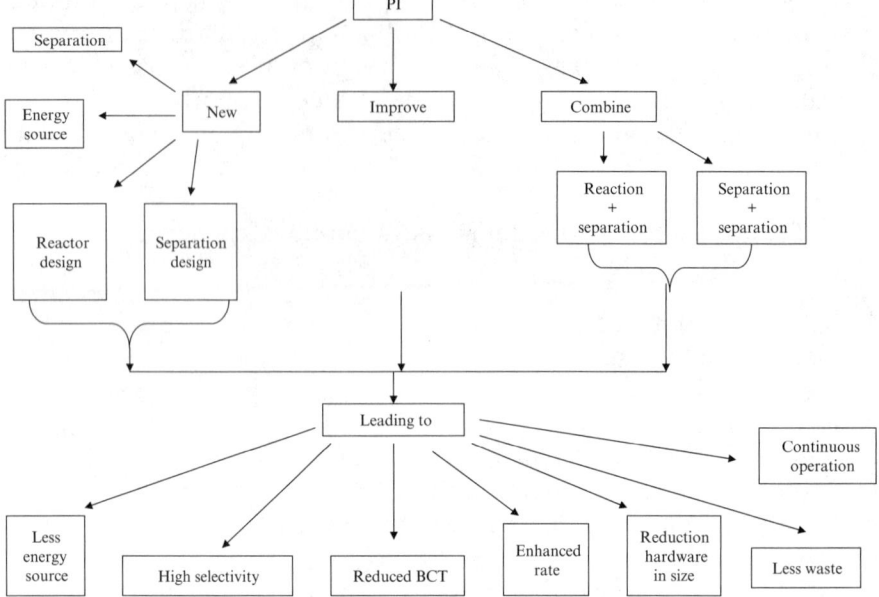

FIGURE 6.2. Process intensification—its features, key principles, and benefits.

1. a reduction in capital cost by 60%,
2. a 99% reduction in impurity amount in the final product, resulting in a significantly more valuable product,
3. a 70+% reduction in energy usage and hence considerable reduction in operating cost,
4. a 93% first-time yield (per pass yield), leading to reduction in downstream processing and less material to recycle back to the reactor,
5. a 99% reduction in reactor volume, leading to inherently safer operation, reduced capital cost, and also less storage of inventory.

Delphi is a forecasting technique based on answers to questionnaires. The technique is widely used to facilitate formation of a group judgment without permitting interaction and biasing that normally happen during a group discussion. It is a method for achieving a structured anonymous interaction between selected experts by means of a questionnaire and feedback. Delphi is commonly used

1. for creative exploration of ideas,
2. to collect suitable information for decision making,
3. to identify factors influencing the future state of process intensification's development,
4. to develop the time scale of differing aspects of the technology's realized potential.

A Delphi study conducted in 2002 focused on identifying the potential opportunities for Process Intensification Technologies (Nikoleris et al., 2002). The questionnaire was sent to a preselected group of experts in order to obtain individual responses to the problems posed. The Delphi studies came up with important findings with respect to PI, and those findings are listed here:

1. Lower energy use in the shorter term. This is a priority in the short term, but it may not be so important in the long term, since the chemical industry would like to move more than 90% of its energy requirements to renewable sources. Recently, the British Government's Performance and Innovation Unit Energy Review recommended a combination of low–carbon, energy-efficient, and renewable technologies to form the basis of a sustainable future energy policy. Forty percent of those surveyed felt that a fuel switch would have no benefit on the acceptance of PI technologies, since they are mutually exclusive. Of course, combining process intensification and renewable technology has a significant role to play in tackling the present and future environmental issues.
2. Improved safety, inventory reduction, and plant physical size reduction. Next to reducing energy, achieving safer plants is seen as the most important goal. PI would lead to reduction in inventory, which would once again lead to safer plants.
3. Switching from batch to continuous processes will help the chemical industries improve their public image. Sixty-two percent of the experts surveyed expressed interest in transforming chemical processes from batch to continuous processes in the short, medium, or long term. Plant accidents and safety studies conducted pointed out that 75% of runaway reactions occurred in batch vessels. A move away from batch would improve safety, since controlling the continuous process would be easier. Except during startup and shutdown, the reactor operating conditions would be constant. PI technology based on continuous processes provides an improvement in heat and mass transfer, lower residence time, lower

inventories, better control, and better-quality products. Continuous-stirred tanks and tubular reactors have lower volume when compared to a batch reactor for the same conversion, Three main areas that have been identified as having the most potential for continuous processing are reactions, separations, and heating.

4. Smaller plants. The experts felt that a 50% plant size reduction target by 2010 and a two-order-of-magnitude reduction by 2010 are desirable. Further reductions may not be practically achievable. Intensified plants, with their lower capital costs, would allow smaller-scale plants to compete economically with their conventional large counterparts. In the 1960s and 1970s, the chemical industries were able to compete due to their larger scales of operation. A small-sized plant was not as economical as a large-scale plant (examples of large-scale plants include petrochemicals, polymers, fertilizers, and sulfuric acid). Intensified plants are expected to have smaller land space. The main advantages associated with plant-size reduction may include
   - lower capital cost of equipment,
   - increased safety,
   - lower maintenance requirements,
   - lower operation costs,
   - less waste to dispose of,
   - reduced time to market,
   - decentralized production.

5. Seventy-eight percent of the respondents considered environmental improvements as a priority in the short term but not a top priority in the long run.

6. Whole plant intensification has not been found attractive or feasible, but it is restricted to specific unit operations and reactions. New processes are expected to be intensified.

7. The compact heat exchangers (CHE) are a good example of PI technology, which now forms the basis of very small reactors, as well as being routinely used for intense heat transfer, in many demanding applications.

8. The majority of experts felt that a major heat and mass transfer intensification opportunity exists, centering around the use of enhanced acceleration in a rotating system (high-G; more details about high-G are given later), with a view to create a fluid-dynamic environment that will allow complete intrinsic kinetics. Such plants are expected to be feasible by 2015.

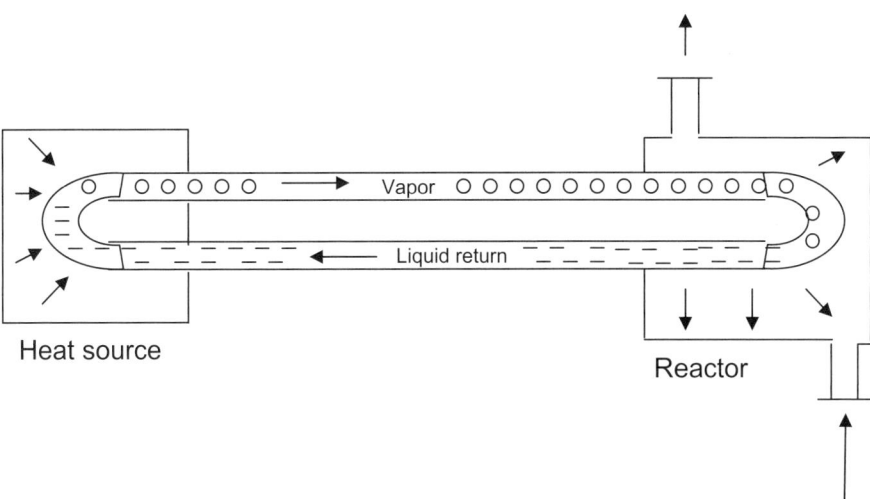

FIGURE 6.3. Heat pipe design.

9. Fifty-seven percent of the respondents believed in the feasibility of the heat pipe technology by the year 2010. Heat pipes are sealed vessels that transfer heat through the evaporation and condensation of a working fluid (see Fig. 6.3). Total heat transferred will also be high during phase changes. For example, the heat of vaporization of water is 240 J/kg. In addition, heat pipes can have thermal conductivity that is 10× or 100× that of copper, with no moving parts and therefore low maintenance requirements, making them reliable and desirable for many applications. Lack of awareness, from both the manufacturers' and end users' points of view, and conservatism are the main causes for it not being used preferentially to conventional heat transfer equipments.

10. Improvements in performance of existing conventional heat exchangers with substantial energy cost savings can be realized through active or passive techniques, such as incorporating fins or tube inserts or electrostatic fields, by adding rotating parts, or through surface enhancement. The goal of enhancement techniques is to achieve a reduction in size, an increase in the capacity of existing equipment, or a reduction in the *approach temperature difference* (the approach temperature difference is the difference in temperature between the two fluids at the entry of the heat exchanger).

Once again there are several barriers for accepting and adapting these enhanced designs by the industries. Such technological and business related barriers include

1. Conservatism in the user industries. Conservatism has been identified as the major barrier for PI implementation, and it is from within an organization, namely from management's attitudes or lack of awareness.
2. Everyone wants to be the "second" to try something, thereby avoiding positive risk.
3. Lack of industrial and academic awareness. Academics are not taking the initiative to drive the advantages of PI to the industrial community.
4. Loss of the buffering effect of large volumes.
5. Lack of codes of practice, unlike the conventional systems, which have the complete data.
6. Limited choices.
7. Concern about fouling of the hardware and handling of solids.
8. Lack of supporting tools.
9. Many industries move their manufacturing operations to other countries to achieve lower production costs. In these countries, issues such as safety, effluent, and waste are not concerns.
10. The experts believe that all the technologies studied such as heat pipes, rotating equipment, and enhancement devices may approach the growth stage in the medium term around 2010 to 2015.

Reactions

PI in reactions has led to several new designs and techniques. Miniaturization can be developed based on the categorization of reaction kinetics. Lonza and Visp have used a kinetically oriented ranking for judging the suitability of chemical processing for microflow operations and used a standard industrial classification, which divides the chemical reactions according to their reaction time, such as Types A, B, and C, where

1. Type A reactions are very fast, reaction times are of the order of less than 1 s, and the overall reaction is mixing controlled.

2. Type B reactions are rapid, where the reaction times are of the order of 1 s to 10 min. These reactions are kinetically controlled, but temperature control is very essential.
3. Type C reactions are slow, where the reaction times are of the order of greater than 10 min. These reactions are well suited for batch processing, but safety and quality are issues that need to be handled.

They found that nearly half the reactions that are practiced in industries bear microreactor potential. However, only about 20% of these reactions can be addressed through the microreactor, while the other 30% of the reactions contain solids or are multiphase in nature. Both types were considered as not easily addressable using today's microreactor technology (Wiley-VCH, Lonza and Visp). It was also concluded that Type A reactions could be handled with micromixers (possibly with integrated or subsequent heat exchangers). Hybrids of microstructured and smart conventional flow-through components could be used for Type B reactions, for example, designs such as micro mixer-tube reactors. Conventional flow-through reactors could be used for Type C reactions, such as tube reactors with stage-wise increase in diameter to counterbalance performance and pressure loss. Also, a microstructured component could be inserted into a conventional batch plant for Type B and C reactions, for instance, tube reactors.

Reactor Designs

PI has led to the design of a variety of new and innovative reactor designs to overcome mass and heat transfer limitations that are normally encountered in large-scale vessels (Semel, 1997). With these designs it is possible to carry out highly exothermic reactions, speed up the rate of reaction by several orders of magnitudes, totally eliminate side reactions and hence waste formation, combine reactions with unit operations, and telescope several steps into a single step.

Microreactor

The ubiquitous batch reactor can be used to carry out polymerization reactions in the laboratory, but the recipe to be used on the plant scale has to be changed to match the relatively poor heat transfer and mixing performance of a larger-diameter vessel. Many

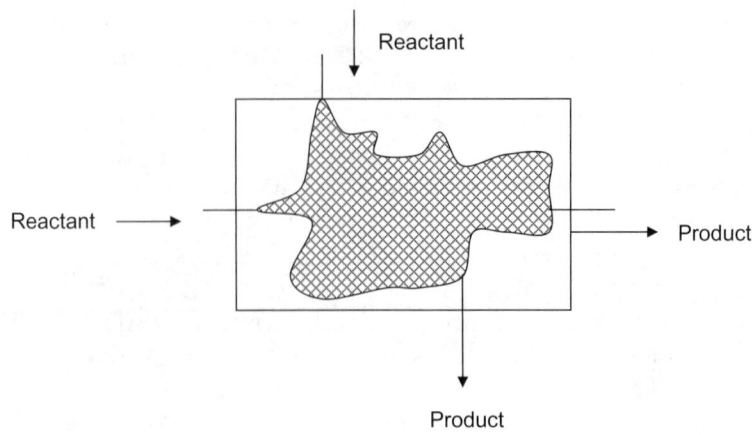

FIGURE 6.4. Reactor on chip (multichannel).

chemical reactions are exothermic, so the rate of reaction is proportional to temperature and the reaction rate is adjusted to match the heat removal rate (Doble et al., 2004). During scale-up instead of tuning the hardware to match the process, many times the process is matched to a particular hardware. A plate reactor, a microchannel reactor (see Fig. 6.4), or a microreactor, on the other hand, has much better heat removal and much shorter residence times.

Microreactor rigs also allow for high-pressure (20 bar) operations, which can help in maintaining the reactants in liquid phase even at high temperatures (~100°C) (Skelton et al., 2001; Hessel et al., 2004). At that pressure low-boiling chemicals and solvents are maintained in liquid phase, thereby achieving high rates of reaction at higher temperatures. Typically, high temperatures may lead to more side and consecutive reactions. But efficient mixing (if the process is kinetically controlled) and shortened residence time reduce side product formation and make the reaction more efficient with no side products. If carried out this way, the reaction time of aqueous-based Kolbe–Schmitt synthesis with resorcinol and phloroglucinol can be decreased by several orders of magnitude, from hours to some 10 s (decreased by a factor of 2000).

The bromination of thiophene involves the use of pure thiophene and bromine flowing below room temperature (sometimes as low as −10°C). In a conventional, stirred batch reactor, the yield of 2,5-dibromothiophene is of the order of 50–70%, while in a microreactor the yield is 86%, with nearly complete conversion

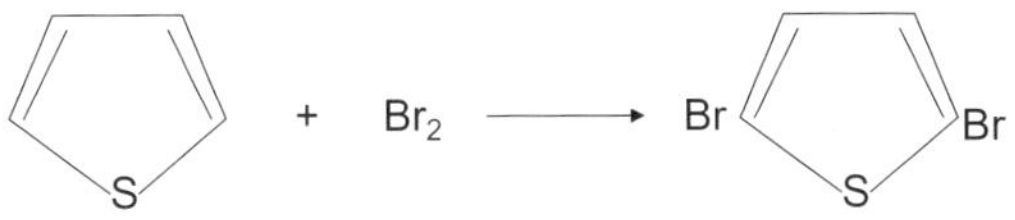

FIGURE 6.5. Bromination of thiophene.

(see Fig. 6.5). Using pure feeds and operating at higher temperatures, the reaction time can be decreased from about 2 h (for the batch process) to less than 1 s (for the micromixer reactor).

Pure hydrogen and oxygen mixtures are highly explosive. Reactions that involve such mixtures are carried out safely in microchannel reactors. For example, the direct preparation of hydrogen peroxide is obtained with a special catalyst, avoiding the circuitous anthraquinone process, used at the industrial scale. Calculations of explosion limits clearly demonstrate that there is a considerable shift when explosive reactions are carried out in microchannels. The safety of the process is not only due to the avoidance of thermal runaway (because of large surface-area-to-volume ratio), but also due to the fact that radical chains are broken down due to the increased wall collision in the small channels of the reactor.

It is well known that nitroglycerin manufacturing plants have several safety features since the probability of explosions is very high. A continuous nitroglycerin pilot plant with microstructured mixer-reactors has been installed at the Xi'an site in China and is operated at a production rate of 15 kg/h, meeting all specs without any of the problems mentioned earlier. The quantity of nitroglycerin present in the plant at any point of time is very small since the whole process consists of the microreactor. The nitroglycerin is manufactured for use as a medicine for acute cardiac infarction.

By suitably coating the microreactor's channel walls (see Fig. 6.6) with catalysts, moderators, promoters, surfactants, and so forth, the unit can be upgraded for specialty applications. A reaction that required the addition of a base to activate a homogeneous catalyst was carried out in a microreactor without the addition of the base. The function of the base is provided in the microchannels by groups such as deprotonated hydroxyls anchored on the surface. Since the specific surface area in a microreactor is large and diffusion distances are short, the surface becomes part of the reaction. Such a phenomenon is not observed in conventional reactors. A

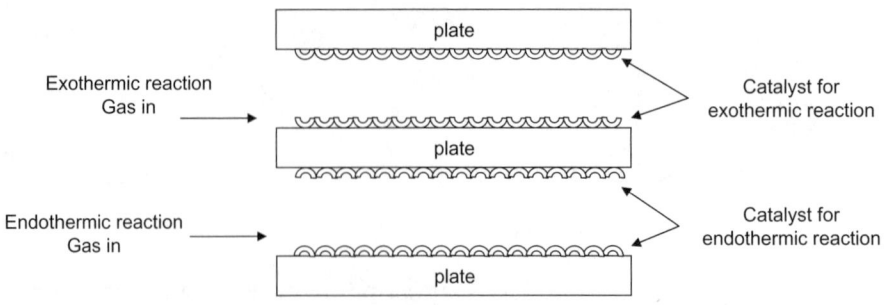

FIGURE 6.6. Catalytic plate reactor—combined exo and endo reactions.

base-free Suzuki–Miyaura coupling using tetrakis (triphenylphosphine) palladium (0) as catalyst and an acid-free esterification of pyrenyl-alkyl acids using only the acidic functions of the microchannel surface are reported to be carried out in plate reactors.

5-hydroxybenzofuran decomposes with a large energy release at relatively low temperatures, limiting the direct synthesis of 5-(2-carboxy-pyridine-2-yloxy)-benz-oxadiazole, using the former as the starting material. Thus, commercially indirect synthesis routes are being practiced. The direct route was recently published by Pfizer using a microreactor technology. Due to short residence time and large surface-area-to-volume ratios, the decomposition problem was completely eliminated. Two well-known methodologies, the Jacobsen asymmetric epoxidation and the Sharpless asymmetric dihydroxylation followed by epoxidation, were scaled up successfully for the preparation of a chiral dihydrobenzofuran epoxide by Bristol-Myers Squibb (USA) using a microreactor.

Through the use of commercially available Cytos® microreactors that are $100 \times 150\,mm$ in size, several commercially important processes have been scaled up. Some examples include

1. the six-stage synthesis of ciprofloxacin,
2. the nitration of toluene with highly explosive acetyl nitrate,
3. the nitration of pyridine-N-oxide at high temperature,
4. the nitration of 2-methylindole.

All these processes are very difficult to scale up, but they have been carried out safely in microreactors. Six Cytos® microreactors running in parallel are equivalent to a miniplant for the synthesis

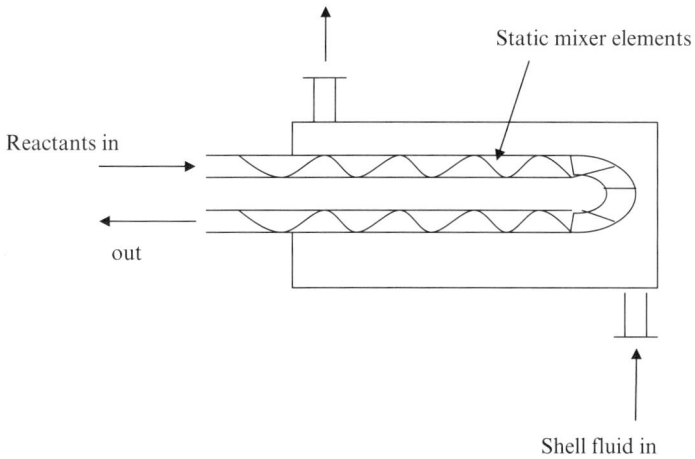

FIGURE 6.7. Flexible reactor with static mixer elements (residence times 25–30 min, 20 bar, ~250°C, single-ended or double-ended design).

of a dye with a capacity of 30 tons/year and a footprint of no larger than that of an office table.

Micro Mixers

The Caterpillar micro mixer is an example of a simple, pilot-sized micro mixer-tube reactor that has an internal microstructure with a ramp-like shape inducing recirculation flow patterns, whereas the overall channel structure can be kept fairly large. The volume varies between 1 L/hr to several hundred liters per hour. Several hundreds of precision-engineered, star-shaped platelets can be stacked, and the flow can be distributed.

A PI reconfigurable reactor developed by the BHR Group (USA) has led to flexibility, a wide range of operating conditions, and easy construction (see Fig. 6.7). The reactor contains static mixer elements (see Fig. 6.8) packaged into a flexible "shell and tube" geometry, which gives a flexible design capable of operating over a wide range of conditions. Generally, static mixers are not flexible, but this design incorporates flexibility. It can be scaled up by adding more length.

Production of glycidyl ethers by reaction between alcohol and chloroalkylepoxide in the presence of a catalyst is traditionally carried out in a batch reactor, and then the products are separated

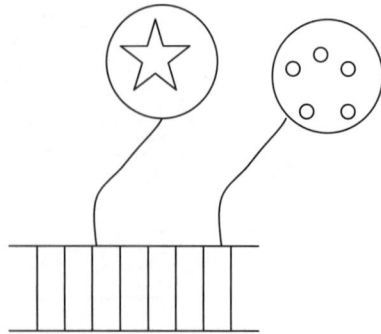

FIGURE 6.8. Micro-structured mixer elements.

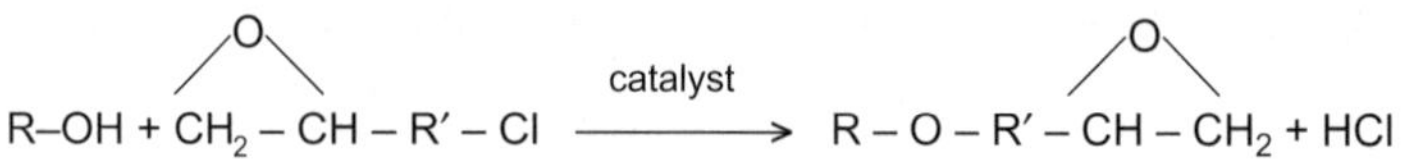

$$R\text{-}OH + CH_2\text{-}CH\text{-}R'\text{-}Cl \xrightarrow{\text{catalyst}} R\text{-}O\text{-}R'\text{-}CH\text{-}CH_2 + HCl$$

FIGURE 6.9. Glycidyl ether synthesis.

by distillation. The latter is also performed in batch mode (see Fig. 6.9). The reaction is highly exothermic, and the rate of raw material feeding is limited by the heat generated. The byproducts formed by side and consecutive reactions reduce the selectivity and increase the energy usage in the distillation columns. Using PI, the stoichiometric feed of components is mixed in a static mixer before the reactor. The reactor consists of two twisted tubes in a helix format that creates strong internal circulation, and radial mixing promotes heat transfer at a low Reynolds number.

The Aldol–Tishchenko reaction of enolizable aldehydes is a simple and effective way to prepare 1,3-diol monoesters. These esters are widely used as coalescing agents in the paint industry. The new process developed by Helsinki University uses monoalcoholates of 1,3-diols as catalysts giving fast and clean reactions, compared to the previous processes that used several inorganic catalysts. The rapid water-free method using microreactors allows fast preparation of these monoesters with high yields and minimum amounts of side products.

Miniature Bubble Column Reactor

Direct fluorination of toluene using elemental fluorine is not feasible since the heat release cannot be controlled with conventional reactors. So the process is deliberately slowed down. Hence, the direct fluorination needs hours in a laboratory bubble column. It can be completed within seconds or even milliseconds when using a miniature bubble column, operating close to the kinetic limit. The Bayer–Villiger oxidation of cyclohexanol to cyclohexanone with fluorine and aqueous formic acid (5% water) is done in miniature bubble column reactors at 60% conversion at 88% selectivity.

Chip Reactor and Library on Chip

Reactors on chip (Fig. 6.4) have reactants flowing in parallel to one another, crossing in a complex and multileveled network, allowing continuous synthesis of organic chemicals (Kikutani et al., 2002). The advantages of using a chip reactor include

1. Minimum raw material and product inventory, particularly important for expensive raw materials.
2. Ability to carry out multistep syntheses in a single unit with a small footprint (floor space).
3. Ability to vary process parameters in a short time. Large, conventional reactors require longer times to move from one steady operating point to another during process changes. Chip reactors can be also extended to quench flow applications (fast reaction followed by quenching of the reaction by either decreasing the reaction temperature or neutralizing the pH in milliseconds).
4. Ideal for collecting kinetic data.
5. Transient operation can be carried out in pulses (Fletcher et al., 2002).

Researchers at the University of Tokyo prepared a small (2 × 2) library in a chip microreactor by reacting 3-nitrobenzoyl chloride and 3,5-dinitrobenzoyl chloride, each with dl-1-phenylethylamine and 4-amino-1-benzylpiperidine using a phase-transfer catalyst. GlaxoSmithKline (GSK) generated a 2 × 2 library for a domino reaction, which consisted of a Knoevenagel condensation that gave an intermediate that immediately underwent an intramolecular hetero–Diels–Alder reaction with inverse electron demand. GSK published the synthesis of a 3 × 7 library using the Knorr reaction of diverse 1,3-dicarbonyl compounds and hydra-zines under ring closure to pyrazoles. The Knorr synthesis is of interest for drug applications as products with a wide range of biological activity and can be generated in this way.

Libraries of a similar size were used for screening of homoge-nous catalysis under multiphase conditions and for the extraction of kinetic data for the asymmetric syntheses, thereby increasing the depth of information that was gathered. Steric, solubility, and electronic effects on reactivity were monitored for a substrate library for the isomerization of allyl alcohol derivatives conducted under liquid–liquid conditions.

Novel Processes and Routes

Most of the known pilot processes in the field of fine chemicals use a microstructured mixer with a connected tubing or micro heat exchanger. The advantages of this over conventional batch processing are

1. compactness,
2. low capital cost,
3. low energy consumption and other operating expenses,
4. negligible wear and no moving parts, thereby minimizing maintenance,
5. lack of penetrating shafts and seals (providing a closed-system operation),
6. short mixing time,
7. well-defined mixing behavior,
8. narrow distribution of product quality.

Microreactor processing can intensify existing chemical pro-cesses in the industry and may even lead to bottom-line returns. Merck has successfully converted a batch production process involving an organometallic reaction using five miniature mixers for about 5 years, until the end of the lifetime of the fine chemical product, increasing the yield by about 20%. The former conven-tional process was run under cryogenic conditions, while the new process was conducted at room temperature without any loss in selectivity. The Clariant Company (USA) has established a pilot process for the synthesis of phenyl boronic acid through the Grignard reaction. The process is sensitive to local concentration changes, thermal overshooting, feed mismatch, improper mixing or temperature control, decreasing the yield of this cryogenic process to 10–20%. The yield from the batch process was only about 65%, while that obtained using a microstructured mixer is about 90%. Scale-up from a laboratory micro mixer to a pilot microstructured mixer did not lead to any decrease in yield. As in the Merck case, the formerly cryogenic process (~−50°C) could be performed at room temperature. This reduced the energy costs of the process as well as led to a higher yield. Clariant published its work on the production of an azo pigment using micro mixers (Kim et al., 2002). Due to the improved mixing characteristics, pigment products with better optical properties in terms of brightness, color strength, or transparency were synthesized. This mainly stemmed from the possibility to produce smaller particles in the microreactor. Since less pigment needs to be incorporated into the commercial dye matrix, there has been an increase in the profit margin.

A direct hydrogen-peroxide process developed by Immtech Pharmaceuticals (IMM) (USA) on the laboratory scale using UOP's proprietary catalyst and process know-how is currently being tested at the pilot scale at UOP. The process is operated at a pressure of 30 bar and a temperature of 50°C, at an O_2 to H_2 ratio of 3.0, and at a space velocity of $1.8\,g\ H_2/(gCATh)$. The H_2O_2 concentration is $1.7\,wt\%$, and the yield is $2.0\,g\ H_2O_2/(gCATh)$. The key to achieving high selectivity toward hydrogen peroxide is to have a noble metal catalyst at a partially oxidized state. At low selectivity, water is the main product.

Solvent-Free Conditions

Michaelis addition of α, β unsaturated carbonyl compound and dimethylamine at a temperature below 55°C (i.e., the boiling point of the amine) in the presence of a solvent is about 48 hr, whereas when the reaction is carried out at 200°C under high pressure and solvent-free conditions, it leads to a reduction in processing time from 48 hr to a few minutes. A high-pressure operation keeps the amine in liquid conditions, obviating the use of a solvent and carrying out the reaction under concentrated reactant conditions, which increases the reaction rate by several orders of magnitude.

Unit Operations

Heat Exchanger Designs

Heat exchangers generally are based on cylindrical pipes that have a minimum surface area per unit volume through which heat is transferred (Kern, 1990). Plate heat exchangers are much more

effective since they provide a higher heat transfer area. The min-
iaturized compact heat exchangers offer a very high heat transfer
coefficient due to their small diameter and high fluid flowrates.
Conventional heat exchangers suffer from fouling, leading to
a decrease in the heat transfer coefficient with time. Klarex's
(Netherlands) self-cleaning heat exchanger consists of a design that
reduces the surface area by approximately 20% (see Fig. 6.10). The
technology features a "shell and tube" heat exchanger in the vertical position. Chopped wire of the same material of construction
as the heat exchanger (if the heat exchanger tubes are made of 316
stainless steel, then the chopped wire is also made of 316 stainless
steel) is circulated through the unit, scouring the tubes clean. The
wire exits the unit at the top and flows back to the inlet through
an external downcomer. In a smaller size, this technology can be
used as a reaction cum heat exchanger system as well. This self-
cleaning system has been successfully used in the petrochemical,
fine chemical, and pharmaceutical industries. Because these units

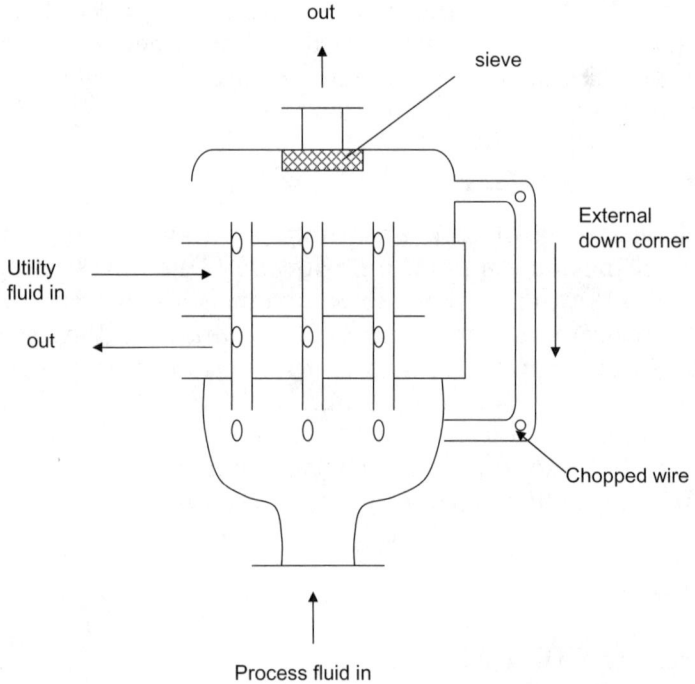

FIGURE 6.10. Self-cleaning heat exchanger.

are self-cleaning, there is no need to oversize the heat exchanger in order to compensate for fouling.

Incorporation of static mixer elements by a Spain company in the shell-tube heat exchangers has led to a 55% increase in the heat transfer area. The volume of the heat exchanger was reduced to 16% in the shell side. In another case the area was reduced by half and the volume by 6.5% on the shell side. There are similar reductions, especially where viscous liquids are involved (Qi et al., 2003).

Distillation Columns

Distillation is predominantly about achieving high gas–liquid mass transfer rates, and the key parameters in achieving this goal that need attention are (McCabe and Smith, 1990; Treybal, 1990)

well-mixed liquid and gas phases,
high interfacial surface area,
thin liquid film,
countercurrent operation.

It has been observed that the overall pressure drop inside a packed distillation column comprises of three major components:

1. gas–liquid interaction at the interface along the flow channels,
2. flow direction changes and losses associated with entrance effects at the transitions between packing layers,
3. gas–gas interaction at the plane separating crossing gas-flow channels.

The latter two, which are responsible for 80% of total pressure drop in packings with a corrugation angle of 45°, do not contribute greatly to mass transfer. This means that reducing these pressure drop components could lead to increased capacity without affecting the mass transfer efficiency.

A dividing wall column (DWC) is a variation of the so-called Petlyuk column. Its major feature is that it allows substantial energy savings while separating in a single body a three-component mixture into pure products (see Fig. 6.11). A conventional configuration for separation of a three-component mixture into pure components requires two distillation columns and associated peripherals (see Fig. 6.12). The dividing wall column represents the most compact configuration, which allows both considerable energy and

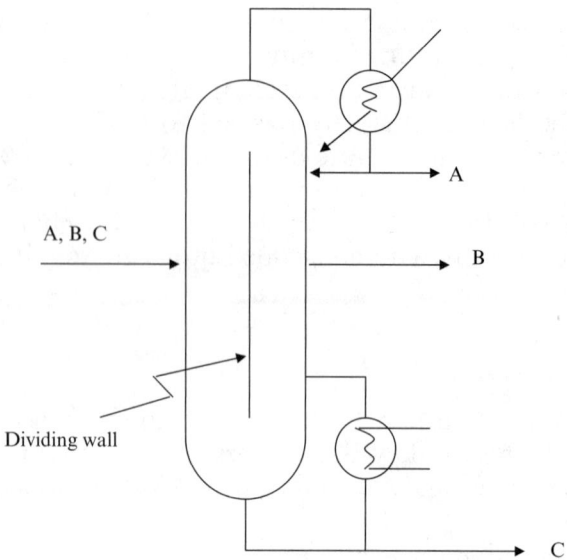

FIGURE 6.11. Dividing wall distillation column.

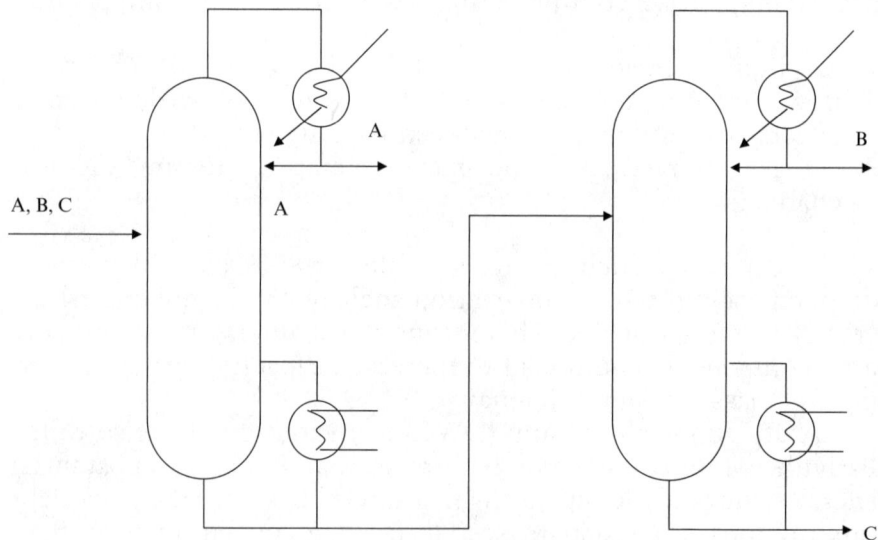

FIGURE 6.12. Conventional two distillation columns.

capital saving in the peripherals. It should be noted that the implementation of this concept requires specific constructional solutions. Energy savings with respect to conventional two-column arrangements are in the range of 30%.

J. Montz GmbH (Germany) designed and built the BASF divided wall columns. The early column designs provided a fixed dividing wall, which was welded on both sides to the column shell. Later, in order to provide better flexibility of the internal configuration and to reduce the design accuracy requirements of the system, a free, movable wall system was employed in the new designs. For larger-diameter columns, the dividing wall is built by assembling it in the column from specially designed, easy-to-install, manhole-size segments. Engineering design details such as edge seals and other construction details are the proprietary know-how of Montz.

In general, gases mix well, and low-viscosity liquids form thin films over the packing. Smaller and finer packing gives high surface area, but a column with very fine packing with countercurrent gas flow and a liquid film flowing down the bed is problematic when the liquid film thickness is around the same as the clearance between the packing. Liquid downflow stops and the column floods. The keys, therefore, are the thickness of the liquid film and identifying what controls it. The higher the applied gravity, the thinner the film and the smaller the packing can be, which would give large mass transfer surface area per volume. In order to increase gravity, the centripetal effect of rotating the packing was realized in a "high-G" machine that resulted in an order of magnitude reduction in liquid film thickness per given size. A significant increase in the liquid side and gas-side mass transfer coefficients resulted in a reduction in the HETP from 25–50 cm in conventional packed beds to about 1–2 cm in a rotating packed bed (RPB) (see Fig. 6.13). The volume of the RPB is very small and can be housed either in the reboiler or in the condenser.

Removal of monomer after polymerization is usually done by flash devolatilization at high vacuum, but residual levels of monomer will be of the order of 200 to 500 ppm. Steam devolatilization can reach lower values, but it is an expensive process, is energy–intensive, and can lead to side reactions. In a pilot RPB having a capacity of 450 kg/hr, the residual concentration of monomer is brought down to 150 ppm. The centrifugal force pushes out the polymer through a rotating packing and a circumferential set of die holes in the rotor, while N_2 gas flows inward in a countercurrent fashion (Fig. 6.13). After being extruded from the rotor,

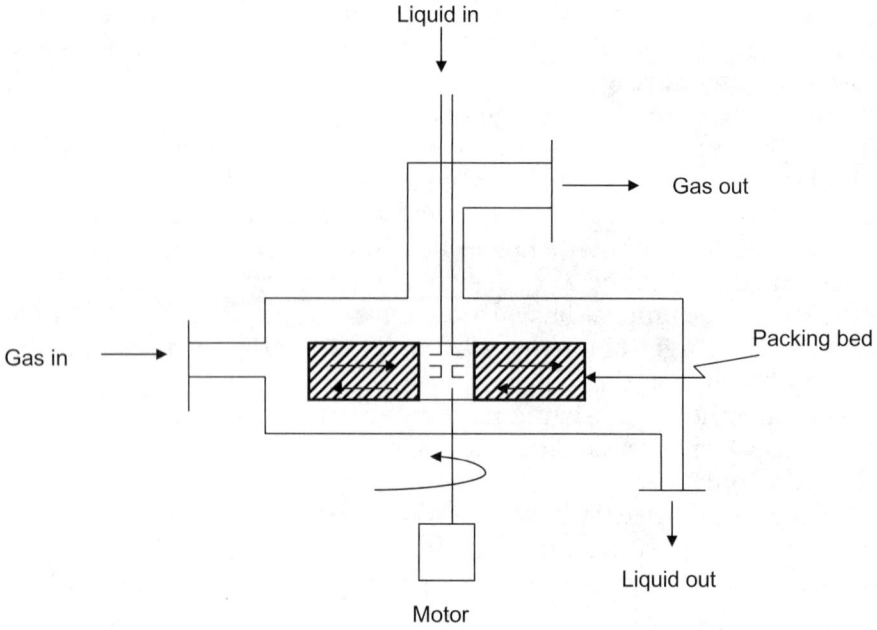

FIGURE 6.13. Rotating packed bed reactor.

the polymer strands are cut into pellets as the rotor rotates past a slowly moving, continuous blade.

Deoxygenated water is required for oil well injection to enhance oil well production and is also used as boiler feed water. Deaeration of water to achieve an oxygen content of less than 50 ppb is not possible through conventional means. Generally, it is done in a packed tower working under vacuum. Oxygen from an inlet concentration of 6 to 14 ppm can be decreased to a concentration of 500 to 1000 ppb. But in order to further decrease the amount, either a chemical is added or a second vacuum tower is used. A high-G deaeration unit of capacity of 10 ton/hr can achieve the desired O_2 levels. A high-G packed column has been there since its invention in the 1980s and has been successfully used in China in deaeration of flooding waters for oil wells and by Dow Chemicals for stripping hypochlorous acid from brines. The high-G reactor replaces 50–60-ft-tall towers and can process up to 250 tons of water/hour with 6-ft-tall equipment. They are small enough to be located on oil well platforms.

Conventional distillation column for concentrating H_2O_2 from 37% to 70% consists of trays, an external reboiler, and a condenser leading to large buildup of component, which can decompose and form an explosive mixture. Intensification is achieved by using a climbing wall reboiler connected to the bottom of the column, a direct condenser, and a structured packed column. Absence of an external reboiler or an external condenser leads to very little buildup, and hence a conventional distillation column is safe to operate. Since all equipment has a low liquid content, it operates at very low temperatures and driving force, is energy-efficient, and leads to a low decomposition rate.

Dehydration of natural gas is traditionally carried out by absorbing H_2O and CO_2 using a solvent such as glycol-water. The solvent is regenerated in a series of two distillation towers [see Fig. 6.14(a)]. Better separations can be achieved with membrane technology, which has several advantages such as passive with no moving parts, low operator intervention, capacity and quality can be incrementally added, limited pressure drop, lower energy usage, and removes water to very low concentrations (e.g., 99% removal possible) [Fig. 6.14(b)]. Separation of a hydrocarbon–benzene–cyclohexane mixture requires two distillation columns (see Fig. 6.15), while the columns could be replaced by a membrane (see Fig. 6.16).

The main disadvantage in membrane technology is methane co-permeation. Another approach that has been developed is to allow the gas to expand adiabatically in a nozzle where the temperature drops from an inlet value of 30°C to –50°C, causing condensation of water and heavier hydrocarbons. Separation of water droplets is achieved by a high-G swirl initiated by a wing (see Fig. 6.17). Twister isentropic efficiency is about 90%, and it is more efficient than a Joule Thomson valve and a Turbo-expander.

Absorption Towers

Conventional packed towers are used for different mass transfer operations such as distillation and absorption. The tower consists of a vertically positioned cylindrical shell filled with packing. Liquid enters at the top of the packed column and flows downward, and gas enters at the bottom and flows upward through the packing. Thus, a countercurrent process takes place, and the packing provides for a high area for mass transfer contact (see Fig. 6.18). The efficiency of the packed tower for mass transfer depends upon the specific area of packing and liquid irrigation rates. It has

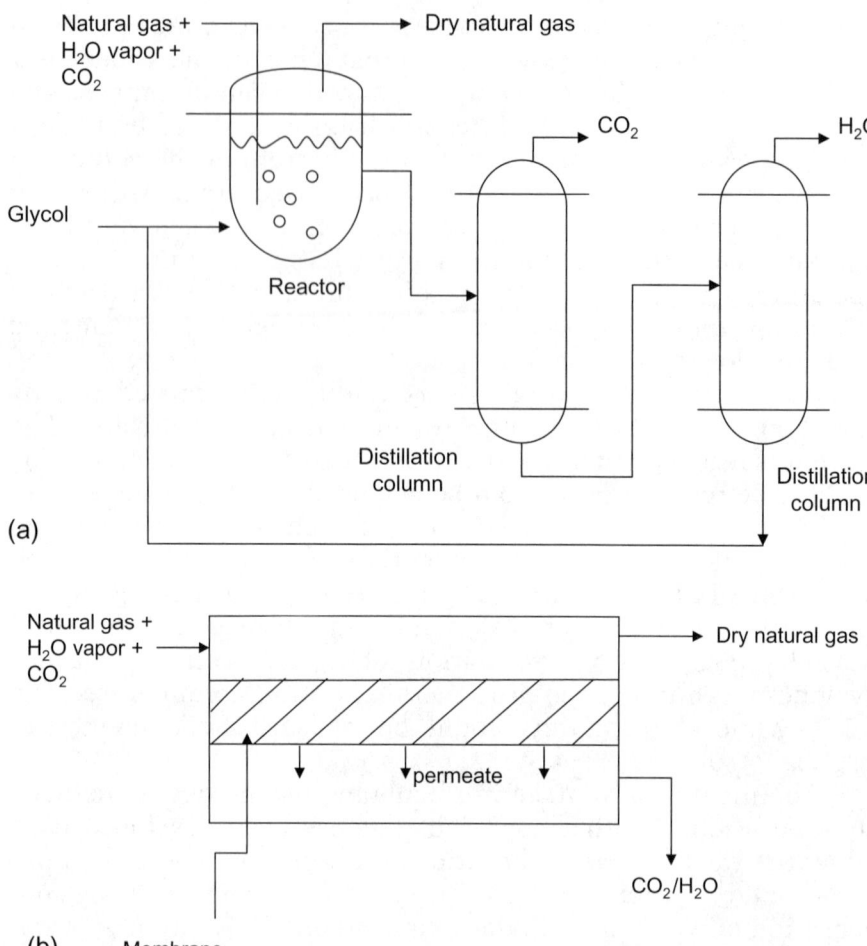

FIGURE 6.14. (a) Conventional process for drying of natural gas. (b) Membrane technology for separation.

long been found that when a liquid flows down an inclined plane, the instability of the film causes waves to form, and the mass and heat transfer coefficients under the waves of a fluid flowing down an inclined plate are very high, which is attributed to turbulence within the waves.

An RPB or high-G mass transfer unit has a torus (or "doughnut-shaped") rotor, which is mounted on a shaft and filled with high specific area packing (Fig. 6.13). Gas enters under pressure from the edge, flows radially inwards, and passes through the rotating

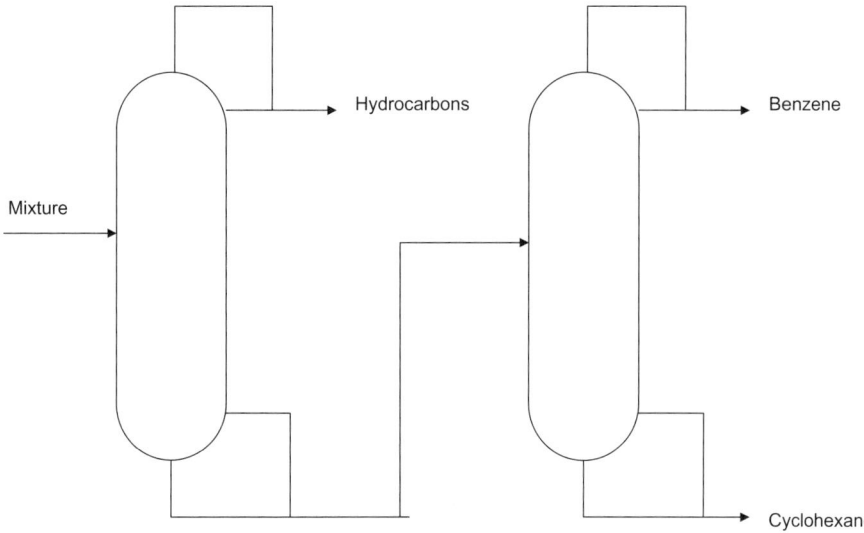

FIGURE 6.15. Traditional distillation assembly for hydrocarbon, benzene, and cyclohexane separation.

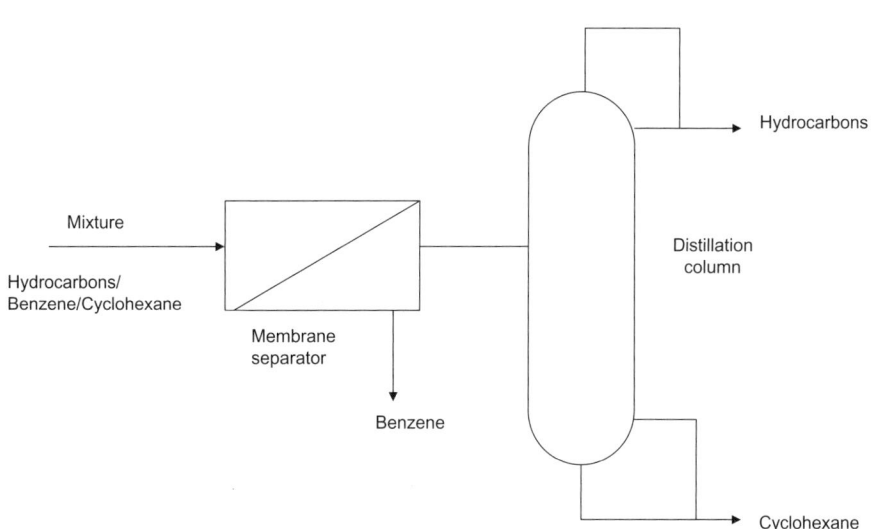

FIGURE 6.16. Hybrid membrane + distillation system for the hydrocarbon, benzene, and cyclohexane separation.

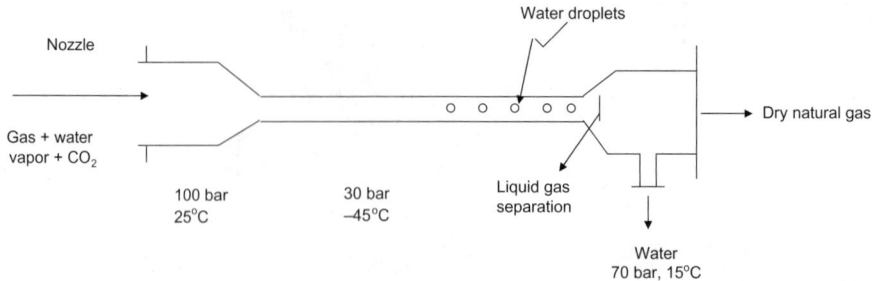

FIGURE 6.17. Adiabatic nozzle for water removal.

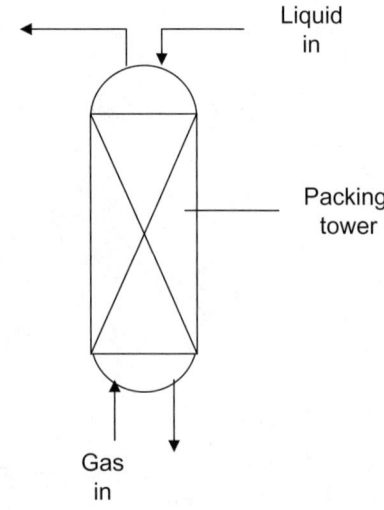

FIGURE 6.18. Conventional gas–liquid absorber tower.

material before exiting through the center. Liquid is sprayed through a nozzle into the center of the packing and propagates radially outward through the packing under the influence of centrifugal forces. The liquid, moving countercurrently to the vapor, exits at the periphery. By selecting a specific rotational speed, one can control both the residence time and the thickness of mass transfer film. The height of a mass transfer unit within an RPB is of the order of a few centimeters.

Removal of n-butanol from water is carried out using activated carbon in a fixed bed. Improvement in efficiency is made by countercurrent operations in a moving packed bed and multistage fluidized bed. Particles between 500 μm and 1 mm in diameter result in large beds with large pressure drops. PI involves using centrifugal adsorption. This design uses small particles (in the micrometer range), low space requirements, short contact times, low adsorbent inventory, and 99% efficiency. An industrial-scale sedimentation centrifuge of 20 to 50 m^3/hr has already reached the markets.

Bubble Column

Bubble columns with vibrating internals and columns with low-amplitude pulsations are currently under investigation in an attempt to improve the column performance. Vibrating internals consist of helical springs leading to 140% gas hold, which is much higher than the ones without internals. Sauter mean bubble diameter is estimated around 0.3 cm for the new type of bubble columns.

Membrane Processes

Membrane processes have the potential to replace energy-intensive processes such as distillation and evaporation, and they are particularly desirable for heat-sensitive products. The membranes are either of plate and frame design [Fig. 6.14(b)] or of tubular design (see Fig. 6.19). Of the membrane processes, reverse osmosis and pervaporation are seen to have the greatest potential. In reverse osmosis, the feed stream flows under pressure through a semipermeable membrane and the solvent flows through the membrane, leaving behind a concentrated solution of the solute. Pervaporation is a combination of membrane permeation and evaporation (see Fig. 6.20). Liquid components that diffuse through the membrane vaporize due to the partial pressure on the permeate side. The process requires only low temperatures and pressures and hence leads to considerable savings in operating cost when compared to distillation. In addition, it is an elegant method for separating constant-boiling mixtures. Pervaporation is particularly suitable for dehydration of organic solvents and removal of organics, such as methanol and acetone, from aqueous streams. It is estimated that savings in primary energy because of the use of membranes (reverse osmosis and pervaporation), melt crystallization,

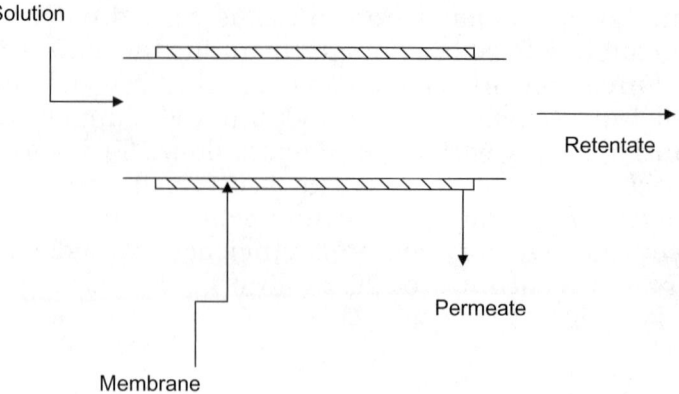

FIGURE 6.19. Tubular membrane design.

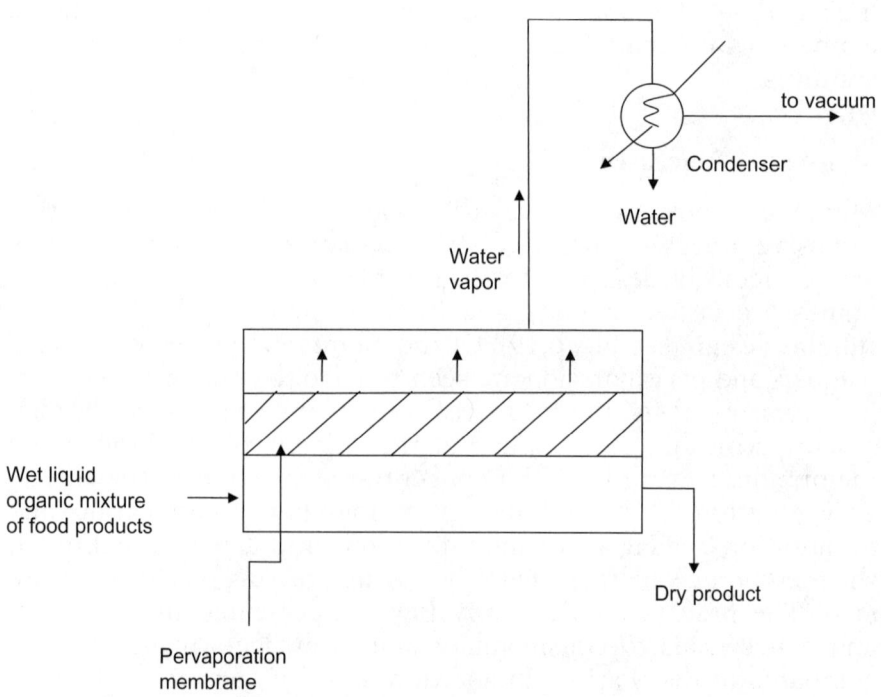

FIGURE 6.20. Principle of pervaporation.

and freeze concentration are of the order of 40% to 55%. Membrane demand is expected to go up in the United States from $10 billion in 2001 to $16.5 billion in 2011. Membranes are now viewed as an alternative to distillation.

Ammonia oxidation for the manufacture of NO (an intermediate in nitric acid production) is carried out in an oxidation reactor incorporating a Ca- and Sr-substituted lanthanum ferrite perovskite membrane. NO selectivity of the order of 98% is achieved by the membrane, while N_2 is rejected completely. The membrane reactor obviates the need for expensive noble metal catalysts and does not produce environmentally harmful N_2O.

Liquid–Liquid Centrifuges

Difficulties that often arise in separation of immiscible liquids include poor or slow phase separation, formation of emulsion layer, and poor process control in batch systems. Centrifuges accelerate separation processes by enhancing the specific gravity differences. Liquid–liquid dispersions requiring hours to separate at 1 G will proceed faster at 100 to 1000 G. They are small in size and have rapid and efficient operations. Chemical processes requiring extraction and washing (or neutralization) as well as separation can be performed in one step with liquid–liquid centrifugal contactors.

Continuous liquid–liquid extraction is carried out in a centrifugal contactor at low rotor speeds to achieve 90% efficiency. Mixing and disengaging times range from 10 to 30 s, respectively. Efficient two-phase mixing is achieved in the annulus between the spinning rotor and fixed housing.

In 1997, a commercial installation at Great Lakes Chemical at El Dorado, Arkansas (USA), replaced a 4000-gal decantation tank by an annular centrifuge to efficiently separate brominated polymer product from the aqueous waste at the rate of 45 L/min. The increase in efficiency led to a 3% improvement in product recovery, which represented an extra profit of $400,000 per year.

The ability of a centrifuge to thoroughly mix two phases in the annular zone prior to separation in the rotor broadens its scope. The annular centrifuge has an upright design in which the vertical rotor pumps, thereby feeding itself. Two immiscible liquids of different densities are fed in the annular space between the spinning rotor and stationary housing. The mixed phases are directed toward the center of the rotor bottom by radial vanes in the housing base. As the liquids enter the central opening of the rotor, they are accelerated toward the wall and separation begins as the liquids are displaced upward by continued pumping. The annular centrifugal contactors operate at low rpm and moderate gravity (100–1000G). Normally, 4000 to 12,000G/s of force is adequate to efficiently separate two immiscible liquids. For separations where the specific gravity differences are slight, centrifugal forces of the order of 50,000G/s can be obtained by merely slowing the feed rate solidus to the contactor.

Extraction Columns

In conventional extraction columns (see Fig. 6.21) for a heavy or light solvent column, flooding places an upper limit on the maximum throughput. If the density difference between the two phases is low, separation becomes a problem. When a suitable solvent is immobilized in a porous medium, absorption and extraction can be carried out like conventional pressure swing adsorption. Consider that a solvent such as heavy oil is immobilized in the pores of the particles of the bed. When a gas mixture of CH_4 and H_2 is introduced into the bed, CH_4 is absorbed by the solvent immobilized in the bed. Pure H_2 escapes through the bed. After breakthrough, CH_4 can be removed from the oil using a vacuum.

Microchannel devices are also successfully used for gas–liquid and liquid–liquid contacting operations such as extraction and

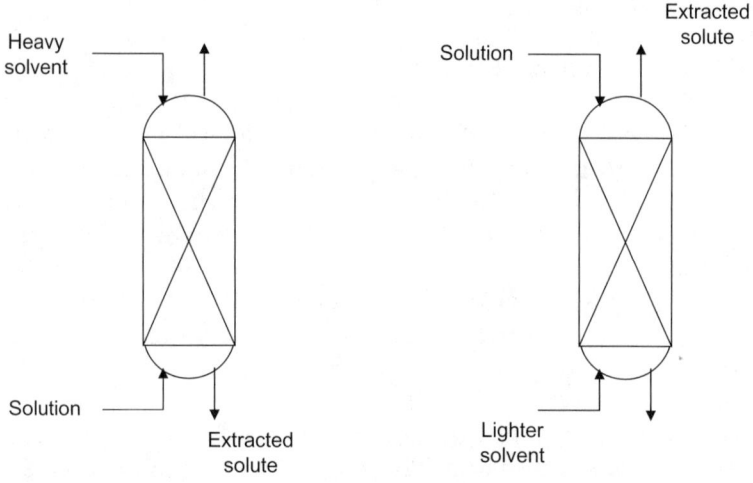

FIGURE 6.21. Conventional liquid–liquid extraction columns.

absorption. Several orders of reduction in mass transfer resistances have been reported with these microchannel contactors (TeGrotenhuis et al., 1998, 1999).

Reactions with Separation Operations

Examples of PI where reactions are combined with separation operations include

1. Reactive distillation. To avoid conversion of product into unwanted components or to drive equilibrium reactions forward, reactive distillation is used to distill out one of the products immediately as it is formed. A conventional equilibrium process consists of a reactor and a distillation column. One of the products is removed in the column, and the bottom is recycled (see Fig. 6.22). In reactive distillation the column is located on top of the reactor, obviating the need for a reboiler (see Fig. 6.23). Reaction of an organic chlorine compound with an aqueous caustic solution can be carried out in a reactive distillation column, which will combine functions such as dissolution of

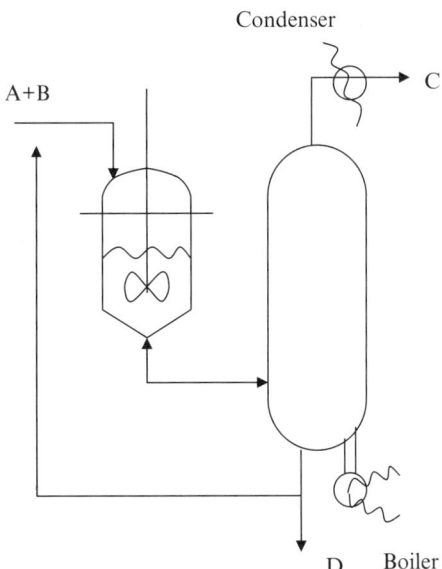

FIGURE 6.22. Conventional reaction cum separation.

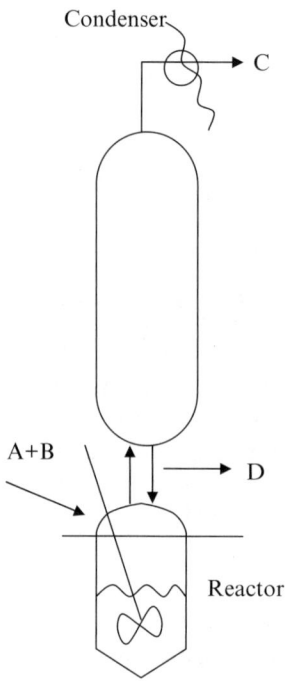

FIGURE 6.23. Reactive distillation.

    alkaline solution, reaction, separation of product from the reaction mixture, through evaporation into the gas phase.
2. Precipitative cum evaporative reactor. This design combines several functions such as reaction, solvent removal by evaporation from the reaction mixture, and separation of solid product by precipitation from the reaction mixture. This design eliminates byproduct formation and controls the crystal size distribution accurately.
3. Liquid–liquid extraction reactor. This combines the functions of liquid–liquid extraction and reaction. The process involves alkaline transfer to the reaction phase, reaction, and extraction of salt to the aqueous phase.
4. Reaction cum pervaporation. This technology can be used for dewatering of organics, which is the removal of organics from waste water in place of the conventional distillation operation. A membrane is the heart of the operation. A combination of reaction and pervaporation, the latter of which occurs when placed in a loop around the reactor, replaces a condenser

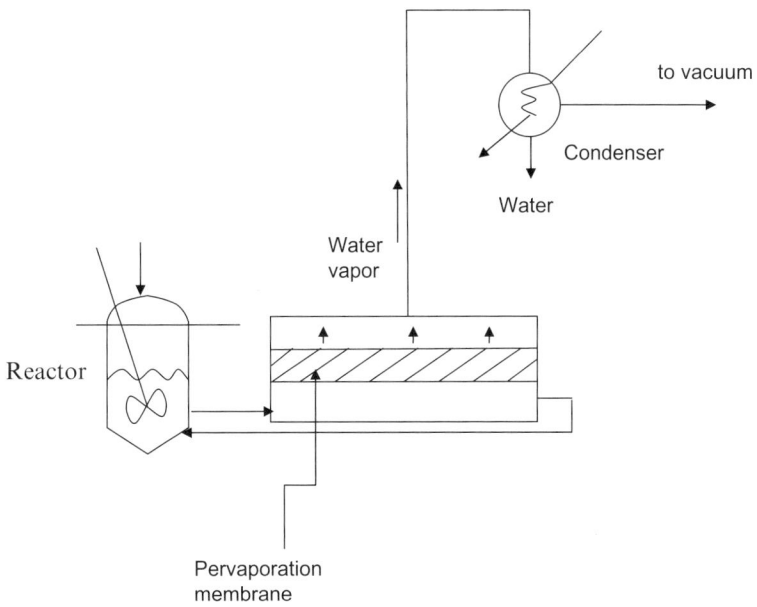

FIGURE 6.24. Reaction cum pervaporation.

assembly (see Fig. 6.24). Akzo Noble (Netherlands) has developed 1-m² ceramic membranes for use in its pervaporation process. The pervaporator could be a plate or tube type. Sulzer (Switzerland), Chemtech (USA), and ECN (Netherlands) have built demo units of such systems at ~180°C and ~10 bar. Fouling of the membrane and its long-term stability are two main issues that need further improvement.

5. Methyl acetate. This is prepared by reacting methanol and acetic acid using an acid catalyst to yield the acetate and H₂O. The reaction is slightly exothermic and has two problems: (1) methylacetate and water form a homogeneous azeotrope, and (2) separation is achieved either by distillation or by decantation. Methylacetate and methanol form a homogeneous low-boiling azeotrope, and removal of acetate removes methanol and lowers conversion. The conventional process consists of a reactor followed by four distillation columns and associated peripherals. An intensified process consists of carrying out the reaction in an 80-m-tall column with a reduced number of components. The process is 80% less energy-intensive than the former.

6. Styrene–butadiene rubber. This is produced by coagulation of latex. This is followed by various unit operations such as washing, extrusion, dewatering, and drying to give a product in the form of crumbs. Usually, all these operations are carried out in separate equipment; however, these operations have been combined into a single piece of equipment (www.ccdcindia.com). The intensified plant is used for several different products, such as ABS, SBR, NBR, and CR. The other advantages are reduced water usage and waste generation; possibilities of recycling of solids and water; recovery of unused monomer; lower energy consumption, utilities usage, and manpower requirements; and an environmentally friendly product. Plants have been built with capacities of 100–7000 kg/hr. Typical 2000-kg/hr main equipment occupies a floor area of 25 m², which is very small when compared to the 400 m² taken up by a traditional plant.

Novel Hydrogenation Processes

A conventional hydrogenation process is carried out in a three-phase slurry reactor (see Fig. 6.25) in a batch stirred vessel. Gas-phase hydrogenation is generally carried out in a fixed-bed reactor containing a noble metal catalyst (see Fig. 6.26). The reaction is fast, with a small reactor size, but its short residence time gives very few byproducts (<0.1%) at reasonable operation pressure. Energy consumption is high, 500 GJ/kton product, since the total

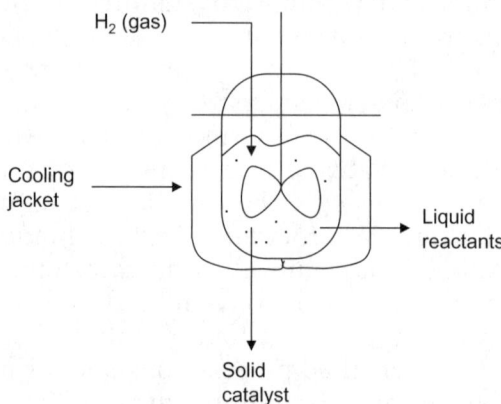

FIGURE 6.25. Batch hydrogenation process.

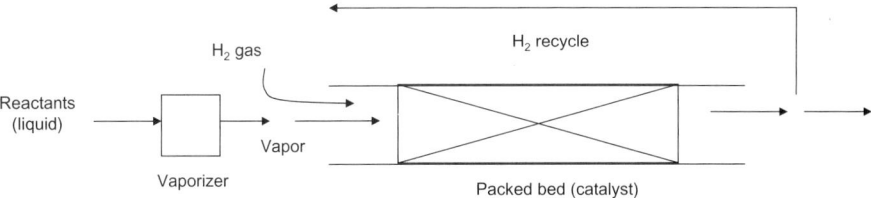

FIGURE 6.26. Gas-phase hydrogenation process.

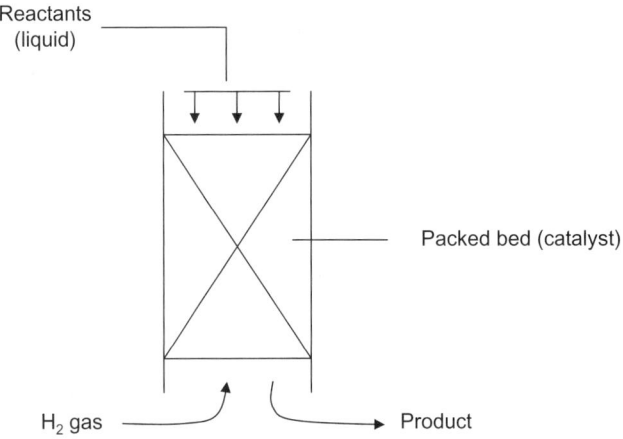

FIGURE 6.27. Trickle bed hydrogenation process.

reactor feed needs to be evaporated. Excess H_2 gas needs to be circulated by blowers. This technology also needs heat exchanger for vaporization of the reactant.

Liquid-phase hydrogenation does not need vaporization and is simple. But a conventional trickle bed reactor (see Fig. 6.27) is generally very large, needs large amounts of catalyst, and has to be operated at a higher pressure. The residence time is long, leading to byproduct formation (~5%). Absorption of sparingly soluble H_2 into the liquid is the first step and is rate-limiting. The reaction is mass transfer controlled, and the catalyst is only partially wetted. The apparent reaction rate is 10–15% that of Langmuir–Hinselwood kinetics, so better wetting is needed to improve the rate of reaction. A high-G, rotating, packed contactor with the reaction in liquid phase leads to lower energy usage and better gas–liquid mass transfer rate (Fig. 6.13).

The rate of hydrogenation of α-methyl styrene over a rotating string of catalyst particles was 5 to 7 times faster than that of the rate achieved in a gravity flow reactor. A 60× faster rate is achieved compared to a trickle bed of alumina beads that are 2.5 mm in size. An industrial trickle bed reactor of $60 \, m^3$ can be replaced with a 1-m^3 rotating bed reactor.

Liquid-phase catalytic hydrogenation involves dissolving H_2 gas in a solvent such as propane or dimethyl ether at a super critical pressure to achieve a single phase (see Fig. 6.28), thereby reducing the number of resistances (see Fig. 6.29).

The liquid-phase hydrogenation of citral using 2.5% Pd metal supported on a clinoptilolite-rich natural zeolite catalyst leads to high selectivity (90%) to citronella. The process has been developed by the İzmir Institute of Technology (Turkey).

The reaction rates are high, and the product quality is good. Heat removal with the new technology is very easy. It can be used for the hydrogenation of organic compounds, fats, oils, and polymers. C. Costello & Associates (USA) reached an agreement with Härröd Research (Sweden) to act as licensing agent for Harrod's Supercritical Single-Phase Hydrogenation Technology. The plant has a very small reactor, and the company claims its installed cost is 25% that of traditional three-phase hydrogenation processes.

Narrow channeled microreactors are good for rapid heterogeneous liquid–liquid reactions. Narrow channels can be manufactured

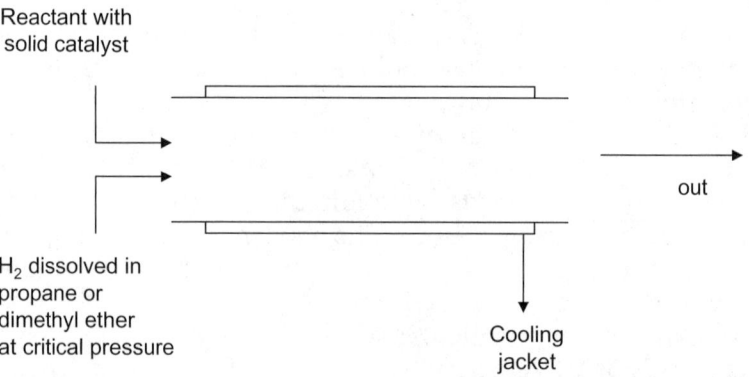

FIGURE 6.28. Liquid-phase hydrogenation.

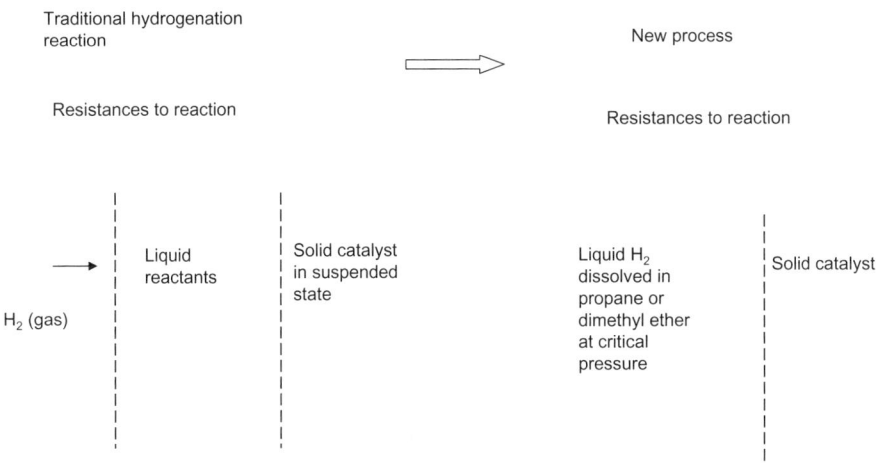

FIGURE 6.29. Resistances in hydrogenation reaction.

by a variety of methods to give a channel width of the order of 10 to 1000 mm. The channels may be etched (mechanically or chemically) into metal, glass, or plastic sheets. A plain sheet is placed on top of this etched plate to enclose the channel. It is possible to incorporate microelectronic components (e.g., miniature pumps, valves, and sensors) within the channels creating a miniature manufacturing process, commonly known as "lab-on-a-chip" (Kobayashi et al., 2004).

Innovative Processes

A general esterification reaction consists of reacting an alcohol with an acid in the presence of a catalyst (such as sulfonic acid) to produce the ester and water. This is an equilibrium reaction and leads to low conversion. The catalyst is usually neutralized with inorganic base after the completion of the reaction. If carried out in a countercurrent reactor under two-phase conditions, this reaction has many benefits. For example, conversion of maleic anhydride to dialkyl maleates or fatty acids to fatty acid esters is performed in a column packed with a solid catalyst. Liquid (acid) flows down the column from the top. Alcohol vapor flows upward from the bottom and absorbs water that is formed and carries it up (see Fig. 6.30). The removal of water by the alcohol drives the

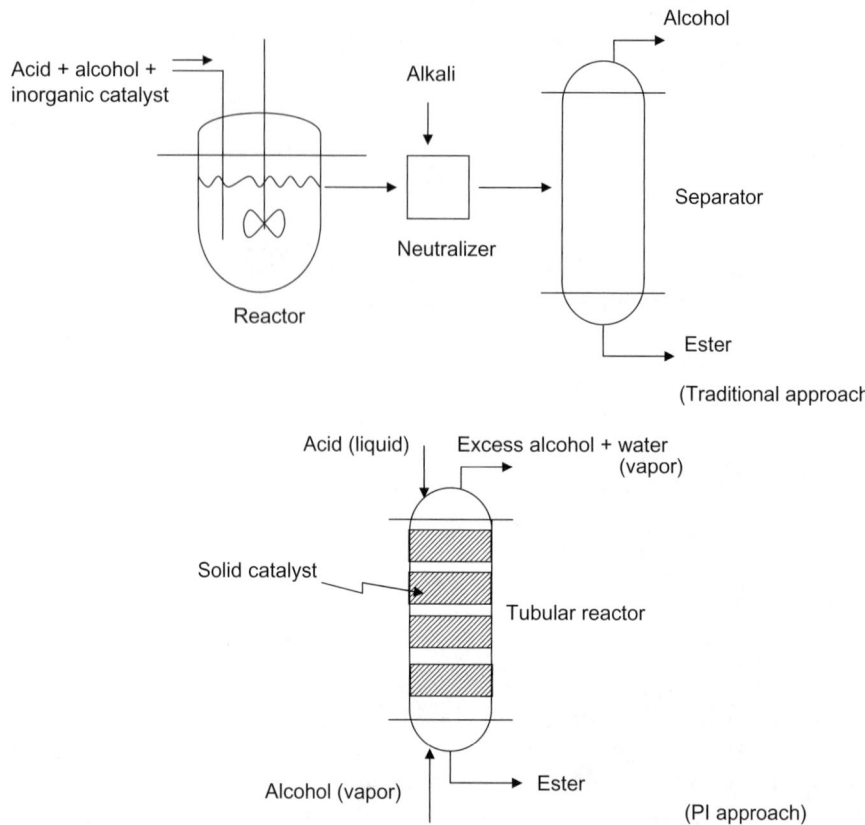

FIGURE 6.30. Esterification reaction.

equilibrium to the right, achieving very high conversion. Other advantages are that the solid catalyst enhances the reaction rate, there is no catalyst removal from process streams, the reactor is small, a high-purity ester is used, and thus no downstream treatment is needed.

The traditional production of hydrogen peroxide from anthraquinone is carried out in two reactors. In the first reactor anthraquinone is hydrogenated, which the second reactor oxidizes to give hydrogen peroxide. This second reactor is a bubble column where air is bubbled and the hydrogen peroxide is extracted from the organic (anthraquinone) medium by water. This unit operation is followed by purification and concentration of the product (see Fig. 6.31). The process improvement consisted of performing the oxidation reaction in a countercurrent tubular reactor where the reaction is carried out in the first half of the tube, and the product

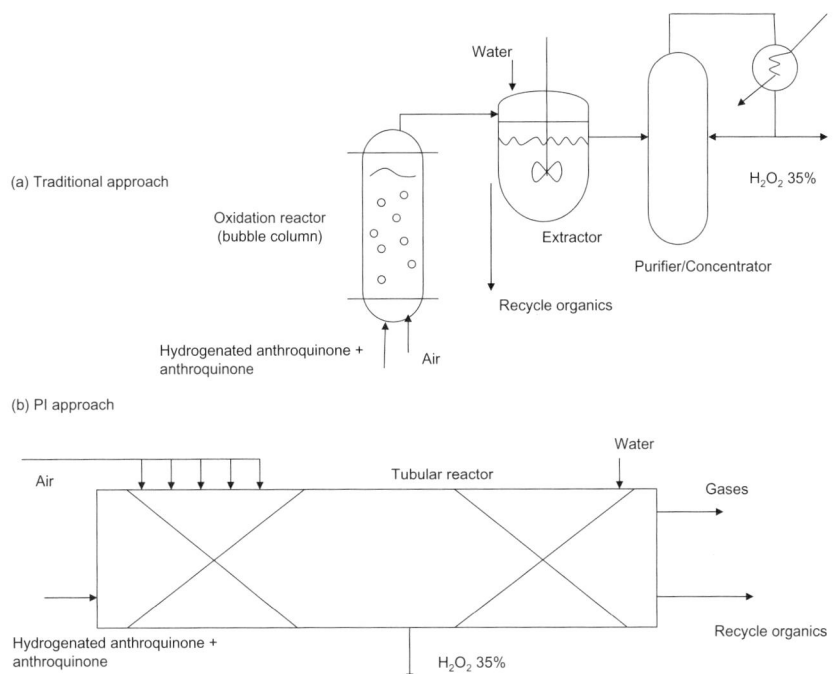

FIGURE 6.31. Hydrogen peroxide production by traditional and PI approaches.

separation using water as the extraction solvent is performed in the second half of the tube.

Linoleum production consists of two unit processes, namely the esterification and oxidation, and several unit operations, namely the mixing, calandering, drying, and finishing. Generally, the production is performed in batch mode, which means extended process time, poor quality control, large energy consumption, and time lost due to reactor cleaning. The process uses several raw materials: linseed oil, tall oil, natural resins, wood sawn, cork, limestone, and pigments. Tall oil esters and linseed oil esters are prepared before the oxidation step to reduce time needed for oxidation. Later a mixture of tall oil ester, linseed oil, and resins is sequentially sent to the oxidation reactor while air is sparged in, which acts as an oxidant. There are four phases in the oxidation process, requiring different operating conditions:

1. the induction step (heating up to 130°C to remove antioxidants, which hamper the oxidation),

2. peroxide formation (at 85°C; the reaction is exothermic),
3. peroxide breakdown due to the addition of the catalyst (which creates active groups or radicals),
4. polymerization (which happens because of the reaction between the radicals).

The reaction is moderately exothermic, and the total residence time is 16 to 30 hours. The viscous mass is discharged to curing tanks, where additional oxidation takes place for about 10 days.

The new PI process is performed in three continuous stirred tank reactors placed in series with air flowing countercurrently. As mentioned, the various process functions are activation of linseed oil in the first CSTR (where linseed oil preoxidation takes place autocatalytically at a lower temperature, thereby deactivating the antioxidants), oxidation in the second reactor (mixing of tall oil and activated linseed oil creates a medium wherein components are oxidized and the polymerization is also initiated), and polymerization in the third reactor. A pilot unit with a reactor volume of 2.5 L was built and tested by Forbo Linoleum, Netherlands. The product was converted into linoleum, and its properties were found to be in the acceptable range.

Oscillatory Flow Mixing Reactor

An oscillatory flow mixing reactor (OFM) consists of baffles placed perpendicular to the axis to form several compartments. The fluid is oscillated between 0.5 and 15 Hz—at an amplitude of typically 1 to 100 mm (see Fig. 6.32). The oscillating fluid motion interacts with each baffle to form vortices, and the resulting fluid motion gives efficient and uniform mixing in the space between the two baffles. Oscillations lead to significant enhancement of heat and mass transfer. Mixing is controlled entirely by the oscillations and not by the throughput. The observed enhancements are effective at low volumetric flow rates (Reynolds number <300). Also, long residence times require only short tubes. Intensification of a batch process could be achieved in a tubular reactor of long residence time without resorting to high tube velocities. It was reported that for a given throughput the heat transfer increases with increasing oscillatory Reynolds numbers. At high throughput, the oscillatory curves approach the steady flow curve. The ideal application of oscillatory flow is in a tubular reactor, which would offer plug flow residence time distribution (RTD), high heat transfer, finely

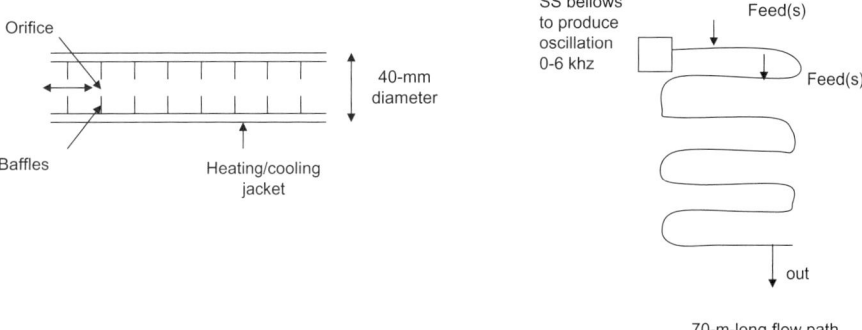

FIGURE 6.32. Oscillating baffled reactor (orifice).

controlled mixing, and small reaction times. NiTech solutions have used the continuous oscillatory baffled reactors (COBRs) for effectively enhancing mixing, dispersion, and controlling the formation of solid particles.

Advanced Loop Reactor

The advanced loop reactor (ALR) is used for highly exothermic reactions, such as that between ethylene oxide and an alcohol to produce glycol ethers. A heat exchanger is included in the loop to remove the heat formed (see Fig. 6.33). The distribution of the glycol ether chain length is influenced by the mixing at the point of the ethylene oxide feed. An efficient mixing reduces localized inhomogeneity and temperature peaks and hence narrows the chain length distribution.

Polyaniline is produced by reacting aniline and an oxidizer such as ammonium persulfate. When carried out in a batch stirred tank, the reaction is highly exothermic and temperature control is a major issue. Also, mixing of the highly unstable oxidizer is a major challenge. These problems can be overcome by carrying out this reaction in a static mixer, which results in more uniform concentration and temperature distributions compared to a stirred tank reactor. Better product quality and lower molecular weight distribution are also obtained. The oxidizer is introduced through several feedpoints that prevent hotspots.

For the manufacture of 2000 tons of P_2O_5/day, a U.S.-based company (www.ccdcindia.com) using the loop reactor was able to reduce the volume by half, power consumption by $1/3$, number of

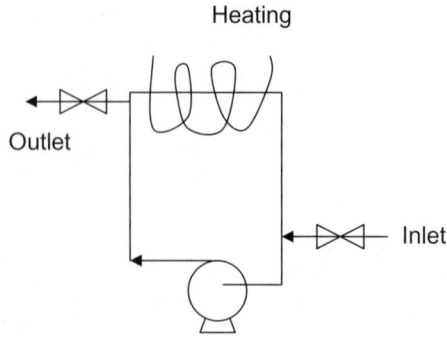

FIGURE 6.33. Loop reactor.

pieces of equipment from 30 to 9, and number of motors from 15 to 3 when compared to the conventional approach. The other advantages claimed are lower environmental emissions and isothermal conditions.

Chlorination using thionyl chloride is mainly carried out in a batch reactor in which the chloride is added slowly to the reaction mass. The process can be made continuous using a loop reactor with a heat exchanger in the loop. The productivity is 340 kg/hr/m^3 of reactor volume, which is several orders of magnitude larger when it is carried out in a batch reactor (which is 10 kg/hr/m^3). Also, 18 m^3 of a glass-lined batch reactor could be replaced by a loop reactor with a volume of 0.5 m^3. A laboratory-scale model would have throughput of about 10 kg/hr, and it could be sufficient to produce 5 to 6 tons of product per month.

Chemical Reactors cum Heat Exchanger

A chemical reactor cum heat exchanger (HEX reactor) design (BHR Group Ltd., UK) is suited for highly exothermic reactions where heat needs to be removed as quickly as it is produced. This design leads to inherently safe processes and also results in significantly improved product yield. The unit is composed of a stack of plates that are etched photochemically to form a series of slots. The plates are stacked such that series of slots form discrete flow paths. Adjacent flow paths are separated by means of intervening solid plates. Two or more separate flow paths may be formed across a group of plates, enabling different fluid streams. Multipoint injection of secondary reactant is permitted, which allows the heat release to be spread across many layers, with each heat-producing layer located adjacent to a coolant layer (www.ccdcindia.com). This design makes it possible to remove the heat of reaction almost as rapidly as it is formed. The reactor is generally made of SS or Hastelloy. Such a reactor can be made of thin, corrugated sheets of polymer such as polyether ether ketone (PEEK) (approximately 100 mm thick), facilitating low conduction resistance. PEEK offers mechanical strength and robustness, can withstand 10-bar pressure, and operates up to 220°C. It also exhibits excellent corrosion resistance and has a very high heat transfer coefficient (~4000 W/m²k). These heat exchangers have applications in the food, aviation, automobile, and fuel cell industries.

Reaction between 1- and 2-naphthol with diazotized sulfanilic acid leads to four different dye products, each formed at rates that differ from one another by several orders of magnitude. A traditional batch reactor leads to a mixture of dyes, whereas when the reaction is performed in a HEX reactor, the product's distribution could be controlled precisely with no side products.

Alfa Laval AB (Sweden) has developed a new reactor system, called Alfa Laval Plate Reactors, which combines the high heat transfer capabilities of plate heat exchangers with the efficient mixing of microreactors into a single unit. The manufacturer claims that the working inventory of potentially dangerous reactants is reduced by up to 95% compared to a conventional stirred tank reactor (Alfa Laval, 2006). This feature makes the reactor especially suitable for performing highly exothermic or explosive reactions. The ART system consists of a reactor sandwiched between heat exchanger plates. The reactor is either a plate with millimeter-deep channels machined onto the surface or a structured plate through which the reactants follow a torturous path to ensure efficient mixing. The system is designed so that reagents can be injected at different locations along the flow path.

Tube Inside Another Tube Reactor

The Holl Technologies Company's (U.S.) STT™ reactor (see Fig. 6.34) consists of a tube inside another tube, where the inner tube spins inside the annular tube, whereby only a very small gap is maintained. The reactants fill the slightly eccentric gap. Immediately upon entry the reactants encounter a very large interfacial contact area leading to extreme rates of surface renewal. Typical shear-rate values are in the range of 30,000/s to 70,000/s. Process rates in reactors are influenced by the minimum length of turbulent eddies and the molecular diffusive mixing time. This reactor

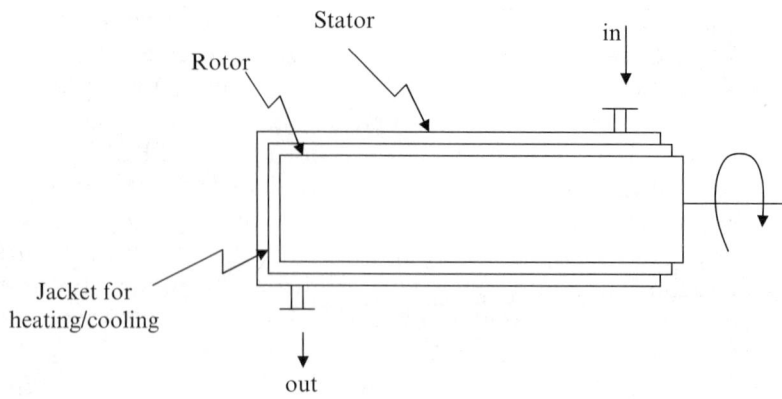

FIGURE 6.34. Holl spinning tube in a tube reactor.

design is capable of creating sub-Kolmogoroff and near-Kolmogoroff eddies and therefore can reduce reaction time. Mixing power requirements also decrease when compared to conventional reactors. The other advantages of a tube inside tube reactors are smaller size (leading to reduced floor space), negligible hold-up volume, continuous operation, self-cleaning, hermetically sealed operation (which reduces hazards), simple construction, and low operating costs. Immiscible materials rapidly interact with each other, achieving perfect homogeneity. It is ideal for emulsification processes and can be used with up to 95% solids. The reactor performance is based on area rather than volume. Heat transfer film coefficients of up to $10,000 \, W/m^2 \, (°K)$ could be achieved in this reactor, reducing unwanted byproduct formation due to wall effects, which normally happens when the temperature gradients across the diameter of the vessel are large, as is the case in large diameter vessels. According to the manufacturer, this design has been successfully used for chemical reactions such as rearrangement, addition-elimination, substitution, condensation, polymerization, catalysis, electrochemistry, and fermentation. It has also been used for photochemical reactions by using stators made of quartz.

Rotating Packed Bed Reactor

A rotating packed bed (RPB) or "high-G" equipment is very different from a conventional packed tower design (Fig. 6.13). Due to high acceleration forces upon the fluids, thinner mass transfer

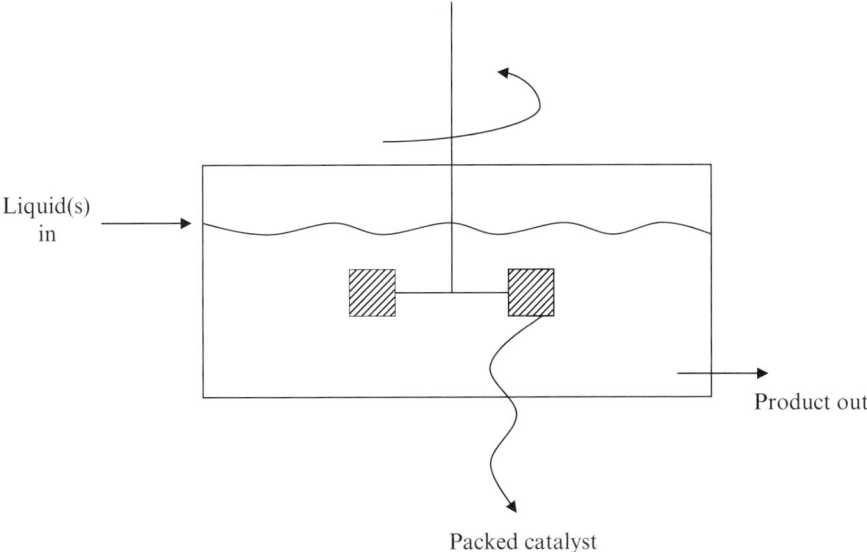

Liquid(s) in

Product out

Packed catalyst

FIGURE 6.35. Spinning basket reactor.

films are formed, leading to higher rates of gas–liquid mass transfer. In addition, high voidage and high specific surface area of the packing material also lead to high mass transfer rates. This design is several orders of magnitude more compact than the conventional packed bed. Another variation of this design is the spinning basket reactor (see Fig. 6.35), where the catalyst placed in a basket rotates at high rpm in a pool of liquid, creating very high liquid–solid mass transfer.

HOCl is produced by reacting chlorine and aqueous caustic soda and subsequently has to be stripped from the brine solution, which is the side product. The many challenges here are as follows:

- Kinetics of the desired reaction is fast, and decomposition products are formed at high residence times.
- HOCl has a low vapor pressure, making it more difficult for it to be stripped from the reaction mixture.
- A solid salt product is formed.
- The energy requirement is high.
- Cl_2 needs to be recycled.
- Part of the process operates near the lower explosion limit of Cl_2O.

The original process yield was 65–80% but improved to 75–80% at Dow Chemical Company (USA) by incorporating a spray distillation assembly with absorption of HOCl in water. The latter unit operation required a large gas to liquid ratio in order to carry out the operation at low gas velocities for minimizing the liquid entrainment and carryover. This required large-diameter equipment, leading to high capital costs. A rotating packed bed (RPB) gas–liquid contactor produces low-chloride HOCl, where stripping of HOCl from sodium chloride brine is achieved completely. The RPB is a small size, producing higher HOCl yields, and uses lower volumes of stripping gas. The commercial application of RPB technology for the production of low-chloride HOCl solutions has been successfully operating for more than several years at Dow Chemical Company (USA).

GasTran's (U.S.) rotating packed bed units use specially engineered materials to shear an incoming fluid stream into ultrafine droplets. The continuous shearing and coalescing of the liquid expose the surface area to the gas medium, enhancing the efficiency of chemical processes.

The rate of hydrogenation of α-Me styrene with palladium as catalyst in the rotating beds of spherical particles and metal foam support was in the range of 30 to 40 times in a centrifugal force field of approximately 450 times the gravitational force field. A trickle bed industrial reactor of $60 \, m^3$ could be replaced with a rotating bed that is less than $1.5 \, m^3$ in volume. Industrial trickle bed reactors are bulky because of poor interfacial transport.

Catalytic Plate Reactor

A catalytic plate reactor (CPR) has metal plates with channels or grooves appearing in a crisscross manner. These channels typically have a height in millimeters and catalyst thickness measured in microns. The grooves are coated with a suitable catalyst and can be arranged so that exothermic and endothermic reactions take place in alternate channels (Fig. 6.6). Heat transfer is by conduction from the exothermic to the endothermic region. The advantages of CPR designs over conventional reactors arise due to the facts that heat transfer rates (through conduction) are very high and that intracatalyst diffusion resistances are minimal. These advantages lead to reactors that are smaller in size, with lower-pressure drop than conventional alternatives. The potential savings on capital cost are enormous. Catalytic plate reactors (marketed by Protensive, UK) can intensify a number of industrial gas-phase reactions as well as promote in-situ reforming of fuel feedstock for commercial fuel cell applications.

The steam reformation process is demonstrated in the CPR by coupling endo- and exothermic reactions. Steam reforming of methane is a highly endothermic reaction and is carried out with the energy being provided by the catalytic oxidation of methane, which is an exothermic process. The CPR for steam reforming suffers from two major disadvantages: (1) it is difficult to replace the catalyst when it is exhausted; (2) since the rate of heat generation decreases as the fuel is depleted (rate approximately of the order of $[CH_4]^{0.76}$), the last section of the reactor contributes very little to the overall conversion of the fuel. Another important application of CPR is the production of syngas, which is the feedstock for many industrial processes.

Scale-up in CPR is achieved by adding more plates in the stack rather than by changing the size of the plates. This leads to reactor performances invariant with scales, thus reducing the time required for scale-up and for moving from lab to commercial production. The replacement of the homogeneous combustion used in conventional reactors by the catalytic process leads to several advantages, including a lower temperature, meaning fewer constraints for materials of construction of the reactor. CPR does not produce any NO_x, and since it is a flameless process, long radiation paths needed in conventional fired furnaces are replaced by channel dimensions of 1 or 2 mm, leading to very compact reactors.

The hydrocarbon product spectrum produced by a Fischer–Tropsch (FT) catalyst is highly dependent upon the catalyst temperature and the rate of diffusion of reactants into the catalyst matrix. The reaction is highly exothermic and, when it is carried out in a fixed bed reactor, if rates of heat removal from the catalyst are not high, hotspots will form, resulting in variation in the product spectrum and also catalyst deactivation. Thin catalyst coatings coated to heat transfer surface areas in a CPR have been found to greatly enhance the yield of the desirable products per unit volume as compared to conventional fixed bed reactors. The reduction in reactor volume and reduction in associated equipment and low-pressure drop make the FT-CPR an attractive technology.

Hot Finger Design

"Hot finger" reactor design (see Fig. 6.36), which has been patented by Protensive (U.K.), consists of two annular tubes (an inner and an outer). It addresses the problem of catalyst deactivation in CPR

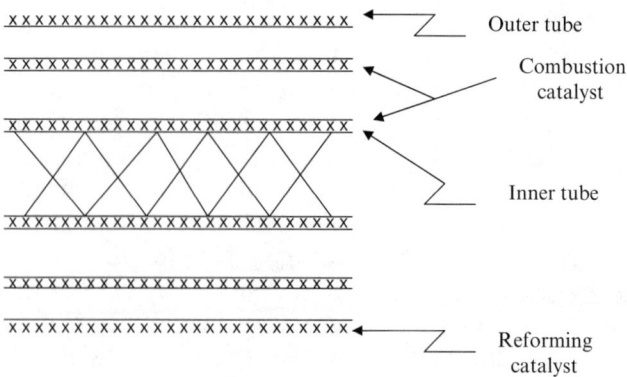

FIGURE 6.36. Hot finger reactor.

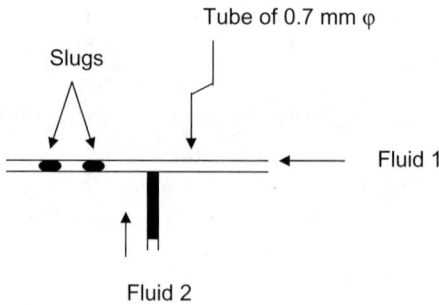

FIGURE 6.37. Slug flow—narrow channel tube.

by providing several easily removable sections, which can be dismantled individually and replaced with sections containing active fresh catalyst. Similar to CPR, it has the provision for carrying out exothermic and endothermic reactions in the two annular tubes.

Slug Flow Reactor

Slug generation within the narrow channels of glass or polymer chip reactors are generally achieved by the continuous pumping of two or more liquid phases into a "T" or "X" channel configuration (see Fig. 6.37). These channels are about 0.38 mm wide and deep. Nitration of benzene and toluene in a slug flow design is used to provide high conversion to nitrobenzene and nitrotoluene,

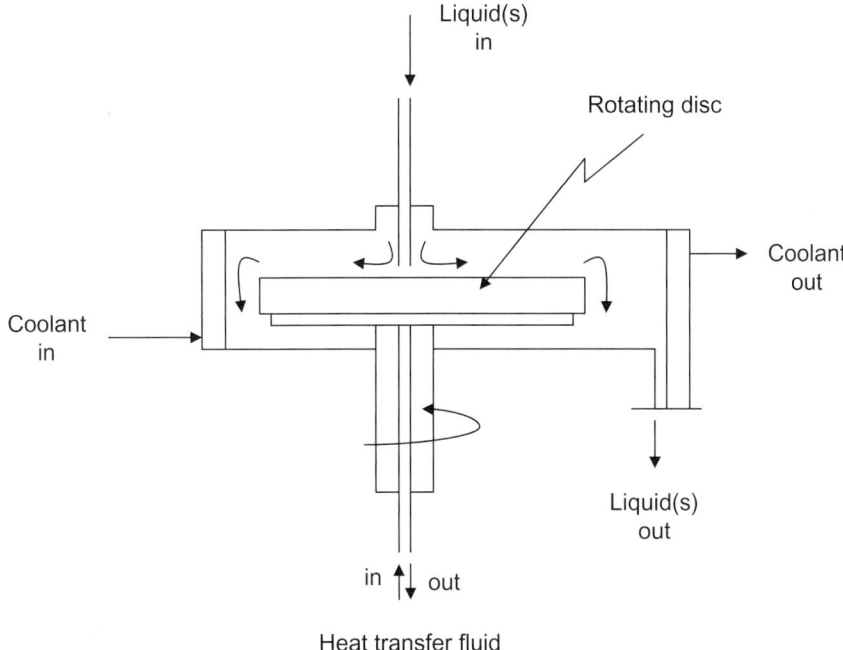

FIGURE 6.38. Spinning disc reactor.

respectively, in seconds using coiled PTFE reactors that are 30 to 180 cm long with a 0.15-mm inner diameter. Control of highly exothermic processes is easier in such small slugs, where each slug acts as a well-mixed homogeneous mass.

Spinning Disc Reactor

Spinning disc reactors (SDR) are capable of operating horizontally or vertically and are mounted on a rotating axle (see Fig. 6.38). Liquid is fed near the center and flows across the surface of a spinning disc under the influence of centrifugal force. This force stretches and spreads the film. The thin liquid film allows for high rates of mass transfer so that it favors unit operations such as absorption, stripping, mixing, and reactions. Residence time on the disc is in the range of 0.1 to 3 s. Both film thickness and residence time are dependent on fluid physical properties, rotational speed, and radial location of the fluid. On exiting the periphery (edge) of the disc, the liquid is thrown onto the enclosing wall and then drains away. The wall is heated or cooled depending upon the process requirements. SDR uses a 100-mm-diameter disc rotating at 600 to 1200 rpm and heated to 150°C using a heat transfer fluid circulated through a chamber below the rotating disc. This leads to very high heat and mass transfer coefficients. There is no back mixing. The generation of films of typically fractions of a milli-meter down to a few microns, through controlled flow and disc speed, can deliver surface to volume ratios of the order of 1000 m^2 per m^3 for high-viscosity materials (such as polymer melts) to 100,000 m^2 per m^3 for low-viscosity systems (such as most of the organic chemicals). Typical mass transfer coefficient values are in the range of 0.01 to 0.03 cm/s for low-viscosity liquids, which gives a kL$_a$ value of 10 s^{-1}, enough to perform mass transfer-limited pro-cesses in fractions of a second (e.g., CaCO$_3$ production from CO$_2$ absorption, completed within 1 s). The order of kL$_a$ value for agi-tated vessels is 0.001–0.1 s^{-1}.

In conclusion, the prominent characteristics of a spinning disc reactor include

- high heat and mass transfer coefficients,
- plug flow,
- intense mixing capability,
- short residence times,
- low fouling,
- very high surface area to volume ratio.

Rapid heating of the process liquids for a short period of time can be achieved in this design, which is essential for sterilization of food products, milk, etc. At low temperatures, the liquid is cooled as soon as it comes in contact with the reactor's walls. This design allows for rapid, continuous processing of temperature-sensitive food and pharmaceutical liquids under process conditions that cannot be attained in conventional stirred tank vessels. Pasteuriza-tion of milk and orange juice, for example, requires controlled rapid heating followed by quick cooling. The heating kills bacteria, but fast cooling is a prerequisite in order to prevent the degradation of the product. The rapid heating or cooling and short residence times provided by the SDR allow for food to be pasteurized effi-ciently. It can be used to concentrate sugar solutions. As this liquor loses water, it becomes increasingly viscous, but the cen-trifugal action of the spinning disc pushes the film radially outward, allowing water to be removed continuously. Complete water removal in a stirred tank would be impossible, however, since the mechanical agitator would find it difficult to stir a highly viscous solution.

In the ice cream making process, rapid crystallization of the water leads to the formation of smaller ice crystals that produce smoother ice cream. By pumping a chilled fluid through the underside of the spinning disc, rapid cooling followed by rapid crystallization can be attained and hence make small ice crystals. SDR is also used for the preparation of custard powder (predominantly corn starch granules). The slurry is pumped over the rotating disc (at 600–1200 rpm) maintained at 70°C at a flow rate of 2 to 3 ml/s, which produces a properly cooked product. At the end of the process a simple wash water jet is introduced at the center of the disc to flush remaining custard out of the system, leaving a clean heating surface. Different dyes could be introduced into the stream to produce a well-mixed product of the desired color. Due to the low inventory within the system, the choice of color could be changed quickly with little product wastage between various choices.

Photo-initiation has been recognized as an excellent method for initiating free radical polymerization; however, due to limited penetration depth of the UV radiation, its commercial application is limited. The film thickness of an SDR is generally less than 500 microns, so radiation-induced polymerization can be performed effectively, even on the large scale.

The supersaturation in a batch vessel determines the crystallization process and hence the particle-size distribution. Low supersaturation leads to larger particles. Controlled and slow addition of a counter-solvent and co-feeding the two components into a product heel are the best techniques to produce precisely sized larger particles. Both techniques suffer since the crystallization environment changes, leading to a large particle-size distribution. The solvent in the solution could be stripped in a controlled manner in the SDR to achieve precipitation and hence control the maximum particle size and the particle-size distribution.

Chain growth addition polymerization based on active carbene (e.g., styrene, vinyl carbazole, isobutylene) is a fast reaction that is most difficult to control since termination by combination or disproportionation is not possible. The initiator for cationic polymerization is a proton from a donor such as water, in the presence of a Lewis acid or sulfuric acid. Polymerization tends to be uncontrollable when undiluted monomers are used, and so these reactions are carried out at low monomer levels, usually in chlorinated hydrocarbon solvents at low temperatures. The reaction would be explosive, above a concentration limit of 30% wt. Exotherms greater than 50°C per minute are observed when the monomer concentration is 48% wt. In contrast, monomer concentrations of 75–80 wt% could be polymerized with ease in an SDR at ambient temperatures. Monomer conversion of 8% per pass can be achieved at a residence time of 1 to 2 s. Molecular weights and polydispersities achieved at these high monomer concentrations in the temperature range of 20 to 40°C are comparable to those found for polymers synthesized at lower concentrations at temperatures below 0°C in a conventional batch vessel.

A polymer dissolved in a 1:1 ratio of toluene when passed over a hot, 30-cm-diameter SDR at 5 ml/s, with hot nitrogen gas flowing in a countercurrent manner, can strip 99.9% of the toluene from the polymer in a single pass. By feeding this vapor into a second, 20-cm-diameter SDR, toluene can be condensed onto the cold, polished surface of this disc. This demonstrates the flexibility of the SDR equipment as both an evaporator and a condenser (Protensive, USA).

Annular Tubular Reactor

Endothermic and exothermic reactions are combined in an annular tubular reactor to achieve tremendous energy savings (see Fig. 6.39). Dehydrogenation of ethyl benzene to form styrene is an endothermic reaction ($\Delta H = 188\,kJ/m$). Side products such as CH_4, ethylene, benzene, and toluene are also formed during the process. The catalyst for dehydrogenation is Fe oxide promoted by K_2O and CrO. This reaction is carried out in a radial flow reactor where the core of the packed bed has an oxidative catalyst. Steam and oxygen are introduced at the center and will react with H_2 to generate heat that flows out of the core to the annulus. The shell side has the dehydrogenation catalyst, where the conversion of ethyl benzene to styrene takes place, capturing the heat flowing out.

Conducting endothermic and exothermic reactions in parallel in the same reactor can be extended to the manufacture of synthesis gas from natural gas. The reaction $CH_4 + H_2O \rightarrow CO + 3H_2$ is endothermic ($\Delta H = 206\,kJ/mol$), while $CH_4 + 0.5\,O_2 \rightleftharpoons CO + 2H_2$ is exothermic ($\Delta H = -38\,kJ/mol$). When these two reactions are coupled in a radial flow, considerable savings in energy can be achieved.

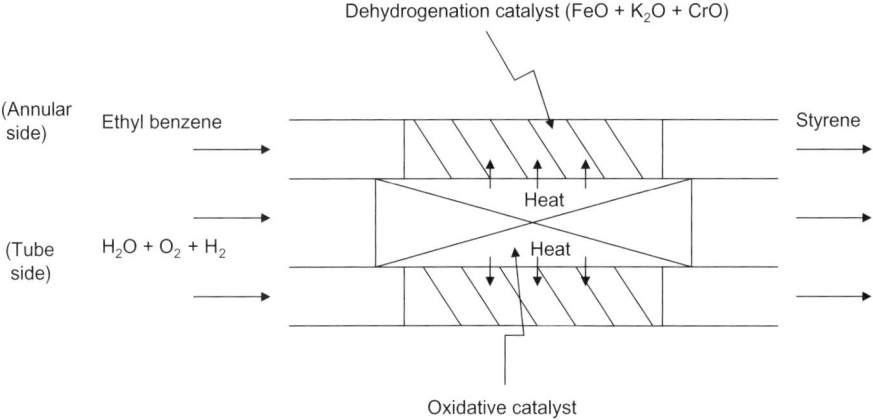

FIGURE 6.39. Annular tubular reactor (radial heat flow reactor).

Segmented Flow Tubular Reactor

The synthesis of crystal product in a batch reactor with a controlled shape, narrow particle-size distribution, and crystal polymorph is a major challenge for many industries. The segmented flow tubular reactor (SFTR) overcomes many of these problems (Bowen et al., 2004). A multichannel SFTR was developed in the European Community Fifth Framework Programme, which involved seven partners. The SFTR is composed of a mixer-segmenter and a tubular reactor (see Fig. 6.40). A supersaturation is created in the mixing chamber, inducing precipitation of the particles. This precipitating suspension is then segmented into identical small volumes by an immiscible fluid. The SFTR achieves a quasi-plug flow when compared to a large batch reactor. The microvolumes created in the SFTR are more homogeneous, and hence the precipitated product is uniform with narrower particle-size distribution. Scale-up of SFTR for commercial-scale production is achieved by multiplying the number of tubes running in parallel instead of scaling-up by increasing the size of the tube. Nanostructured calcite and nanosized barium titanate were prepared successfully on a pilot scale using this SFTR technology.

Products are achieved using continuous stirred tank reactors, a jet implement reactor (see Fig. 6.41), or a loop reactor (Fig. 6.33).

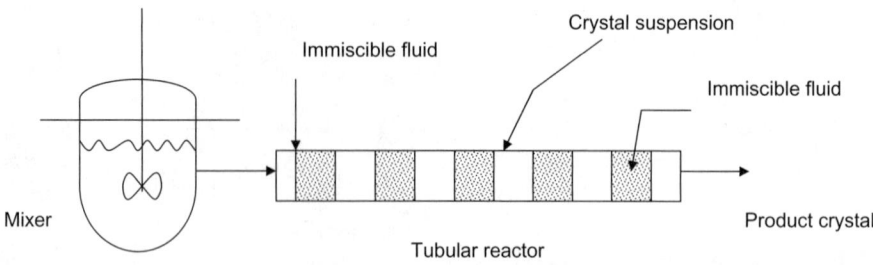

FIGURE 6.40. Segmented flow reactor for a crystallization process.

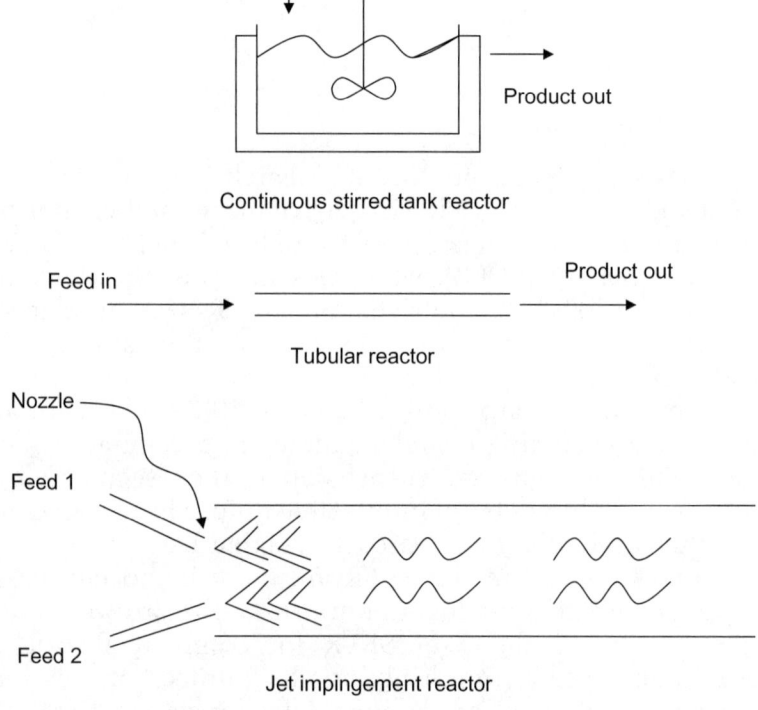

FIGURE 6.41. Continuous process operation using different types of reactors.

From Batch to Continuous Processes

Nitrated products such as nitroglycerine were manufactured traditionally in large batch reactors. Now modern nitration plants use either small, continuously stirred tank reactors that provide intense mixing and large heat transfer areas or jet reactors to deliver the intense mixing and rapid contacting of reactants, leading to short residence time and reduced inventory. The latter point leads to safe operation. Noram Engineering & Constructors Ltd. (Canada) has built jet impingement reactors for producing nitrobenzene, with a 10-fold increase in reaction rate. This reactor contains reactants in the form of high-speed jets impinging on each other to create complete mixing and a large contact area.

The Buss loop reactor (Fig. 6.33), which has been around for quite some time, has been successfully applied for hydrogenation, amination, and sulphonation types of reactions. In a conventional process, the production of phosphorus oxychloride involves reacting phosphorus trichloride with oxygen or air in a batch reactor. About 500 tons per month are produced in three reactors, each of which is $13\,m^3$ in volume. In a continuous process, 700 tons per month of this material are manufactured in a reactor volume of $0.5\,m^3$, leading to a 95% increase in productivity. About 15–25% excess oxygen is used in the batch process, whereas only 5% excess oxygen is used in the new continuous design. Other benefits include more uniform load on the utilities such as chilled and cooling water. This has also led to a decrease in the size of the utility plants since the peak loads have decreased tremendously.

Monobromobenzaldehyde, an intermediate for the manufacture of the pesticide m-phenoxybenzaldehyde, is produced in a batch reactor. Due to side reactions, the productivity is about $15.5\,kg/m^3/hr$. If made in a continuous reactor, the same process productivity increases to $34.5\,kg/m^3/hr$.

The process for the manufacture of ethyl acetate involves the esterification of methanol with acetic acid in the presence of catalyst. The other steps include removal of water of reaction, distillation of product, and recovery and recycle of excess reactants. As many as six distillation columns achieve these operations. Eastman Chemicals has completely modified the process, replacing the distillation columns with a single multifunctional unit, leading to a reduction in the number of reboilers, condensers, pumps, etc. The heat input and rejection occur at only two points. Sulzer (Switzerland) has similarly changed the process of hydrogen peroxide distillation, while Degussa, Creavis Technologies and Innovation (Norway) has intensified the process of manufacturing of

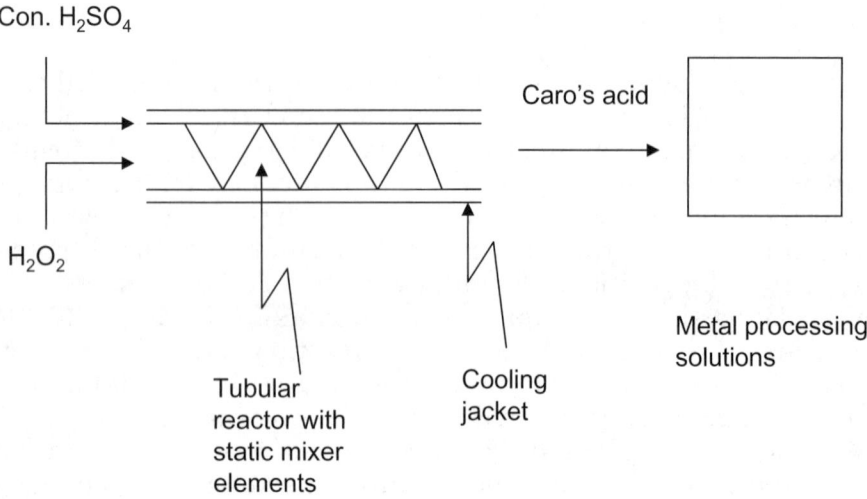

FIGURE 6.42. Process for the manufacture of Caro's acid.

hydrogen peroxide by combining oxidation and extraction in a single step.

Caro's acid, used in metal processing, is a powerful oxidizing agent and decomposes readily. It is made by reacting concentrated sulfuric acid with hydrogen peroxide. A process was developed to manufacture 1000 kg/day of Caro's acid in a tubular reactor with a volume of only 20 ml and a residence time of less than 1 s, with product being mixed immediately with the solution to be treated so as to avoid any intermediate storage and safety problems (see Fig. 6.42).

Phosgene is a very toxic chemical, and storage of it in large quantities poses a serious safety problem. A continuous tubular reactor is developed to make this chemical for immediate consumption. So an inventory of only 70 kg of gaseous phosgene has to be maintained when compared to an inventory of 25,000 kg of the liquid in the storage in the old batch process. Methyl isocyanate (MIC), the infamous chemical that was released at Bhopal, can be generated and immediately converted to the final pesticide in a tubular reactor that will contain a total inventory of less than 10 kg of MIC.

A process developed by Hitachi for the production of the polymer polyethylene terephthalate from ethylene glycol and polyethylene terephthalic acid by esterification and polycondensation is performed in three reactors, whereas the traditional process requires two reactors and four mixers for the esterification reaction and three reactors and three mixers for the polycondensation reaction process. In the Hitachi process the cost per unit power of the main reactor is approximately one sixth that of the traditional process. The number of units requiring maintenance has decreased by a factor of seven.

Other New Reactor Designs

Crude vermiculite ore produces the basic material used in the manufacture of fire protection and industrial insulation products. The process equipment consists of three rotary furnaces for heating and reaction. Torftech Ltd. (UK) replaced these with a single toroidal fluidized bed (Torbed) furnace of 1-m diameter with a capacity of 2 ton/hr. This led to a reduction in overall energy consumption and in vermiculite wastage and maintenance costs. The new process resulted in a payback period of 16 months. Now 11 plants are operational in Europe with this Torbed technology. This technique is also commercialized to produce silica from rice husk and roasting of sulphide ores, etc.

Milling and grinding have traditionally been done in ball, roll, or grind mills. Dispersed media mill operates in liquid media with small-sized steel, ceramic, or glass balls that are circulated in a vessel consisting of rotating members. A large reduction in power consumption and infrastructure was achieved by this innovative design. The new design has been applied to the paint and ink industry, red phosphorus, ferrite powder, pigments, etc.

PI can reduce the Dow Hazard and Dow Toxicity indices considerably. Mono- and dinitration of xylenes can be carried out in small-sized equipment with high heat transfer rates, thereby eliminating the need for inert diluents. A pesticide intermediate is prepared by chlorination at $-40°C$ since considerable exothermic heat is evolved. The process could be carried out in novel mixers and heat exchangers, so that an ordinary coolant system is used instead of cryogenic systems, thereby reducing operating costs. Combinations of static mixers and heat exchangers can be used for dilution of concentrated acids and alkalis; for reactions with high heats of reaction such as neutralization, nitration, sulfonation, etc.; in dye industries; and for catalytic nitration of aromatics.

A novel fluidized bed reactor with a submerged membrane is developed for the production of ultrapure H_2 from CH_4 with less than 10 ppm CO. The H_2 produced is used online in the polymer electrolyte membrane fuel cells for small-scale applications. A high degree of process intensification is achieved by integrating permselective Pd metallic membranes for H_2 removal operating at 500–600°C inside a fluidized-bed reactor along with selective O_2 addition through dense perovskite membranes operating at 900–1000°C. Incorporating both membranes within a single reactor has the clear advantage of producing ultrapure H_2 and pure CO_2, circumventing expensive CO_2 sequestration. The membrane-assisted fluidized bed reactor consists of a partial oxidation bottom section and a steam reforming/water gas shift top section. Higher yields are obtained since the thermodynamic equilibrium is shifted to the right.

Process Integration

Process intensification involves making fundamental changes to processing technologies to yield improved product quality, throughput, and energy efficiency. It is a management and design tool used to optimize energy resources in process plants employing conventional technologies.

Pinch analysis is the most common process integration tool. It involves the use of heat exchanger networks to optimize heat energy by linking hot and cold process streams in the most thermodynamically advantageous way. Exergy analysis is another tool used for process integration that takes into account all energy flows.

The potential for energy savings from process integration depends on the individual plant and applications. The major short-term benefits are expected in the food and drink and pulp and paper industries. Estimates of the potential energy savings and the payback period for various applications, identified by process integration studies in the EU (only due to fuel use, excluding feedstock in the chemical industry) are given in Table 6.1. The application of PI differs widely per EU member state.

The main barriers to the further use of process integration in the EU include

- the need for capital expenditure for additional hardware,
- the absence of more flexible process integration design tools,
- the lack of understanding and knowledge about the technique in many industries,

TABLE 6.1
Expected Savings and Payback Period in European Union Countries Due
to Use of Process Integration

Country	Application	Savings (%)	Expected Payback Period (Years)
U.K.	Oil refineries	10–20	1–2
	Chemical industries	10–20	2–7
	Food and beverage industries	24–40	1–5
Netherlands	Synthetic fibers	2	1
Germany	Oil refineries	5	<3

Sources: Caddet, 1993, 1996; De Beer et al., 1994; ETSU, 1994.

- impact of changes carried out on plant reliability, flexibility, and maintenance,
- the risk of disturbances to production,
- the long payback in certain industries.

Microfluidic Reactors

A microreactor that contains a number of different catalysts fixed in different compartments connected via a microfluidic network or reactor modules connected by microtubing may well be the optimal chemical production unit. Each compartment may operate simultaneously, which leads to efficient use of the microreactor. Different combinations of steps in cascade reactions may give a whole range of products (Fletcher et al., 2002; Song et al., 2003).

Fabrication of an SU-8–based microfluidic reactor on a polyether ether ketone (PEEK) substrate sealed by a "flexible semisolid transfer" (FST) process is described by Song et al. (2004). A continuous-flow, polymeric microfluidic reactor utilizing SU-8 as a photoresister on a PEEK substrate was fabricated by standard UV lithography. Embedded multilayer structures were fabricated between the substrate and the inlets and outlet of the microfluidic reactor that facilitated fabrication of the entire microfluidics using SU-8, resulting in improved bonding between the substrate and the pattern. An FST process, based on a reduced exposure dosage, was developed to seal the microfluidic channels. Scanning electron microscopy (SEM) images and photographs revealed no trace of blockages in channels due to the sealing process. The maximum pressure drop without any leakage was found to be 2.1 MPa. The microfluidic reactor withstood temperatures as high as 150°C and was found to be suitable for carrying out wet chemical synthesis (Song et al., 2004).

Directed enzyme-prodrug therapy involves administering a drug designed only to be toxic after being chemically activated by an enzyme that is also delivered into the tumor cell. To develop the prodrugs, researchers at the Georgia Institute of Technology created a microfluidic reactor that could screen a large number of chemicals for their ability to be activated by the delivered enzyme. First the enzyme nitrobenzene nitroreductase (NbzA) is linked to a water-soluble polymer poly(ethyleneimine). A silica shell is grown around this polymer-enzyme construct, producing porous silica beads with diameters ranging from 500 to 1000 nm. After this has been filled into the channels of a microfluidic device, the activation of several compounds by the enzyme is measured.

Conventional microfluidic reactors have drawbacks such as the reactants diffusing slowly and the particles not moving through the channels at the same speed; those in the middle move faster than those alongside the channel walls, resulting in nanocrystals spending different amounts of time in the reactor. These two phenomena lead to quantum dots with a wide range of diameters. To overcome these problems, a two-phase microfluidic system in which gas bubbles divide the stream of liquid in the channels into individual and very regular segments was designed. Within these segments, back mixing resulted in a constant exchange of material between the walls and center of the channels, and so all particles spend equal time in the reactor. In order to accelerate the diffusion of the reactants, the mixing zone of the channel is made with tight curves. With their microfluidic reactor, the researchers could prepare quantum dots of uniform size in significantly higher yields than with previous microfluidic techniques.

Pacific Northwest National Laboratory's (USA) microchannel reactor unit consisting, in part, of a combustor/evaporator made of stainless steel with an overall size of $41 \times 60 \times 20$ mm, with micro-machined combustor channels of $300\mu \times 500\mu \times 35$ mm, is used to perform methane partial oxidation reaction at 900°C to produce carbon monoxide and hydrogen. Methane conversion efficiencies were more than 85% and 100% with 11 and 25 ms residence times, respectively.

The channels are separated by a micro-machined contactor plate of a 25-μ-thick Kapton substrate that has a matrix of uniform holes 25 μ in diameter. The solvent and feedstreams can be operated co-currently or countercurrently, bringing in intimate contact the two immiscible fluids as they flow through very thin channels smaller than the normal mass-transfer boundary layer.

Automotive fuel processing is another application where the potential exists for significant market demand for microtechnology. The fuel cells will generate electricity to drive the electric motors that move the vehicles, which will carry liquid hydrocarbon fuels, plus a fuel processing plant to produce hydrogen for the fuel cells. The proton exchange membrane (PEM) fuel cell is the electricity generator and is being developed by the U.S. Department of Energy. The fuel processor produces hydrogen-rich streams from gasoline or methanol fuel using a heterogeneous, catalytic microchannel chemical reactor. It is a multistep process involving fuel vaporizer, primary conversion reactor to produce synthesis gas, water gas shift reactor, and CO cleanup reactor. The fuel processor will have a volume of less than 0.3 L to support a 50-kWe output from fuel.

Photochemical Reactor on Chip

Photochemical reactions proceed via a free-radical mechanism. The radicals, which are formed near the light source if they do not diffuse quickly to react further with other species, will recombine, generating excess heat instead of a productive reaction. Large-scale photochemical reactions are usually performed with macro-scale lamps immersed in the reaction vessel. Issues involved in such design are scalability of light sources, heat and mass transfer in the processes, and safety concerns (e.g., explosions caused by excess heat). Radical recombination reduces the quantum efficiency of the overall process. By PI miniaturization the diffusion length is reduced, leading to an increase in frequency of collision with other molecules to produce the desired product.

By deep reactive ion etching (DRIE) of a silicon substrate channel that is 500 mm wide and 250 mm deep, serpentine-shaped long channels are fabricated with large surface areas (see Fig. 6.43). A layer of CVD oxide (1.5–2 mm thick) is first deposited and densified on the Si wafer. A mask with the patterns for the flow channel and inlet/outlet ports is used to pattern a layer of photoresist. The patterns are transferred into the oxide layer using a buffered oxide etch (ammonium fluoride and hydrogen fluoride in water). The photoresist is then removed using a sulfuric acid and hydrogen peroxide mixture, and then a new layer is coated. A second mask

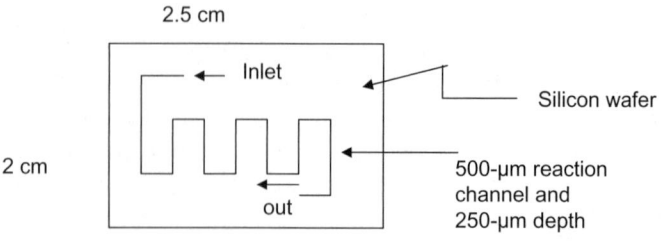

FIGURE 6.43. Photochemical reactor chip.

that has only the inlet and outlet is then used to pattern the new layer of photoresist. The photoresist layer is used to mask the exposed silicon to produce the inlet/outlet patterns (approximately 250 mm thick) in the first DRIE step. The wafer is then cleaned with acid solution, exposing the patterned oxide. In the second DRIE step, the oxide layer is the mask for etching the channel and the inlet/outlet, producing throughholes at the inlet/outlet areas and a 250-mm-deep channel. After the residual oxide layer is removed by buffer oxide etch, the Si wafer is anodically bonded to a Pyrex wafer. Finally, the bonded wafer is diced to produce individual reactors.

Microinstruments

Online or inline process analytics (UV-Vis, IR, Raman, NMR, GC, MS) have a crucial role for providing information on the nature of species and their dynamic changes. The instruments should be able to fit into microreactors and should have low dead volume and fast dynamic changes. Such smart, commercial analytical instruments are entering the PI market.

Microreactors also require microinstruments for detection of the products and side products, process measurements, and control. Some of the microinstruments that have been developed are

- micro-LC,
- surface tension detection,
- gas chromatography,
- fringe field sensors,
- vapochromic sensors,
- Raman,
- grated light reflective spectroscopy,

- reflectometry,
- flow-through particle analyzer,
- surface plasmon resonance,
- mini-NMR,
- ultrasound.

Conclusions

Process intensification encompasses the philosophy of green chemistry: waste reduction, improved safety, reduced energy usage, etc. (Anastas and Warner, 1998). Different processes might have different rate-controlling steps, for example, mass transfer, mixing, diffusion, etc. Often the controlling step is obvious, sometimes it is completely unknown, and sometimes there are different rate-controlling steps during the course of the reaction. It is of fundamental importance that some understanding of the rate-limiting steps is obtained. Oscillating columns offer moderate residence times with heat transfer coefficients better than the batch vessels. Plate heat exchanger-type reactors (Hex reactor from BHR) are good for clean, high-heat transfer duties. When performed in high-G or microreactors, diffusion-controlled reactions lead to several orders of magnitude increase in the overall rate of reaction. Spinning disc reactors offer very high surface-area-to-volume ratios, which can be used effectively to perform fast polymerization reactions and concentration of temperature-sensitive materials. Continuous reactors, including simple plug-flow pipe units, tubular models containing static or other mixing devices, and various jet devices, have been used to efficiently produce toxic materials for immediate consumption in downstream processing with little or no inventory. Unsafe and highly exothermic reactions could be performed in such designs safely under undiluted conditions.

The unit operations and processes that are modified are reactive distillation, reactive extraction, membrane separations, oscillating flows in reactors, membrane reactions, fuel cells, etc. (Stankiewicz and Moulijn, 2000). Examples of new equipments that are available due to PI include compact heat exchangers, structured packed columns, static mixers, high-G column, spinning disc reactor, oscillating flow reactor, loop reactors, spinning tube in tube reactor, heat exchange reactor, supersonic gas–liquid reactor, static mixing catalysts, microchannel reactors, and microchannel heat exchangers. Equipment such as sonochemical reactors, microwave reactors (see Fig. 6.44), ceramic cross-flow

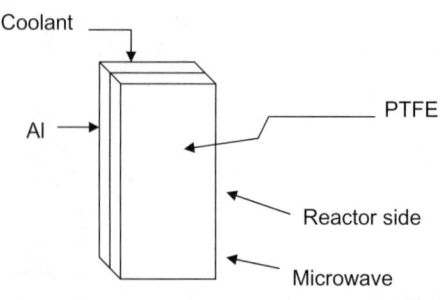

Reactor volume = 0.27 cc

FIGURE 6.44. Microwave radiation reactor.

heat exchangers and reactors, gas lift reactors, and membrane reactors are other designs that can lead to PI.

References

Akay, G., *Process Intensification and Miniaturization, Principles and Applications in Biological, Chemical and Environmental Technologies*, Elsevier, New York, 2004.

Alfa Laval AB, edlinks.che.com/5830-531, 2006.

Anastas, T. and Warner, J. C., *Green Chemistry, Theory and Practice*, Oxford University Press, Oxford, 1998.

Benson, R. S. and Ponton, J. W., Process miniaturisation—a route to total environmental acceptability?, *Trans. IChemE*, **71**, part A, 160–168, 1993.

Bowen, P., Testino, A., Legagneur, V., Donnet, M., Hofmann, H., and Cobut, N., Precipitation of nanostructured & ultrafine powders: Process intensification using the segmented flow tubular reactor—still in search of the perfect powder, 2006 Spring National Meeting, Orlando, Florida, 2004.

Caddet, 1993, Proceedings IEA Workshop on process integration, International experiences and future opportunities, Sittard, The Netherlands.

Caddet, 1996, CADDET register on energy efficiency demonstration projects, Sittard, The Netherlands.

De Beer, J. G., Van Wees, M. T., Worrel, E., and Block, K. 1994, "ICARUS-3, The potential of energy efficiency improvement in the Netherlands from 1990 to 2000 and 2015," Dept of Science, Technology & Society, Utrecht University, Utrecht, The Netherlands.

Doble, M., Kumar, A., and Gaikar, V., *Biotransformations and Bioprocesses*, Marcel Dekker, New York, 2004.

Ehrfeld, W., Löwe, H., Michel, F., Lohf, A., and Hofmann, C., Hybrid, modular concept, see, e.g., WO patent 00/62918, microreactor module, issued Oct. 26, 2000.

Energy Technology Support Unit (ETSU), 1994. An appraisal of UK energy research, development, demonstration and dissemination, HMSO, London, UK.

Fletcher, P. D. I., Haswell, S. J., and Zhang, X., Electrokinetic control of a chemical reaction in a lab-on-a-chip micro-reactor: Measurement and quantitative modeling, *Lab Chip*, **2**: 102–112, 2002.

Green, A., Process intensification magnifies profits, *Chem. Eng.*, p. 66, 1999.

Hendershot, D. C., Process minimization: Making plants safer, *Chem. Eng. Prog.*, pp. 35–40, 2000.

Hessel, V., Hardt, S., and Löwe, H., A comprehensive overview of micro-reactor technology, *Chemical Micro Process Engineering*, Wiley-VCH, Weinheim, Germany, 2004.

Kern, D. Q., *Heat Transfer Principles*, McGraw-Hill, New York, 1990.

Kikutani, Y., Horiuchi, T., Uchiyama, K., Hisamoto, H., Tokeshi, M., and Kitamori, T., Glass microchip with three-dimensional microchannel network for 2×2 parallel synthesis, *Lab Chip* **2**: 188–192, 2002.

Kim, H., Saitmacher, K., Univerdorben, L., Wille, Ch., Adler, H. J., and Potje-Kamloth, K., *Macromol. Symp.*, **187**: 631–640, 2002.

Kobayashi, J., Mori, Y., Okamoto, K., Akiyama, R., Ueno, M., Kitamori, T., and Kobayashi, S., A microfluidic device for conducting gas-liquid-solid hydrogenation reactions, *Science*, **304**: 1305–1308, 2004.

McCabe, W., Smith, J., and Harriott, P., *Unit Operation of Chemical Engineering*, McGraw Hill, New York 1990.

Mullin, R., Fine Chemicals—Chronicles of Chemistry II—Enterprise of the Chemical Sciences, 2004.

Qi, Y., Kowaguchi, Y., Christensen, R. N., and Zakin J. L., Enhancing heat transfer ability of drag reducing surfactant solutions with static mixers and horney combs, *Int. J. Heat and Mass Transfer*, **46**(26): 5161–5173, 2003.

Semel, J. (ed.), *Process Intensification in Practice: Applications and Opportunities* (BHR Group Publication 28), Wiley, New York, 1997.

Simons, K., Intensive progress, *Eur. Chem. News*, p. 22, 2004.

Skelton, V., Greenway, G. M., Haswell, S. J., Styring, P., Morgan, D. O., Warrington, B. H., and Wong, S. Y. F., The generation of concentration gradients using electroosmotic flow in micro reactors allowing stereoselective chemical synthesis, *The Analyst*, **126**: 11–13, 2001.

Song, Y., Kumar, Challa, S. S. R., and Hormes, J., Fabrication of an SU-8 based micro-fluidic reactor on a PEEK substrate sealed by a flexible semi-solid transfer (FST) process, *J. Micromech. Microeng.*, **14**: 932–940, 2004.

Song, H., Tice, J. D., and Ismagilov, R. F., A microfluidic network for controlling reaction networks in time, *Angew. Chem. Intl. Ed.*, **42**: 768–772, 2003.

Stankiewicz, A. I. and Moulijn, J. A., Process intensification, *Ind. Eng. Chem. Res.*, **41**: 1920–1924, 2002.

Stankiewicz, A. J. and Moulijn, J. A., Process intensification: Transforming chemical engineering, *Chem. Eng. Progr.*, 22–34, 2000.

TeGrotenhuis, W. E., Cameron, R., Viswanathan, V. V., and Wegeng, R. S., Solvent extraction and gas absorption using microchannel contactors, *Proc. 3rd Intl. Conf. Microreaction Tech.*, 1999. Also available online, at http://www.pnl.gov/microcats.

TeGrotenhuis, W. E., Cameron, R., Butcher, M. G., Martin, P. M., and Wegeng, R. S., Micro channel devices for efficient contacting of liquids in solvent extraction, *Separation Sci. Tech.*, 1998. Also available online, at http://www.pnl.gov/microcats.

Treybal, R. E., *Mass Transfer Operations*, 3rd ed., McGraw-Hill Higher Education, 1980.

van den Berg, H., Application of PI in the process industries, review and analysis of 20 cases, 5th International PI Conference, Maastricht, Oct. 13–15, 2003 (www.technology.novem.nl/nl/process-intensification/pin/documents/20 PI.pdf).

Wegeng, R., Call, C., and Drost, M. K., Chemical system miniaturization, *Proc. AICHE 1996 Spring National Meeting*, 1996. Also available online, at http://www.pnl.gov/microcats.

Wess, G., Urmann, M., and Sickenberger, B., Medicinal chemistry: Challenges and opportunities, *Angew. Chem. Intl. Ed.*, **40**: 3341–3350, 2001.

CHAPTER 7

Alternate Energy Sources

One form of energy can be converted into another form with a certain loss of efficiency. The various forms of energy currently available and the methods to generate them appear below. Some of these energy forms are renewable, and others are nonrenewable, as outlined in Fig. 7.1. The nonrenewable energy forms have been practiced traditionally for several hundreds of years and are currently expected to disappear completely in a few hundred years or less.

Heat
   by burning fossil fuels
   solar radiation
   warm air, water, subsurface water, and ocean
   nuclear energy
   earth's core (hot springs)
   electricity passing through wires
Light
   sun
   fluorescent and incandescent lightbulbs
   LED
   laser
   burning fuels
Electricity
   photovoltaic
   dynamo generators

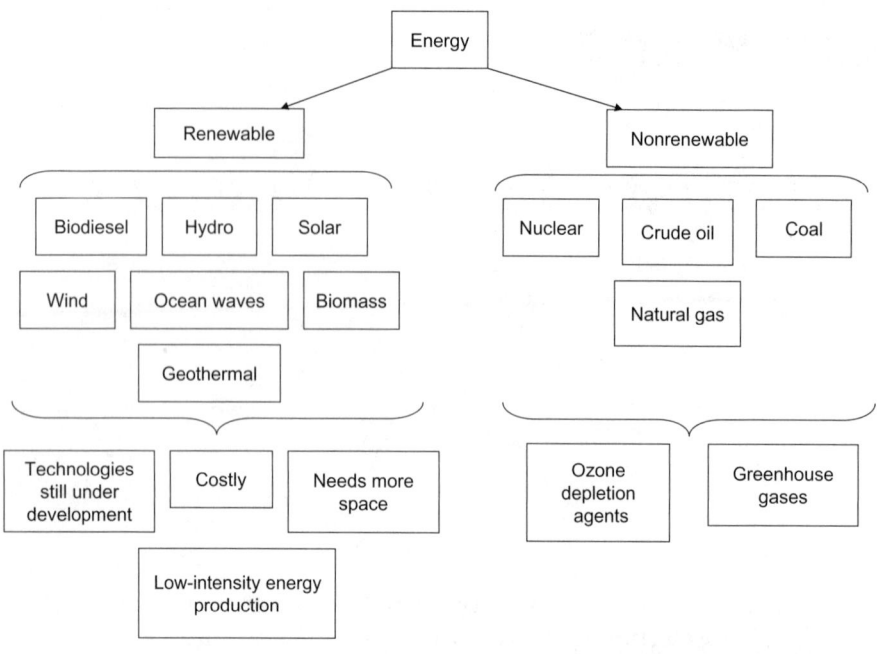

FIGURE 7.1. Energy sources.

batteries
hydrogen fuel cells
static (friction, lightning)
Radio waves
radio transmitters
Mechanical
hydroelectric facilities
Sound
vibrating surfaces (microwave)

The two most common forms of energy used by humans are heat and electricity.

Greenhouse Gases

When sunlight strikes the earth's surface, some of it is reflected toward space as infrared radiation (heat). Many chemical compounds found in the earth's atmosphere allow sunlight to enter the atmosphere freely, while certain gases and vapors absorb this infrared radiation and trap the heat in the atmosphere (these are known as *greenhouse gases*). Over a period of time, the cumulative effect of these greenhouse gases is a slow increase in the earth's surface temperature. Some of the gases found in nature, such as water vapor, carbon dioxide, methane, and nitrous oxide, and others, which are exclusively human made, like gases used for aerosols and chemicals such as chlorofluorocarbons (CFC), exhibit greenhouse properties.

During the past 20 years, about three quarters of human made carbon dioxide emissions were due to the burning of fossil fuels. Based on 2004 data, next to the United States, China is the biggest producer of greenhouse gases (Jia, 2004). China mainly relies on coal and oil for energy. Between 1996 and 2003, oil imports in China increased from 20 to 90 million tons. Energy consumption in China is expected to continue rising significantly as it aims to quadruple its GDP by 2020. According to China, in 1994 it emitted 2.6 billion tons of carbon dioxide, 34.3 million tons of methane, and 850,000 tons of nitrous oxide. In the United States, greenhouse gas emissions arise mostly due to energy requirements (which represent 82% of total U.S. human made greenhouse gas emissions) (Energy Information Administration, Office of Integrated Analysis & Forecasting, 2002). Energy requirements include fuel used for electricity generation and that used for heating and cooling needs of the homes and offices. The United States currently produces about 25% of global carbon dioxide emissions from burning fossil fuels but is projected to lower its carbon intensity by 2025 and remain below the world average (National Energy Information Centre, EIA, Washington, DC (www.eia.doe.gov/environment. html).

The Kyoto Protocol is an international treaty bringing many of the world's developed nations together in an effort to limit greenhouse gas emissions and reduce the effects of global warming. Russia ratified the treaty in February 2005, while, as of this publication, the United States and Australia were among the few countries that had not ratified it. All companies operating in ratifying countries must comply with the protocol regardless of where they are based. Six specific gases that contribute toward global warming have been identified: carbon dioxide, methane, nitrous oxide, hydrofluorocarbons (HFC), perfluorocarbons (PFC), and sulphur hexafluoride (SHF). The global warming potential of these gases with respect of CO_2 are 21, 310, 140–11,700, 6500–9200, and 23,900, respectively (Dore et al., 2003). Although carbon dioxide has the lowest global warming potential potency, it is released in far greater amounts than any of the other gases due to human activities and is responsible for 82% of all the global warming caused in the United Kingdom. The half-life of some manmade chemicals is so long that they persist longer than the natural global warming gases.

A *sink* is a reservoir that uptakes a chemical element or compound from another part of its cycle. For example, soil and trees tend to act as natural sinks for carbon, as billions of tons of carbon each year in the form of CO_2 are absorbed by oceans, soils, and trees. The EU-15 countries had agreed to cut, by 2012, 8% of the 1990 values, but the data collected in late 2006 by the European Commission predict that the values will be only 0.6% below the base year levels by 2010. Worse still, Austria, Belgium, Denmark, Ireland, Italy, Portugal, and Spain may even exceed their individual limits.

The next most prominent greenhouse gas is methane (9%), which comes from landfills, coal mines, oil and gas operations, and agriculture. Nitrous oxide (5% of total emissions) is emitted from burning fossil fuels and from the use of nitrogenous fertilizers and industrial processes. Manmade gases such as HFC, PFC, and SHF (2% of total emissions) are released as byproducts from industrial processes and through leakage from cooling systems.

World carbon dioxide emissions are expected to increase by 1.9% annually between 2001 and 2025. Much of the increase in these emissions is expected to occur in developing nations such as China and India (which will be above the world average at 2.7% annually between 2001 and 2025). Data collected from Antarctic ice cores, before the industrial emissions began, led to the conclusion that atmospheric CO_2 levels were about $280\,\mu L/L$. The concentrations stayed between 260 and $280\,\mu L/L$ throughout the 10,000 years preceding the beginning of the years of industrial emissions. Since the beginning of the Industrial Revolution (1850), the concentrations of many of the greenhouse gases have increased. Most of the increase in carbon dioxide occurred after 1945. The current amount of CO_2 is 364 ppm (increase of 31% over pre-Industrial era), and CH_4 is 1745 ppb (increase of 150% over pre-Industrial era).

Human activities have raised the levels of greenhouse gases primarily by releasing carbon dioxide, methane, and other gases (see Fig. 7.2). These activities include burning of fossil fuels and deforestation, leading to higher carbon dioxide concentrations; livestock and paddy rice farming, land use and wetland changes, pipeline losses, and covered, vented, landfill emissions, all of which lead to higher methane release; use of CFCs in refrigeration systems; use of halons in fire extinguishers; and various manufac-

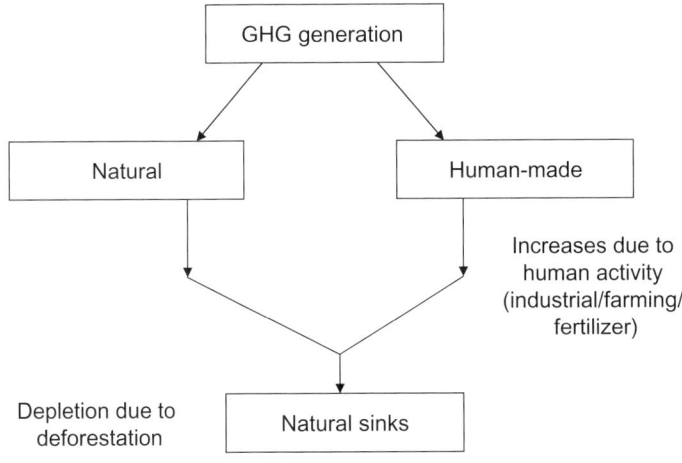

FIGURE 7.2. Greenhouse gas (GHG) generation and consumption.

turing processes. Water vapor is a natural greenhouse gas that accounts for the largest percentage of the greenhouse effect. Water vapor levels fluctuate regionally, but in general humans do not have a direct effect on its levels. The current amounts of CFC-11, CFC-12, CFC-113, CCl_4, and HCFC-22 are 268, 533, 84, 102, and 69 ppt, respectively (IPCC, 1998).

The concentration of greenhouse gases in the atmosphere keeps changing dynamically due to several reactions and processes, such as (Archer, 2005; Caldeira and Wickett, 2005)

1. condensation and precipitation of water vapor from the atmosphere,
2. oxidation of methane by hydroxyl radical and water vapor,
3. mixing and interchange of atmospheric gases into oceans or other regions,
4. chemical reactions of gases in atmosphere with other compartments of the planet, such as reduction in CO_2 amount due to photosynthesis of plants or, after dissolving in the oceans, formation of carbonic acid and bicarbonate and carbonate ions,
5. dissociation of halocarbons by UV light releasing $Cl^{\bullet}$ and $F^{\bullet}$ as free radicals in the stratosphere with harmful effects on ozone (such as "ozone depletion" or "ozone hole"),
6. dissociative ionization reactions caused by high-energy cosmic rays or lightning discharges (e.g., lightning forms $N^{\bullet}$ atoms from N_2, and the former then reacts with O_2 to form NO_2).

A material balance for the accumulation of CO_2 in the environment consists of CO_2 generation due to natural causes and human activities. The depletion of CO_2 is due to its absorption in the natural sinks, as shown in the following model:

$$\frac{Vd[CO_2]}{dt} = r^h{}_G + r^n{}_G - r_S,$$

$r^h{}_G$ = rate of generation of CO_2 due to human activity
   = [rate of burning of fossil fuels]$[f_1]$ + Σ [rate of production of industrial chemicals/manufacturing operations]$[f_2]$,
$r^n{}_G$ = rate of generation of CO_2 due to natural causes,
r_S = rate of consumption of CO_2 due to absorption in various sinks
   = Σ[amount of various sinks] $[g_1]$,
V = volume of the biosphere,
f_1, f_2 = stoichiometry of CO_2 production,
g_1 = efficiency of absorption of CO_2.

The various sinks for CO_2 absorption have been depleting because of deforestation (leading to a decrease in rate of CO_2 absorption), while the rate of manufacturing operations has been increasing. This leads to an overall increase in CO_2 concentration ($[CO_2]$) with time.

Lifetime of Greenhouse Chemicals

Recovery from a large input of atmospheric CO_2 from burning fossil fuels will result in an effective lifetime of tens of thousands of years (Archer, 2005). Methane has an atmospheric lifetime of 12 to 15 years. The methane is degraded to water and CO_2 by chemical reactions in the atmosphere. Nitrous oxide has an atmospheric lifetime of 120 years, while CFC-12 has an atmospheric lifetime of 100 years. HCFC-22 has an atmospheric lifetime of 12.1 years, and tetrafluoromethane has an atmospheric lifetime of 50,000 years. Sulfur hexafluoride has an atmospheric lifetime of 3200 years.

Oil

The crude oil production (measured in thousands of barrels per day) in various countries (Nov. 2006 data) (Development report on the oil markets in 2006, Ministry of Energy, State of Kuwait (www. moo.gov.kw) is given in Table 7.1.

TABLE 7.1
Crude Oil Production

Algeria	1,343
Indonesia	872
Iran	3,803
Kuwait	2,383
Iraq	2,002
Libya	1,713
Nigeria	2,258
Qatar	802
Saudi Arabia	8,857
UAE	2,448
Venezuela	2,488

In 1956, Hubbert proposed that over time crude oil production in a country would follow a bell-shaped curve, with a peak production at a particular time and a decrease with further time. Nine major large oil-producing countries, including the United States, Great Britain, Venezuela, and Norway, had reached their peak global oil production in 1998, and their production volume has been slowly decreasing since then. Factors that influence the peak date include worldwide recession, military or political factors, etc. The cumulative depletion among this group is now about 1.5 million barrels per day (Mbpd). Thus, presently the current demand for oil is growing at about 3.5%, which is 82 Mbpd worldwide. The peak production is estimated to be above 90 Mbpd. But the International Energy Agency (IEA) as well as the OECD are optimistic, and authorities claim that production will be 110 Mbpd within the next 10 to 14 years. This is based on the assumptions that significant future discoveries will be made and that technologies for the use of nonconventional sources such as shale oil and oil sands will become competitive and hence will be practiced.

Solar Energy

Solar energy can be converted directly into heat by passive or active systems. The passive systems use a thermo siphon and have no pumps. The thermo siphon operates only when the fluid is hot. Other space heating systems use a thermal diode to achieve similar effects. Passive solar water distillers may rely on capillary action to pump water. Active solar systems use additional equipment such as circulation pumps, air blowers, or tracking systems that aim the solar arrays or collectors at the sun. These mechanisms are typically powered by electricity. A wide range of power technologies exist to convert the solar energy. A few of these include

1. Photovoltaic cell produces electricity directly from solar energy.
2. Hydroelectric power stations produce indirect solar power.
3. Sunlight is concentrated onto a thermal collector, and the surface is heated up. The heat is carried away by a fluid.
4. Sunlight strikes a solar sail on a spacecraft and is converted directly into a force on the sail, which causes motion of the craft.
5. Sunlight strikes a light mill and causes the vanes to rotate.
6. Sunlight is focused on an externally mounted reflective channel that conducts sunlight into building interiors to supplement lighting.

Photovoltaic (PV) devices use semiconducting materials to convert sunlight directly into electricity. Solar radiation, which is nearly constant outside the earth's atmosphere, varies with changing atmospheric conditions (clouds and dust) and the changing position of the earth relative to the sun. Of the total solar energy received, 19% is absorbed by the atmosphere, while clouds reflect 35% of the total energy. The peak power received at sea level is $1000\,W/m^2$. For example, in North America the average power of the solar radiation lies somewhere between 125 and $375\,W/m^2$, meaning at a rate between 3 and $9\,kWh/m^2/day$. Photovoltaic panels currently have an efficiency of 15%, and, hence, a solar panel delivers 19 to $56\,W/m^2$, or 0.45–$1.35\,kWh/m^2/day$ (annual day and night average). A 173-m^2 photovoltaic system in the 30-year lifetime of the system is estimated to prevent $2100\,lb$ of NO_x, $6100\,lb$ of SO_x, and 756 tons of carbon dioxide that will be produced if oil is used to produce the same amount of energy. The average lowest retail cost of a large solar panel declined from $7.50 to $4 per watt between 1990 and 2005. The cost of producing electricity from solar radiation is still not yet competitive.

According to an April 2000 article in the *Electric Power Research Institute (EPRI) Journal*, photovoltaic arrays in a geostationary earth orbit at an altitude of 22,300 miles would receive eight times the sunlight that is received at earth's surface. Such arrays would be unaffected by cloud cover, atmospheric dust, or the earth's day–night cycle. A drawback to concentrated sunlight is that it is hot. If not converted into electricity, radiation that is focused turns into heat and can damage the arrays. Current research is directed toward studying ways to capture waste heat and convert it to electricity by means of thermal voltaic processes and special coatings on the mirrors and lenses that can reject portions of the sun's spectrum that PV arrays do not use, thereby reducing excess heat. Another approach is to convert stored solar energy to microwave radiation and beam it down to a combination rectifier-antenna (rectenna), located in an isolated area. The rectenna could convert the microwave energy to direct current power.

A Solar Updraft Tower

Figure 7.3 shows a low-tech solar thermal power plant where air passes under a very large agricultural glass house (between 2 and 30 km in diameter) that is heated by the sun and channeled upwards toward a convection tower. It then rises naturally and is used to drive turbines, which generate electricity. An energy tower (see Fig. 7.4) is an alternative proposal for the solar updraft tower and is driven by spraying water at the top of the tower. Evaporation of water causes a downdraft by cooling the air, thereby increasing its density, driving wind turbines at the bottom of the tower. It requires a hot, arid climate and large quantities of water (sea water may be used for this purpose) but does not require the large glass house of the solar updraft tower.

A solar pond is a low-cost approach to harvesting solar energy. The pond has three layers of water: the top layer with a low salt content; an intermediate layer with a salt gradient, which sets up a density gradient that prevents heat exchange by natural convection in the water; and a bottom layer that has a high salt content, which can reach a temperature of 90°C. The different densities of the layers prevent convection currents. The heat trapped in the salty bottom layer can be used for different purposes, such

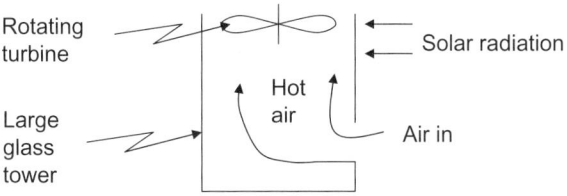

FIGURE 7.3. Solar updraft tower.

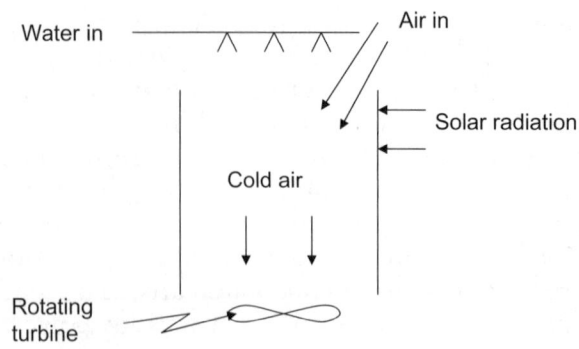

FIGURE 7.4. Energy tower.

as heating of buildings, industrial processes, or generating electricity.

Solar chemical refers to a number of possible processes that harness solar energy by absorbing sunlight in a chemical reaction in a way similar to photosynthesis in plants but without using living organisms. Metals, such as zinc, have been shown to drive photoelectrolysis of water. Transition metal compounds, such as titania, titanates, niobates, and tantalates, exhibit very low efficiency photoelectrolysis of water. Man-made photosynthesis reactions could also convert solar energy and CO_2 into useful chemicals.

The world's largest solar power plant, which is located in the Mojave Desert, in California, consists of $4 \, km^2$ of solar reflectors. This plant produces 90% of the world's commercially produced solar power. The total world peak power of installed solar panels was around 5300 MW as of the end of 2005 (International Energy Agency, 2006). The largest solar plant, SEGS in California, produces 350 MW, and the largest nuclear power plants each generate more than 1000 MW per year.

Ocean Waves

Since water is about a thousand times heavier than air, even a slow-flowing stream of water can yield great amounts of energy. The winds cause waves on the surface of the ocean and on lakes as it transfers some of its energy to the water, through friction between the air molecules and the water molecules. Stronger winds cause larger waves. Wave energy is another form of alternative and sustainable energy. The up and down motion of the waves can be captured to generate power. An experimental wave energy machine installed off the coast of Scotland can generate enough electricity for about 400 homes on a nearby island. The machine has an oscillating chamber made of concrete. As the waves come inside the chamber, the air pocket inside this chamber is compressed. This compressed air is forced through a small hole onto a turbine that generates electricity.

Hydroelectricity

Canada generated 61% of its electricity supply from hydroelectricity in 1999, mostly with large dams (Renewable Energy in Canada; Conference Board of Canada, 2003). Hydroelectric generation does not produce significant greenhouse gases, but it does have other major environmental impacts. The reservoirs often destroy nearby habitat by submerging vast areas of highly productive forest and wildlife habitat. The dams also damage freshwater ecosystems by blocking the movement of fish and other organisms. Pollution from mercury and other contaminants is a problem in many reservoirs in northern Canada. Large dams are also known to cause earthquakes. Hydroelectric power now supplies about 715,000 MWe, or 19% of the world's electricity. Hydroelectric power can be far less expensive than electricity generated from fossil fuel or nuclear energy.

Wind

In 2005 the U.S. wind energy industry installed 2431 MW of new wind power capacity in 22 states, making the total U.S. wind power capacity 9149 MW (Wind & Hydropower Technologies Program; U.S. Department of Energy; Energy Efficiency and Renewable Energy). Commercial wind turbines are now installed in 30 U.S. states, producing enough electricity annually to equal the power used by 2.3 million U.S. households. Wind power developers invested more than $3 billion in new wind turbines in 2005. The AWEA predicted that installations in 2006 would approach 3000 MW of new wind power capacity (U.S. Department of Energy; Energy Efficiency and Renewable Energy, 2006). Texas now challenges California's status as the state with the most installed wind power. FPL Energy and GE Energy supply most of the wind turbine in the United States. China, Germany, and India follow in the capacity of installed wind power generation systems. The current installed capacity of wind farms is 59,322 MW (less than 1% of worldwide electricity use). The total installed wind power capacity in 2005 of the top five countries are

1. Germany, 18,428 MW,
2. Spain, 10,941 MW,
3. United States, 9549 MW,
4. India, 5200 MW,
5. Denmark, 3128 MW.

Small-scale turbines that are approximately 2 m in diameter weigh about 16 kg, and produce 900 W are available. Wind strengths vary and hence do not give a continuous, steady flow of power. The wind speeds are seasonal, and wind farms require a large land area, where no other activity could be performed. Wind turbines may harm birds. Wind mills located at a slightly higher altitude could capture more energy than those located at the ground level.

Geothermal Energy

Geothermal energy is obtained from the earth's internal heat and can be used to generate steam to run a steam turbine, which in turn could generate electricity. Three miles below the earth's surface, the temperature is 100°C, which suffices to boil enough water to run a steam-powered electric power plant.

Instead of drilling three miles beneath the earth's surface, scientists could access this power at *geothermal hotspots*, which are volcanic features found all around the world. At the hotspot the mantle is thin, and excess heat from the interior of the earth is transmitted to the outer crust. A few of the hotspots are the volcanic islands of Hawaii, the mineral deposits and geysers in Yellowstone National Park, or the hot springs in Iceland. Iceland produced 170 MW geothermal power and heated 86% of all houses in the year 2000 through geothermal energy (Ragnarsson, 2000). Geothermal power is generated in over 20 countries at an operational capacity of 8000 MW (equivalent to 17% of its electricity from geothermal sources). These geothermal hotspots can easily be used to generate electricity. Some examples follow:

1. One system consists of pumping hot water into permeable sedimentary hotspots found underground and then using the steam liberated to generate electricity. The used steam is condensed and sent back down to the permeable sedimentary stream.
2. Another system utilizes volcanic magma that is still partly molten at around 650°C to boil water, which generates electricity.

3. The third approach uses hot dry rock, which is just hardened magma, but still is extremely hot. To recover this heat from these rocks, water is circulated through the rock to produce steam.

The first system listed above is not as useful as the other methods because of the acidic nature of the fluids (sulfurous and sulfuric acids) found underground. These acidities would damage the equipment, reducing the economic effectiveness of the system. This is a general problem of geothermal energy systems, making them more expensive than other alternative energy sources.

Hydrogen

Hydrogen is not an energy source but an energy carrier. One of the main reasons for switching to hydrogen is to prevent global warming caused by fossil fuels since the energy produced by hydrogen does not produce acid rain, CO_2, dust, or nitrous oxides. Of course, if H_2 is produced from fossil fuel, then the whole purpose is lost. Hydrogen and oxygen can react in a fuel cell to produce electricity and water as their reaction's byproduct. Currently, 96% of hydrogen is made from fossil fuels. Based on 2004 data, in the United States 90% is made from natural gas, with an efficiency of 72%. Only 4% of hydrogen is made from water via electrolysis. Currently, the vast majority of electricity comes from fossil fuels in plants that are 30% efficient and from electrolysis that is 70% efficient, which leads to a 20% efficient process to create one unit of hydrogen energy. Using renewable energy is much more useful than using fossil fuel to produce hydrogen. Current wind turbines perform at 30–40% efficiency, producing hydrogen at an overall 25% efficiency. The best solar cells available have an efficiency of 10%, leading to an overall efficiency of 7%. Algae can be used to produce hydrogen at an efficiency of about 0.1% (see Fig. 7.5).

Hydrogen can be made from biomass, but this process has several problems:

1. It is a very seasonal process.
2. It contains plenty of moisture and hence requires space to store and energy to dry before gasification.
3. Limited supply of biomass is available for this process.
4. The quantities are not large or consistent enough for large-scale hydrogen production.

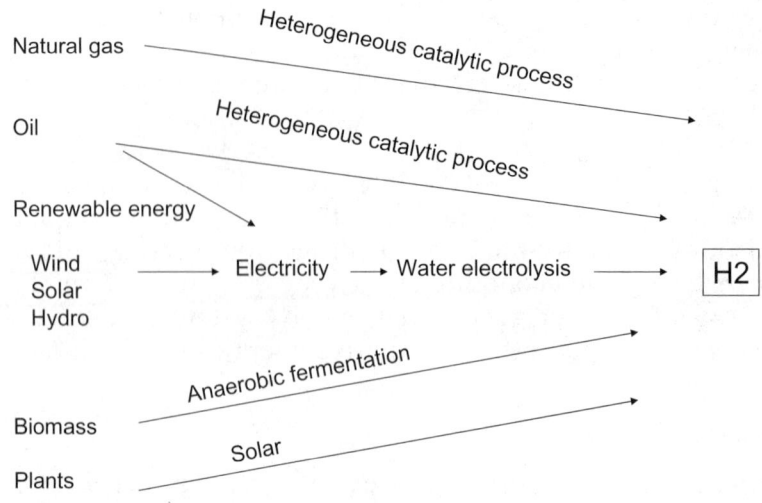

FIGURE 7.5. Various processes for the production of hydrogen.

5. A large land space is required (cultivated biomass in good soil has a low yield of 10 tons/2.4 acres).
6. The soil will be degraded from erosion, and there will be a loss of fertility if stripped of biomass.
7. Any energy put into the land to grow the biomass, such as fertilizing, planting, and harvesting, will add to the energy costs.
8. The delivery costs to the central power plant are high.
9. The process is not suitable for pure hydrogen production.

To be stored, hydrogen must be compressed and liquefied. To compress hydrogen to 10,000 psi in a multistage compressor, 15% of the energy needs to be utilized. Handling and storage require extreme precautions because the hydrogen is so cold (−423°F), requiring cryogenic support systems. As hydrogen pressure in the storage tank is increased, the volume decreases but the thickness of the tank material increases, adding to cost. Transporting hydrogen or sending it through pipelines also requires a lot of care. Hydrogen also tends to make metal brittle due to the formation of metal hydride. Hydrogen has the lowest ignition point of any fuel, 20 times less than that of gasoline, so it can explode easily due to a spark or static.

The numerous problems associated with fuel cells must be addressed before such technology becomes practical:

1. Fuel cells are heavy. A metal hydride storage system that can hold 5 kg of hydrogen, including the alloy, container, and heat exchangers, would weigh approximately 300 kg, which would lower the vehicle's fuel efficiency.
2. Fuel cells are expensive. In 2003, they cost at least $1 million [U.S. Department of Energy's National Energy Technology Laboratory (NETL) and the Electric Power Research Institute (EPRI)].
3. They are currently not reliable.
4. They need a much less expensive catalyst than platinum.
5. They can clog and lose power if the hydrogen has impurities.
6. They do not last more than 1000 hr.
7. They have yet to achieve a driving range of more than 100 miles. They still cannot compete with electric hybrids like the Toyota Prius, which is already more energy-efficient and lower in CO_2 generation than projected fuel cells.

Nuclear Power

Nuclear power plants provide about 17% of the world's electricity. There are 400 nuclear power plants around the world, with more than 25% in the United States. France depends more than most countries on nuclear power for electricity, generating about 75% of its electricity from nuclear power. In the United States, nuclear power supplies about 15% of the electricity.

Nuclear reactor uses enriched uranium in the form of *pellets*, with a 1-in. diameter and length. The pellets are arranged into long rods, which are collected together into *bundles*. The bundles are submerged in water inside a pressure vessel. The water acts as a coolant. In order for the reactor to work, the bundle, submerged in water, must be slightly supercritical. Control rods made of a material that absorbs neutrons are inserted into the bundle. Raising and lowering the control rods allow operators to control the rate of the nuclear reaction. To increase the rate of reaction and produce more heat, the rods are raised out of the uranium bundle. To create less heat, the rods are lowered into the uranium bundle. When the rods are lowered completely into the uranium bundle, the reactor shuts down. The uranium bundle heats the water and turns it to steam. The steam drives a steam turbine, which spins a generator to produce power. In some reactors, the steam from the reactor goes through a secondary heat exchanger to convert another loop of water to steam, which drives the turbine. The advantage of this design is that the radioactive water/steam never comes in contact with the turbine. In some other reactors, the coolant fluid in contact with the reactor core is gas (CO_2) or liquid metal (Na, K); these types of reactors allow the core to be operated at higher temperatures. The reactor's pressure vessel is typically housed inside a concrete liner that acts as a radiation shield.

The main disadvantages of nuclear power are the handling and later disposal of the radioactive waste. A safe method for the disposal of radioactive waste, other than burying it deep in the land in concrete bunkers, has yet to be identified. Also, the reactors need several safety protections and safeguards to avoid human exposure to radioactive radiation and leakage of material. While Uranium-235 fission produces no CO_2 gas directly, the processes of mining, milling, refining, moving uranium, and disposing of radioactive waste (ore concentrations) require energy equivalent to the CO_2 gas emissions of a natural gas-fired power station.

Biodiesel

Biodiesel is a vegetable oil-based (soy or canola oil) fuel that runs in the present diesel engines, without any modifications to the hardware. Biodiesel and biodiesel blends can be used in all compression-ignition (CI) engines that were designed to be operated on diesel fuel. It is cheaper than oil, sustainable, and nontoxic; it does not produce acid rain (absence of sulfur); and it does not contribute as much as fossil fuels do to global warming. Studies have shown it reduces engine wear by as much as 30%, primarily because it provides excellent lubrication (Agarwal, Bijwe, and Das, 2003). Even 2% biodiesel in normal diesel will help achieve this improvement. Biodiesel fuel yields 220% more energy than that required to produce, transport, and distribute it, which is due to the fact that the feedstock crop collects solar energy and transforms it into the biodiesel feedstock oil.

Various biodiesel blends, which include different ratios of biodiesel and diesel from crude oil, can be used in vehicles depending upon the vehicle's requirement and weather conditions. A 20% biodiesel will provide a higher octane rating, superior lubricity, significant emission reductions, and less toxic emissions; will virtually eliminate visible soot emissions; and will have similar fuel consumption, horsepower, and torque. Premium biodiesel is a fuel manufactured from vegetable oils by a transesterification process. Soybean oil is currently the leading source of vegetable oil for biodiesel manufacture in the United States.

Problems with biodiesel are that it is not readily available in large quantities and the amount of NO_x increases by 15%, which contributes to the generation of smog. Another disadvantage is that the viscosity increases at lower ambient temperatures, hence requiring additives for lowering the fuel's gel point.

Renewable Energy

Green power describes electricity produced from renewable sources that are less harmful to the environment than fossil fuels. So energy produced from solar, wind power, geothermal, biomass, and small hydroelectric plants is considered green power. *Renewable energy* is an unending source of energy that quickly replenishes itself. Renewable energy does not cause pollution or release toxic substances into the atmosphere; of course, hydroelectric dams cause damage to flora and land area. Some renewable energy systems have environmental problems. Wind turbines can be hazardous to flying birds, while hydroelectric dams can create barriers for migrating fish—a serious problem in the Pacific Northwest, which has seen the destruction of salmon populations. Biomass and biofuels require large amounts of land area. Renewable energy sources provide relatively low-intensity energy, so in order to convert them into useful energy, the collection needs to be distributed over large areas.

Renewable energy comes from an energy resource being replaced by a natural process at a rate that is equal to or faster than the rate at which that resource is being consumed. Renewable energy is a subset of sustainable energy. Most renewable forms of energy, other than geothermal and tidal power, ultimately derive from solar energy. Energy from biomass is derived from plant material and is produced by photosynthesis using the power of the sun. Wind energy derives from winds, which are also generated by the sun's uneven heating of the atmosphere. Hydropower depends on rain, which again depends on the sunlight's power to evaporate water. Even fossil fuels are derived from solar energy since fossil fuels originated from plant material. Renewable energy resources may be used directly or used to create other, more convenient forms of energy. Examples of direct use are solar ovens, geothermal heating, and water and windmills. Biomass refers to any form of plant or animal tissue, including wood, straw, biological waste products such as manure, and other natural materials that contain stored energy. The energy stored in biomass can be released by burning the material directly or by feeding it to microorganisms that use it to make biogas.

The various types of renewable energy are

1. wind energy,
2. water power,
3. solar energy,
4. geothermal energy,
5. biofuels (including liquid biofuel, solid biomass, and biogas).

Biofuel

Biofuel is any fuel with an 80% minimum content by volume of materials derived from living organisms harvested within the 10 years preceding its manufacture. A drawback with biomass is that it needs to be grown, collected, dried, fermented, and burned. All these steps require resources and an infrastructure. The carbon in biofuels was recently extracted from atmospheric carbon dioxide by the growing plants, so burning it does not result in a net increase of carbon dioxide in the earth's atmosphere. Hence it is considered a renewable source.

Agricultural products, including straw, lumber, manure, sewage, garbage, and food leftovers, are used for the production of bioenergy. Currently, most biofuel is burned to release its stored chemical energy, which is not very efficient. Converting it into electricity using fuel cells is very efficient. Most bioenergy is consumed in developing countries and is used for direct heating. Sweden and Finland supply 17% and 19%, respectively, of their energy needs with bioenergy.

The biomass could be residue from harvesting or crops specifically grown for the task. The biomass could be converted into useful products through the sugar platform. The bulk of the plant material contains cellulose and lignin. The cellulose or hemicellulose is broken down into sugars. Then these sugars can be converted into ethanol or other building block chemicals. Biofuels can be classified as solid, liquid, and gaseous, and several examples exist in each of these classes. Lignin can be burned as a fuel. The second approach involves conversion of biomass to gaseous or liquid fuel by heating it under oxygen-limiting conditions (pyrolysis), which are further converted to useful products. A large number of information resources are available with the U.S. Department of Energy Efficiency and Renewable Energy (http://www1.eere. energy.gov/biomass/for_researchers.html). The Biomass Document Database provides access to most biofuels and many other biomass documents produced by the National Bioenergy Center and its subcontractors since 1980. The National Renewable Energy

Laboratory (NREL) Publications Database contains bibliographic information about publications developed or written by NREL staff and subcontractors (http://www.nrel.gov/publications/). The Department of Energy's (DOE) Information Bridge (http://www.osti.gov/bridge/) contains documents and bibliographic citations of DOE research report literature from 1994. The citations relate to physics, chemistry, materials, biology, environmental sciences, energy technologies, engineering, computer and information science, and renewable energy.

Solid Biomass

Solid forms of biomass are

- wood,
- straw,
- animal waste,
- crops such as maize, rice, peanuts, and cotton,
- dried, compressed peat.

Certain types of biomass have attracted research and industrial attention, since they are available in very large quantities and have low market value. They are algae, bagasse from sugarcane, dried distiller's grain, firewood, hemp, jatropha, maize (corn), manure, meat and bone meal, peat, rice hulls, silage, stover, and whey.

Liquid Biomass

A number of liquid forms of biomass can be used as a fuel:

- Bioalcohols.
- Ethanol produced from sugar cane is being used as automotive fuel in Brazil. Ethanol produced from corn is being used as a gasoline additive (oxygenator) in the United States.
- Methanol, which is currently produced from natural gas, can also be produced from biomass.
- An acetone–butanol–ethanol mixture can be produced by anaerobic fermentation. This mixture can be used in existing gasoline engines.
- Biologically produced oils can be used in diesel engines:
  straight vegetable oil,
  waste vegetable oil,
  biodiesel obtained from transesterification of animal fats and
      vegetable oil.

Liquid biofuel is usually bioalcohol such as ethanol and biodiesel and virgin vegetable oils. E85 is a fuel composed of 85% ethanol and 15% gasoline that is currently being sold to consumers in the United States. The European Union plans to add 5% bioethanol to Europe's petrol by 2010.

Gaseous Biomass

Forms of gaseous biomass include
- biomethane produced by the natural decay of garbage or agricultural manure,
- wood gas,
- hydrogen produced by the electrolysis of water,
- gasification, which produces carbon monoxide.

Many organic materials can release gases, due to the metabolism of organic matter by bacteria under anaerobic fermentation. Under high-pressure, high-temperature, and anaerobic conditions, many organic materials such as wood can be gasified to produce gas. Biogas can be produced from current waste streams, such as paper production, sugar production, sewage, and animal waste.

Future Sources of Renewable Energy

A difference in salt concentration exists between sea water and river water. This gradient can be utilized to generate electricity by separating positive and negative ions by ion-specific membranes. Brackish water is produced. It is predicted that one third of the electricity needs in the Netherlands can be covered with this system. Ocean thermal energy conversion (OTEC) uses the temperature difference between the warmer surface of the ocean and the cold lower recesses to employ a cyclic heat engine. Lake-bottom water is constant at 4°C, and this water can be used for cooling fluids that flow through submerged pipes.

Conclusions

Countries that are currently highly dependent on fossil fuel should focus all their efforts toward harnessing renewable energy sources. Although the cost of production is still high, more effort would bring down this value. Tapping renewable energy sources would also decentralize energy supply to the individual household, ending energy supply's status as a political issue.

Different sources (including waste) are being investigated for extracting energy. For example, Ozmotech, a major Australian company, has developed a pyrolysis process under inert-atmosphere to convert 400,000 tons of plastic waste into 350 million liters of diesel per annum (Ecos, 2006).

References

Agarwal, A. K., Bijwe, J., and Das, L. M., Wear assesment in a biodiesel fueled compression ignition engine, *J. of Engng. Gas Turbines and Power*, **125**(3): 820–826, 2003.

Archer, D., Fate of fossil fuel CO_2 in geologic time, *J. Geophys. Res.*, **110**: C09S05, 2005.

Biodiesel for Oregon, Oregon Environmental Council, Portland OR (www.biofuels4oregon.org).

Caldeira, K. and Wickett, M. E., Ocean model predictions of chemistry changes from carbon dioxide emissions to the atmosphere and ocean, *J. Geophys. Res.*, **110**: C09S04, 2005.

Conference Board of Canada, Renewable Energy in Canada, 2003.

Dore, C. J., Goodwin, J. W. L., Watterson, J. D., Murrells, P., Hobson, M. M., Haigh, K. E., Baggott, S. L., Thistlethwaite, G., Pye, S. T., Coleman, P. J., and King, K. R., National Atmospheric Emissions Inventory, U.K., Emissions of Air Pollutants 1970–2003, 2003.

Ecos, Ozmotech's plastic-to-fuel solution in demand, **130**: 5, 2006.

Energy Information Administration, Office of Integrated Analysis & Forecasting, Emissions of Greenhouse Gases in the United States 2001. Washington, D.C., 2002.

Henley, M., Potter, S., Howell, J., and Mankins, J., Wireless power transmission options for space solar power, *EPRI Journal* (Spring): 6–17, 2000.

Hubbert, M. K., *Nuclear Energy and the Fossil Fuels*, Presented before the Spring Meeting of the Southern District, American Petroleum Institute, March 1956.

IPCC, Radiative forcing report from 1994, updated (to 1998) by IPCC TAR, 1998.

Jia, H. P., The People's Republic of China—Initial National Communication on Climate Change—2004. United Nations Framework Convention on Climate Change (UNFCCC), 24 Nov. 2004; www.scidev.net

Ragnarsson, A., *Geothermal Development in Iceland*, Proceedings World Geothermal Congress 2000, Kyushu–Tohoku, Japan, May 28–June 10, 2000.

U.S. Department of Energy; Energy Efficiency and Renewable Energy, U.S. Wind Power Industry Tempers Its 2006 Forecast Slightly. EERE Network News, November 2006.

CHAPTER 8

Inherent Safety

Industrial processes are prone to hazardous or unexpected events. In their paper, Nicholas D. Anastas and John C. Warner wrote, "Hazard should be considered as a design flaw and efforts need to be made in the designing phase to minimize or eliminate it" (2005). Hazards may arise from many factors, and they manifest themselves in various forms. These various hazardous events could be external or internal fire, confined or unconfined explosion, noxious gas release, pollution, moving object hazard, etc. (USCSHI, 2000). The description of these events and their consequences are listed later in this chapter.

A hazard could lead to the following consequences:

1. External fire. May be prompted by flammable gas or vapor; liquid, solid, metal, wood, or waste material; pyrophoric material and presence of ignition source such as sparks, static, friction, hot spots, welding, lightning, auto ignition, or furnace. Immediate consequence: engulfment, thermal radiation, fire damage, smoke, domino effect (where one incident is the primary cause for several other incidents).

2. Internal fire (inside equipment or a confined space). May be prompted by flammable gas or vapor; liquid, solid, metal, wood, or waste material; pyrophoric material; oxygen, halogen, and presence of ignition source such as sparks, static, friction, hot spots, welding, lightning, auto ignition, or furnace. Since the space is confined, the chances of developing a flammable atmosphere are very high. Immediate consequence: damage to

equipment, domino effect, fumes/gases extending to an external fire.

3. Internal explosion (inside equipment). May be prompted by uncontrolled reaction, equipment testing, filling, purging, or physical overpressure. Immediate consequence: equipment damage, missile/fragment, structural damage.

4. Confined explosion or detonation. May be prompted by flammable gas or vapor; liquid, solid, dust, mist, oxygen, halogen; or explosive or unstable compound. Immediate consequence: damage to equipment, missile/fragment, domino effect, fumes/gases.

5. Unconfined explosion, such as VCE (vapor cloud explosion), gas explosion, or detonation. May be prompted by flammable gas or vapor; liquid, solid, dust, mist, oxygen, halogen; or explosive or unstable compound. Immediate consequences: missile, noise, light, domino effect, fumes/gases.

6. Acute or chronic harmful/noxious exposure. May be prompted by gas or vapor; mist; liquid; acid; alkali; biology; or fume. Immediate consequences: chronic effects on employees and the public.

7. Pollution. May be prompted by escape of gas or vapor; mist; liquid; acid; alkali; biology; fume; algae; or flue gas from storage. Immediate consequences: chronic effects on employees and the public and harm to flora and fauna.

8. Violent release of energy. May be prompted by electrical, kinetic, or potential energy. Immediate consequence: equipment damage, domino effect.

9. Noise. May be prompted by machinery, flares, ejectors, vents, pressure releases, sirens, or mechanical handlings. Noise above a certain decibel is considered a hazard. Immediate consequence: temporary loss of hearing, nuisance.

10. Visual impact. May be prompted by fire or explosion. Immediate consequence: temporary blindness, nuisance.

11. Moving object. May be prompted by elevators, ramps, road vehicles (tankers, articulated lorries), dumpers, fork-lift trucks, stackers, cranes, rail wagons, or robots. Immediate consequence: physical injury to personnel or damage to equipment.

The long-term consequences of a hazard could be fatality/permanent injury to employees or the public, damage to equipment, harm to flora or fauna, bad publicity, and evacuation of site (Edwards, 1995). A hazard may lead to a range of financial losses to the company, from heavy to light, including bad publicity, strong reprimand from the local authorities, or public issue of notice for shutting down the operation or litigation. A heavy financial loss could be similar to that faced by Union Carbide after the toxic gas emission in Bhopal, India, which led to several thousand deaths in the 1980s. The company had to shut down its operations permanently at that site. Noise is also considered a hazard. Several of the real industrial accidents listed later could have been avoided and thousands of lives could have been saved if sufficient thought had been given by the designers, engineers, and managers.

The insecticide carbaryl was made from α-naphthol, methylamine, and phosgene at the infamous Union Carbide plant at Bhopal. Methylamine and phosgene were reacted together to make methyl isocyanate (MIC). The MIC was then reacted with α-naphthol to make carbaryl. At times the intermediate MIC, which is highly unstable, was stored in tanks. In Bhopal on December 3, 1984, about 15 tons of highly toxic methyl isocyanate were released from an intermediate storage tank, killing many thousands. The safe approach would have been to avoid the need for intermediate storage of MIC by matching MIC production and consumption rates. An alternative process consists of using the same three raw materials but reacting them in a different order: α-naphthol and phosgene would be reacted together to give a chloroformate ester, which would then be reacted with methylamine. This process does not produce any MIC. Neither process is ideal, because both involve the use of phosgene, but the alternative process at least avoids producing MIC. Going further, one could think of an even safer approach, such as making an alternative insecticide to carbaryl—that is safer to produce—or developing pest-resistant plants (or making use of natural predators to overcome the problem of pest). Table 8.1 lists the various approaches that could have been followed to avoid the Bhopal accident.

In the accident at Seveso, Italy, on July 10, 1976, 2 kg of dioxin was discharged from a reactor vent which contaminated about $20 \, km^2$ of surrounding land. Although many operational mistakes led to the accident, the main reason was that steam at 300°C was used for a process whose maximum operating temperature was 160°C. The process was expected to have a thermal runaway temperature at about 185°C. The accident occurred because the steam heated the reactor walls and the reaction mass to 300°C, which in turn led to the mass being raised to thermal runaway temperature. An inherently safer design would have been to have the steam pressure controlled and superheated to ensure that the maximum operating temperature of 160°C could not be exceeded.

TABLE 8.1
Different Alternatives That Could Have Been Adopted at Bhopal to Reduce Risk (Moving Down the Table One Achieves Inherent Safety)

Approaches	Solution
Safety	Store less MIC
	Two levels of safety checks
	Better training of manpower
	Special safety checks to avoid water ingress; produce enough to match production rate downstream
	HAZOP, what-if analysis
Inherent safety	Modify process to prevent MIC production
Avoid use of phosgene	Newer raw materials
Avoid use of carbaryl	Use alternate safe insecticide
Avoid use of synthetic	Pest-resistant plant (GM) or
Insecticide	Biopesticide

The Chernobyl (former USSR; now Ukraine) accident in 1986 was the result of a flawed reactor design that was operated by inadequately trained personnel and without proper attention to safety. The resulting steam explosion and fire released 5% of the radioactive material into the atmosphere. Such types of nuclear reactors were known to have unstable operating regions during shutdown.

The chemical plant owned by Nypro (UK) produced caprolactam, a precursor used in the manufacture of nylon. The process involved oxidation of cyclohexane with air in a series of six reactors to produce a mixture of cyclohexanol and cyclohexanone. Two months prior to the explosion in Flixborough, England, a crack was discovered in the fifth reactor, and a temporary 50-cm diameter pipe was installed to bypass that leaking reactor. On June 1, 1974, the temporary bypass pipe ruptured, possibly as a result of a fire in a nearby region. Within a minute, about 40 tons of cyclohexane leaked from the pipe and formed a vapor cloud. Upon coming into contact with an ignition source, it exploded, completely destroying the plant and killing all 18 employees in the nearby control room and 9 other site workers.

Inherent safety as a concept was first promulgated by Trevor Kletz in the late 1970s and is based on common sense, which includes avoiding use of hazardous materials and hazardous activities. An inherent safe process avoids hazard instead of creating situations that will lead to the hazard and then trying to control it. This could be through several approaches. A chemical process can have multiple hazards associated with it. Hazards may arise due to raw materials, intermediates, final products, side and waste products, the nature of the process, the mode of operation, the complexity of the process steps, environmental conditions, experience or level of training of the personnel, etc. Certain guidelines were formulated for improved inherent safety, hygiene, and environmental protection. There are several basic principles and guidelines available for developing inherently safe process or accessing processes for their inherent safety. They are discussed in the subsequent pages of this chapter. There is no general answer to the question of which process is inherently safe, but similar processes could be graded according to their level of inherent safety (or their level of potential for danger).

One problem design engineers face is how to minimize simultaneously the risk associated with all of the hazards in a process. In the real world, the various hazards are not independent of each other, but are closely linked together (Hendershot, 1995; Hendershot et al., 2005). A process modification, which reduces one hazard, will always have some impact on other process steps and hence could have a positive or negative impact on other hazards. The advantages and disadvantages of each option must be compared for a particular case and the choice made based on the specific details of the process, materials used, probability of occurrence of the hazard, and its impact. These are discussed in detail later.

In general, the strategy for reducing risk could involve reducing the frequency of either the risk or the consequence of the potential accidents. The frequency of the risk could be reduced by having several layers of safety checks or controls (see Fig. 8.1). A process could have a slave and a master controller, followed by a set of alarms, and, finally, manual supervision. In such a design, an uncontrolled event would arise only if all the controls and checks failed. If the probability of such failure of each of the event is about 0.02 (i.e., 2%), then the probability of a hazard to occur would be equal to $0.02^4 = 0.00000016$.

The consequence of the hazard could be reduced by isolating the hazard with a protective cover. For example, atomic reactors are covered with a thick concrete shield so that the public and the ecosystem are not exposed to radiation if a runaway reaction occurs. But still, unforeseen human errors led to the venting of radioactive gases in Chernobyl to the atmosphere, causing

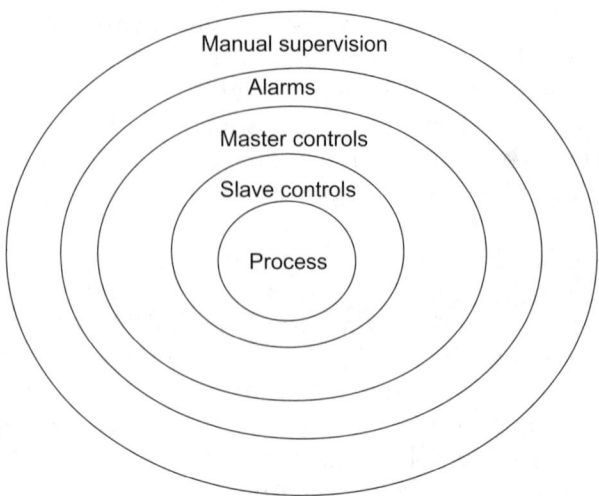

FIGURE 8.1. Various levels of safety checks and controls.

contamination and radiation deaths. The four approaches by which the incidence of hazards could be reduced are listed below in decreasing order of reliability and robustness:

1. **Inherent.** Reducing or eliminating hazards by using less hazardous materials and process conditions. This philosophy of inherent hazard reduction is currently being advocated since it is the surest method for achieving safe working conditions.
2. **Passive.** Reducing or eliminating hazards by process and equipment design features that reduce either incident frequency or consequence without the active functioning of any device.
3. **Active.** Using process controls, safety interlocks, and emergency shutdown systems to detect potentially hazardous process deviations and to take immediate corrective action.
4. **Procedural.** Using operating procedures, administrative checks, emergency response, and other management approaches to prevent incidents or to minimize the effects of an incident. This is followed very rigorously while operating complicated processes such as running an atomic reactor.

Risk control strategies in the first two categories are more reliable and robust than the other two, because they depend on the physical and chemical properties of the system rather than on the

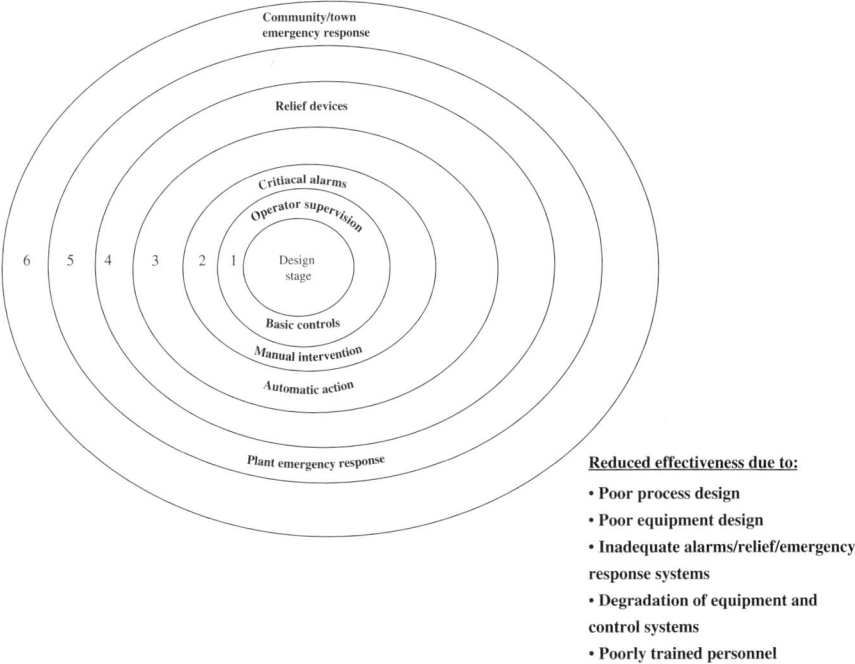

FIGURE 8.2. Safety at the design stage and at various levels.

successful operation of instruments, devices, and operating procedures. The fourth approach depends squarely on human intelligence, training, and experience. So it could fail without any warning. Inherent and passive strategies are not the same and are often confused. A real, and truly inherently safer, process will reduce or completely eliminate the hazard, rather than simply reduce its impact. Bringing in the issues related to the process safety during the design stage is the core to achieving a safe process. All other layers of safety checks such as alarms, supervisory control, and plans can always become ineffective over a period of the process's operating time (see Fig. 8.2).

Key Considerations

The principal considerations in inherent safety are to avoid processing using toxic, flammable, or environmentally hazardous materials; reducing the inventory of hazardous material and the potential for surprise; and separating people from chemicals and solvents. There are several subclassifications to these main principles. Several keywords can be used to develop inherent safety principles:

1. intensification,
2. substitution,
3. attenuation,
4. limitation of effects,
5. containing/enclosing/reinforcing,
6. error tolerance,
7. avoiding domino effects,
8. preventing incorrect assembly,
9. clarifying equipment assembly,
10. easing control.

Intensification, or minimization (principle 1 listed above), consists of reducing the quantities of hazardous chemicals in the plant or combining unit operations (also known as telescoping operations). "What you don't have can't leak" is the bus word. Reduced inventory of methyl isocyante at the Bhopal Union Carbide plant could have reduced the intensity of the accident.

A company that uses ethylene oxide as a raw material traditionally used to purchase it from a vendor. The material was shipped to the plant and stored in a large tank prior to use. A new manufacturing plant was built adjacent to the ethylene oxide plant, and it was received by pipeline, eliminating the need for storage and transportation of large quantities of the hazardous chemical (Orrell and Cryan, 1987). This approach minimized the intensity of possible hazard.

A company in Europe used phosgene to manufacture a fine chemical intermediate; phosgene was manufactured in a separate batch plant, and many tons of it were stored in large intermediate storage tanks. A new, continuous process to manufacture phosgene "on demand" was developed, which reduced the inventory dramatically. When the process needed phosgene, the new process was started, brought quickly to steady state, producing acceptable quality at the required rate, and fed directly to the process with no intermediate storage (Osterwalder, 1996). After the campaign was completed, the phosgene plant was shut down. This approach completely eliminated the storage tank for phosgene.

A chlorination process produced in a batch stirred tank reactor was replaced by a process using a loop reactor with intense mixing and recirculation (see Fig. 8.3). The new reactor was 33% the size

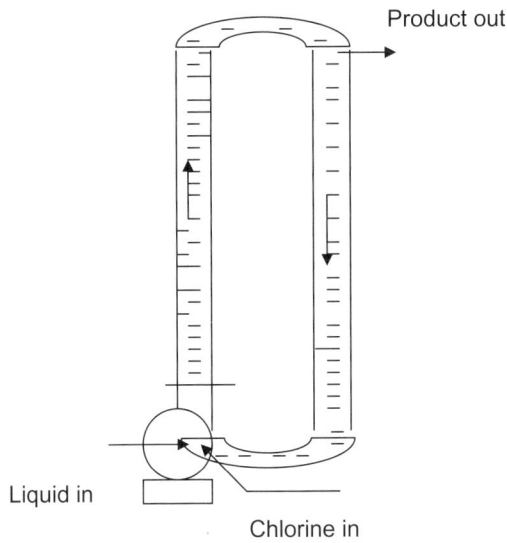

Product out

Liquid in

Chlorine in

FIGURE 8.3. Loop reactor.

of the original one with a batch time 25% that of the original. The chlorine consumption was also reduced by 50% due to better mixing and absorption of the gas in the loop reactor.

Combining unit operations not only reduces the number of equipments but also decreases hazards involved. The chloride route for the production of titanium dioxide pigment was improved by increasing the pressure of the oxidation reactor from atmospheric pressure to 2.5 bar, enabling a chlorine-rich gaseous stream to be fed directly into the chlorinator to recover chlorine. This avoided the need for an absorption stage previously required to boost the pressure of the highly corrosive chlorine-rich mixture using liquid titanium tetrachloride as the absorption medium. Further, a distillation column was also required to remove the chlorine gas from the tetrachloride solution. Inventory of the hazardous titanium tetrachloride was reduced from 1000 tons to about 10 tons.

Safe Reactor Design

Reactors are one of the major contributors to risk and hazards in a chemical manufacturing process, since their operations are generally carried out at high temperatures and pressure, occasionally with toxic and corrosive chemicals. In addition, the reactor has to hold the reaction mass during the course of the reaction. An optimal design of the reactor could be achieved only if a thorough understanding of the reaction mechanism, reaction kinetics, mass transfer, heat transfer, and mixing is known. Slow and inadequate mixing could increase the inherent reaction time although the intrinsic reaction time is small. Such processes require innovative reactor designs, optimum mixers, and perhaps small reactors.

A continuous stirred tank reactor (CSTR) is usually smaller than a batch reactor for a specific production rate (see Fig. 8.4). In addition to reduced inventory, CSTR usually results in enhanced safety, reduced costs, and improved product quality. Tubular reactors offer the greatest potential for inventory reduction since they have the lowest volume for a given conversion when compared to the previous two reactors (see Fig. 8.5). In a fed batch reactor, the reactants are added slowly, thereby controlling the rate of reaction and the exotherm. The design equations for these reactors are as follows:

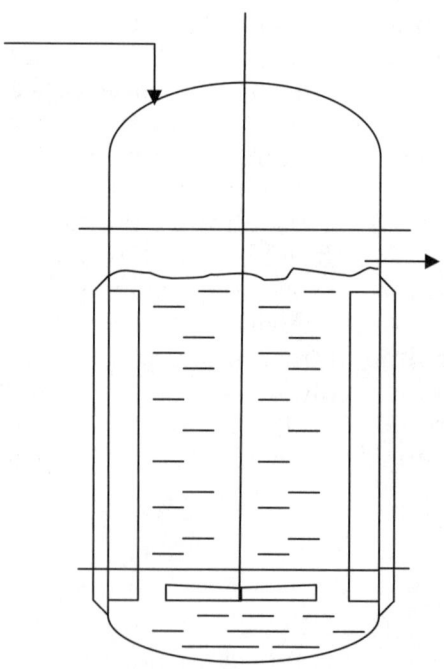

FIGURE 8.4. Continuous stirred tank reactor.

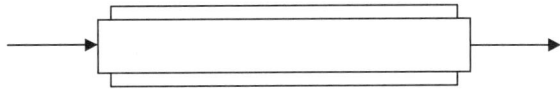

FIGURE 8.5. Tubular reactor.

Batch:

$$\tau = \int_{S_0}^{S} \frac{dS}{-r_S},\tag{8.1}$$

where

τ = reaction time (s),
S = substrate concentration (%),
S_0 = initial substrate concentration,
r_S = rate of reaction (may take any form).
Conversion is given by $(S_0 - S)/S_0$.

CSTR:

$$\tau = \frac{V}{F} = \frac{(S_0 - S)}{S_0},\tag{8.2}$$

where

V = reactor volume,
F = flow rate,
S_0 = feed substrate concentration.

Tubular:

$$\tau = \frac{V}{F} = \int_{S_0}^{S} \frac{dS}{-r_S}.\tag{8.3}$$

Fed batch reactor:

$$V\frac{dS}{dt} + S\frac{dV}{dt} = FS_0 - r_S V.\tag{8.4}$$

In a fed batch reactor, the reactor volume and concentration buildup vary as a function of time. Tubular reactors are usually of a simple design, contain no moving parts, and have a minimum number of joints and connections. The main disadvantage of continuous reactors is that the reactor would produce off-spec material during its startup and shutdown phases. These reactors are ideally suited for preparing large quantities of material and are not suitable for producing different types of small amounts of products, such as specialty chemicals, fine chemicals, pharmaceutical products, or paints. A loop reactor is a continuous tube or pipe connecting the outlet of a circulation pump to its inlet (Fig. 8.3). Reactants are fed into the loop, where the reaction occurs, and product is withdrawn from the loop. Loop reactors have been used in place of batch-stirred tank reactors in a variety of applications, including chlorination, ethoxylation, hydrogenation, and polymerization. A loop reactor is smaller than a batch reactor for producing the same amount of product and can achieve intense mixing. For example, for a polymerization process, a 50 L loop reactor has a capacity equal to that of a 5000 L batch reactor.

Semi-batch operation, or the gradual addition of one or more reactants to a reactor, limit the quantity of reactants inside the vessel and increase safety when compared to batch processes in which all reactants are included in the initial batch charges (see Fig. 8.6). For an exothermic reaction, in a semi-batch process the total energy of reaction available in the reactor at any time is minimized. Gradual addition can help in controlling the rate and

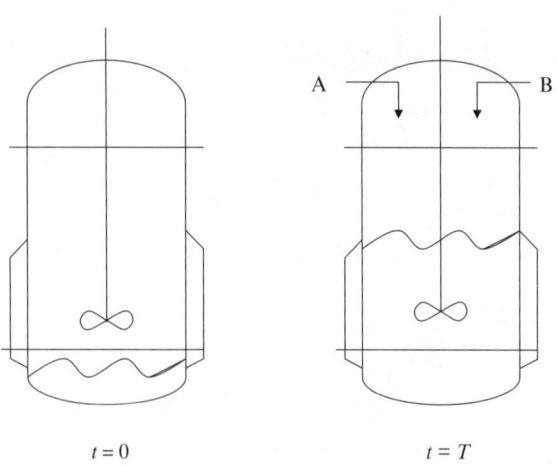

$t = 0$ $t = T$

FIGURE 8.6. Semi-batch reactor operation.

hence the heat of reaction. The main advantage of a CSTR or tubular reactor is that, except during startup and shutdown, the operation is under steady conditions, while in the case of a batch or fed batch reactor, the operating parameters keep changing with time (unsteady) and may lead to unsafe operations.

Elimination/substitution (principle 2) implies the use of a safer material in place of a more hazardous one. It may be possible to replace flammable solvents with nonflammable ones or toxic chemicals with nontoxic ones. It is necessary to evaluate not only the property of the substance but also the volume that is handled. Changing to a safer reagent may be a better solution than using a possible hazardous chemical and incorporating several safety checks or control alarms and trips to prevent the hazards. An example of this principle is the use of air instead of hydrogen peroxide in oxidation reactions or the use of hydrogen instead of hydrazine for reduction reaction. Use of a magnesium hydroxide slurry instead of a concentrated sodium hydroxide solution to control pH is a better alternative since the former is less corrosive than the latter. Inherent safety could be achieved by using milder alkalis or acids in reactions instead of stronger alkalis and acids.

A new generation of paints has been invented that are water-based rather than the conventional solvent-based ones, thus eliminating the need for handling solvents or having expensive VOC treatment facilities necessary when solvent-based paints are used. Also, the hazards from flammable materials are totally eliminated. Risks involved in the transportation of flammable solvents are also eliminated by this substitution.

For each process, the choice of solvent involves a number of considerations, including toxicity and compatibility with other raw materials. Another aspect that needs to be kept in mind is its boiling point. In a particular reaction, acetone was used as the solvent. The heat of reaction of the process was such that an uncontrolled addition of one of the reactants, or loss of cooling, could lead to vigorous boiling of the batch, overpressurization of the reactor, and flooding of the condenser. A simple change of solvent from acetone to toluene produced a reaction mixture with a boiling point sufficiently high to overcome all these possible hazards. The ambient average temperature as well as the highest ambient temperature reached during summer months are important parameters during selecting the solvent, more so in equatorial regions, since a higher ambient temperature leads to higher vapor pressure and hence higher vapor concentration in the atmosphere.

ICI Explosives (Australia) traditionally used to prepare its emulsion explosive in plants located in remote locations, store it, and transport it to the mine site. Such an operation was inherently very unsafe due to the storage and transportation of explosives. Later ICI switched to preparing the emulsion explosive mixtures at the site using mobile mixing plants and directly charging it into the borehole. This "miniplant on wheels" was very safe since transporting individual components is less hazardous than transporting an intact explosive. This philosophy totally eliminated the risk.

Purification of ethylene oxide from water solution can be carried out by heating the solution or injecting live steam. The latter approach avoids the overheating of ethylene oxide since the maximum temperature will equal the steam temperature. In a boiler inadvertent heating above 140°C could lead to internal deflagration. An early ethylene oxide (EO) plant used an inventory of 450 tons of boiling hydrocarbons as coolant to control the reaction of 25 tons of gas, with a high integrity protection system. Containment of the coolant, in the event of leakage or loss of reactor control, required massive capital investment. Pressurized water cooling is now considered both safer and more environmentally acceptable. Leakage of coolant would mean escape of water to the atmosphere.

Substitution means using less hazardous reaction chemistry or replacing a hazardous material with a less hazardous one. Traditionally, acrylate esters were manufactured using the Reppé process, which involves reacting CO with acetylene in the presence of nickel carbonyl and anhydrous hydrogen chloride. There are numerous hazards in this process. Acetylene is reactive and flammable; carbon monoxide is toxic and flammable; nickel carbonyl is toxic, an environmental hazard (heavy metal contaminant), and a suspect carcinogen; and anhydrous hydrogen chloride is toxic and corrosive. Today most acrylate production uses a propylene oxidation process. Although this new process route is not fully safe, it is inherently safer than the Reppé process.

Hydrogen peroxide is a commonly used oxidizing agent that decomposes sometimes violently with the evolution of oxygen. The decomposition process can also happen due to the presence of impurities. Another problem faced while using hydrogen peroxide is that during its addition into the reaction mixture, if the agitator stops due to power cut, unreacted material builds up. When the agitator is restarted, the accumulated hydrogen peroxide will react quickly at a high decomposition rate, generating large exotherm, leading to pressure buildup inside the reactor and ejection of the batch. As an alternative, air can be used as the oxidizing agent, which not only is cheaper but also has no instability problems. A catalytic oxidation process is easily controllable and eliminates the problems of accumulation. Also, switching off the supply of air stops the reaction.

Polyisobutylene succinic anhydride, an intermediate for a large number of surfactants, can be prepared by reacting polyisobutylene and maleic anhydride. Two approaches are possible to achieve this coupling, one by using chlorine gas and the other by using paratoluene sulphonic acid as a catalyst. Although the yield is higher in the former, handling of chlorine gas and hydrogen chloride, which is generated as a byproduct, makes the process inherently less safe than the second one. In addition, the former process requires towers for absorbing excess chlorine and the generated hydrogen chloride gases.

The choice of the manufacturing process should involve consideration of all the materials involved in the total process and not just those involved in the chemical synthesis step alone. A process that used methanol as solvent had an effluent treating facility downstream; a cheap option was treating the waste with sodium hypochlorite solution. However, it was realized that the residual methanol in the effluent would react with the hypo to produce the dangerous impact-sensitive explosive, methyl hypochlorite. So an alternate, methanol-free process had to be developed to avoid the hazard downstream.

Early refrigeration systems used a variety of hazardous refrigerants, which included ammonia (which is toxic and flammable), light hydrocarbons (which are flammable), and sulfur dioxide (which is toxic and corrosive). Later, in the 1930s, chlorofluorocarbon (CFC) refrigerants were introduced to eliminate these chemicals and gases that were inherently hazardous. Recently it has been discovered that CFCs have a major impact on the ozone depletion, and hence their use is being phased out. Home refrigerators now use only about 120 g of isobutane refrigerant. These changes are a good example of using the inherent safety strategies such as "minimize" and "substitute." This example also indicates that continuous improvement is needed as more knowledge is generated with time.

Attenuation, or moderation (principle 3), means reducing the severity of operation. If the use of a hazardous chemical in a process is unavoidable, then one needs to determine whether the severity of its usage could be moderated by operating the process at a lower pressure or temperature. Chlorine and ammonia can be stored as refrigerated liquids at atmospheric pressure rather than at high pressure at ambient temperature in autoclaves. The lower pressure results in lower leak rates, and the lower temperature lowers the rate of vaporization.

Dilution is another approach that can lead to attenuation (e.g., using air as an oxidant instead of using pure oxygen could avoid explosive situations). Use of 28% aqueous ammonia solution instead of anhydrous ammonia can lead to a 10× decrease in the vapor cloud downwind during a reactor leak.

Dynamite is nitroglycerine absorbed onto an inert carrier. Alfred Nobel invented dynamite in 1867; it was safer to use than directly using nitroglycerine. Plastic materials for use in molding and fabrication processes will be safer if they can be handled as pellets or granules rather than as fine powders, since powders can form explosive dust cloud. Examples of materials that have been handled in a dilute form to reduce the risk of handling and storage include aqueous ammonia or methylamine in place of anhydrous material, muriatic acid in place of anhydrous HCl, and sulfuric acid in place of oleum (SO_3 solution in sulfuric acid).

The ammonia manufacturing process has seen a steady improvement in safety. In the 1930s, ammonia plants operated at pressures as high as 600 bar. In the 1950s, process improvements led to reduction in operating pressures to 300–350 bar; by the 1980s, this process was operating at 100–150 bar. Besides being safer, the lower-pressure plants are also cheaper and more efficient.

Limitation of effects (principle 4) is achieved by changing designs or process conditions rather than adding protective equipment that may fail. For example, it is better to prevent overheating by using a lower-temperature fluid rather than using a hotter fluid and relying on a control system to maintain overheating. Steam is safer than oil as a heating medium with respect to overheating since steam is limited by the operating pressure (temperature is directly proportional to pressure) while the latter could be overheated and hence may require controllers to check overheating.

The philosophy of "just-in-time production" was primarily introduced as an industrial engineering concept to avoid unnecessary storage of products in the factory outlets (which signifies locked up, unused operating cost). Just-in-time production can help in dispatching the material as soon as it is needed by the customer, without unnecessarily storing the finished goods, which leads to blocking of working capital. This concept is also well suited for producing intermediates that are toxic or dangerous, for example, if a process requires hydrogen cyanide or isocyanide as an intermediate, a process that produces this dangerous chemical just in time and in the desired amount is much safer than having a facility to produce it in large quantities with an intermediate storage. For example a cyanohydrin plant that requires hydrogen cyanide can take it directly from the top of the distillation column's reflux tank, thereby eliminating hazards involved in its storage.

Instead of manufacturing in large quantities, compressing, liquefying, storing, and vaporizing chlorine, it can be supplied to the customer plant directly as a gas from the compressor discharge, if the rate of production could be exactly matched to the customer's need. This eliminates liquid chlorine piping inventory, in the storage tank, and the inventory in the vaporizer. Of course, during the start of the chlorine plant there will be a small flow of impure chlorine from the plant.

By linking the ethylene oxide manufacturing and purification directly to the consumer plant, the ethylene oxide inventory is reduced from 50 tons to 6 tons for a 300,000-tpa ethoxylate surfactant manufacturing plant.

The principle of simplicity is based on the fact that simpler plants are safer than complex plants, as they provide fewer opportunities for human, equipment, instrument, and control errors and contain a smaller number of equipment that can fail.

An equilibrium reaction of the type $A + B \rightleftharpoons C + D$ can be driven to the right if one of the products is removed continuously. Bringing the reaction to equilibrium followed by removing one of the products by distillation and recycling the unreacted reactants back to the first reactor is not simple (see Fig. 8.7). But by using a reaction cum distillation assembly that combines the reaction and continuous removal, one could achieve reduction in hardware. In the combined system the distillation column is placed above the reactor. A process for the manufacture of methyl acetate by using reactive distillation reduced the number of distillation columns from eight to three and also eliminated an extraction column and a separate reactor. Such a system reduced process inventory and auxiliary equipment such as reboilers, condensers, pumps, flanges, pipes, and heat exchangers, leading to savings in capital and operating costs.

Containing/enclosing/reinforcing (principle 5) could lead to separation of personnel, the general public, flora or fauna from hazardous chemicals or situations. Radioactive waste is generally enclosed in concrete and buried underground, preventing its

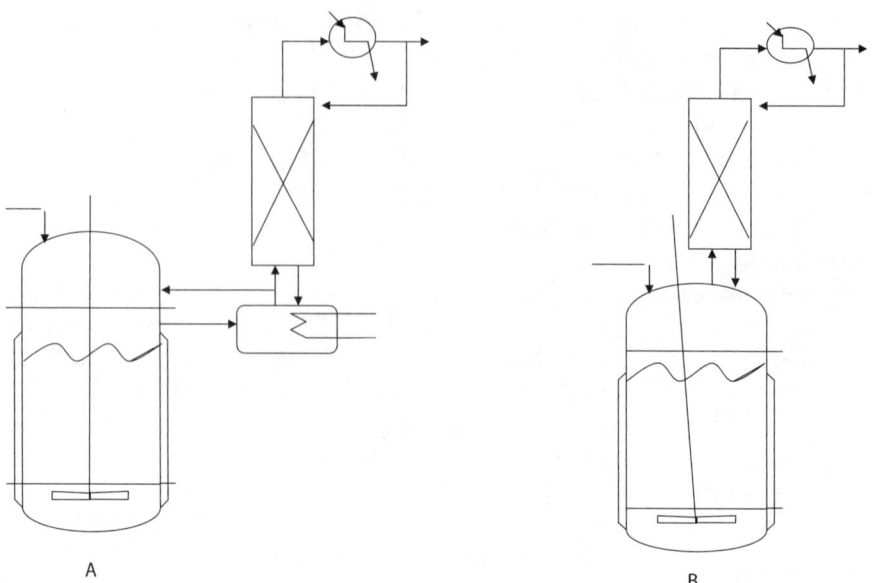

FIGURE 8.7. Equilibrium reaction: (A) traditional approach; (B) reactive distillation.

leakage and exposure to humans. Oil spills in the ocean are generally isolated and prevented from contaminating the rest of the ocean using booms, which are floating barriers to oil (for example, a big boom may be placed around a tanker that is leaking oil, to collect the oil). Superfund sites are contained from the rest of the environment to avoid leakage.

Error tolerance (principle 6) involves more robust equipment, processes that can absorb upsets or deviations in operating temperature, pressure and excess of reactants, or catalysts without leading to runaway reactions or exothermic conditions. A sensitivity analysis of the various operating variables on the process performance could reveal the stability of the system. At times, in order to achieve robust operating conditions, it may be necessary to operate the process at a lower rate of productivity rather than selecting a set of conditions that lead to higher productivity but are very sensitive (see Fig. 8.8) to small changes in the parameter(s).

Domino effects (principle 7) are secondary or incidental effects that arise if a problem in one system leads to problems in other places. A leaky valve that is located in the storage tank of a toxic

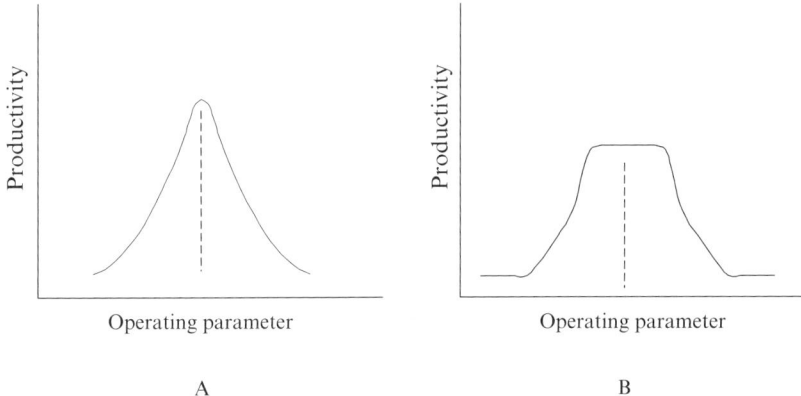

FIGURE 8.8. Process sensitivity: (A) sensitive; (B) stable/less sensitive process.

or hazardous chemical could lead to contamination in other vessels downstream from this storage vessel. Venting of toxic gases could lead to hazards to humans, flora, and fauna, which could be short or long term in nature. Fail-safe shutdown designs, open construc- tions, and ample layout spacing are some of the ways domino effects could be minimized. Multiple collisions on the highway on foggy or rainy days are due to domino effects.

Many hazards arise due to incorrect assembly of components (principle 8), which is due to ambiguity in the subcomponents' designs. This could be avoided using unique assembly components to reduce or eliminate human error during assembly. A nonreturn value placed in the wrong direction is a common problem, which leads to stoppage of flow and pressure buildup.

An example of the principle of clarifying equipment status (principle 9) means that if an equipment or instrument is opened up for maintenance or repair, it is very essential that others clearly know the status, without any ambiguity, so that mistakes are not made. Starting electric power when technicians are still working on the power line—and starting a utility flow before the comple- tion or proper commissioning—are two of the unsafe incidents that arise due to improper clarification of the equipment status. Avoiding information overload to the maintenance, operation, and plant personnel is also an important point to keep in mind. One of the reasons for the radioactive leakage in Chernobyl was due to the information overload faced by the operating personnel, who could not cope with so many alarms going on at the same time, leading them to make the mistake of opening the wrong valves thereby venting radioactive fumes.

Ease of control (principle 10) means less hands-on control and high controllability.

Safety Management and Hazard Identification

Chemical reaction and process hazard identification is an important step before the startup of a manufacturing process. Several of the incidents listed below could have been avoided if hazard identification techniques had been followed by the manufacturer prior to startup.

- An explosion during a blending operation resulted in five fatalities and destruction of much of the manufacturing facility at Lodi, New Jersey (1995).
- An uncontrolled reaction in a phenol-formaldehyde resin plant killed one worker, injured four others, and extensively damaged the plant at Columbus, Ohio (1997).
- A runaway reaction in a batch dye manufacturing process injured nine people in Paterson, New Jersey (1998).
- A decomposition reaction caused by water contamination and a failed pressure controller on steam tracing resulted in a pipe rupture in Ringwood, Illinois (2000).

These types of hazards could have been avoided if good process safety management systems were in place, including consideration of reactive chemistry issues, handling and storage of individual reactive chemicals, and proper understanding of the process operations.

Chemical reaction hazard identification involves consideration of several points during the process scale-up and pilot plant stages:

1. Heat of reaction for the main and other side reactions. All main and side reactions that could occur in the reaction mixture should be identified, and the heat of reaction of each one of these reactions should be estimated. There are a number of techniques available for measuring heat of reaction, including calorimetery, plant heat and energy balances for processes already in operation, analogy with similar chemistry, literature sources, supplier knowledge, and thermodynamic estimation techniques.

2. Stability of the reaction mixture at the maximum adiabatic reaction temperature. Decomposition reactions, which generate gaseous products, are a serious concern because a small mass of reacting condensed liquid can generate a very large volume of gas from the reaction products (22.4 L of gas is produced by a liquid of 1 g mol; for example, when evaporated, 18 g of water will occupy 22.4 L of space at atmospheric pressure). This results in rapid pressure generation in a closed vessel. These reactions will need safeguards such as emergency pressure relief systems, which can quickly vent this excess gas. Understanding the stability of a mixture of components may require a laboratory testing facility.

3. Stability of all individual components of the reaction mixture at the maximum reaction temperature. This might be done through literature information, through supplier contacts, or with the help of a differential scanning calorimeter. The differential scanning calorimeter can determine the temperature at which the material would decompose. It is also possible to collect the gases and vapors liberated during decomposition and analyze them using other analytical instruments. If any component can decompose at the maximum adiabatic reaction temperature, one has to understand the nature of this decomposition, the gases that are evolved, and their quantities and evaluate the need for safeguards, including emergency pressure relief systems, downstream scrubbers, absorbers, or flares. For a compound $C_xH_yO_z$, the oxygen balance is given by $-1600(2x + y/2 - z)/mol\,wt$. All explosive materials have values between -100 to $+40$. So if the raw materials, intermediates, or products have values in this range, one could be sure that it could be an explosive.

4. Know the heat of addition and removal capabilities of the production reactor. A large batch reactor has a jacket and/or coils for heating or cooling. Coils have to be used if the reaction is exothermic in order to control the process. Large-diameter vessels will have a considerable temperature gradient along the radius of the vessel. The reactor agitator also adds up to the heat source, about 2550 Btu/hr/hp. The heat transfer coefficient can decrease due to fouling of the surface and also due to presence of multiphase fluids. With time the reactor's heat removal capabilities may decrease due to fouling. Factors such as reactor fill level, agitation, type of agitator, fouling of internal and external heat transfer surfaces, variation in the temperature of heating and cooling media, variation in flow

rate of heating and cooling fluids, changes in the properties of the heating and cooling media (such as decomposition of heating oil, water ingress into the media, etc.) affect the thermal processes.

5. Identify potential reaction contaminants. Raw materials are never 100% pure, and each raw material brings its own contaminants. Even ubiquitous items present in a plant environment (e.g., air, water, rust, oil, and grease) could affect the safety. Trace metal ions such as sodium, calcium, and others commonly present in process water can act as a catalyst initiating unwanted reactions. These may also be left behind from cleaning operations. In the pharmaceutical and paint industries, a large number of products are made with the same set of hardware, and products from the previous batch could become contaminants for the next batch. Nickel present in SS vessels could catalyze hydrogenation reactions. Acid glass linings could catalyze acid-catalyzed reactions (such as esterification and addition reactions).

6. Scale-up leads to higher temperature gradients. Heat addition, removal, and agitation will be less effective in a plant reactor when compared to small pilot vessels. The temperature of the reaction mixture near the jacket wall or coil wall has to be higher (for systems being heated) or lower (for systems being cooled) than the bulk mixture temperature in order to achieve the same heat transfer rate. This also means that the process fluid is exposed to higher temperatures near the heat transfer surface, which may lead to its decomposition if it is thermally labile. The temperature may also be higher near the addition ports because of poor mixing and localized reaction at the point of reactant contact. The location of the reactor temperature thermometer (or thermo couple) relative to the agitator and to heating and cooling surfaces may impact its ability to provide correct information about the actual reactor temperature. These problems will be more severe for very viscous systems, slurries, large-diameter vessels, and foul chemicals. A high temperature could result in a higher rate of chemical reaction or decomposition. A low temperature near a cooling coil could result in slower reaction there and a buildup of unreacted material, increasing the potential for sudden release of chemical energy if the conditions in the reactor change. If the chemical energy release is accompanied by gas evolution, it could lead to buildup of pressure inside the reactor. If the relief valve is not able to cope with this sudden rise, it could lead to explosion.

7. Possibility of vapor phase reactions, which include combustion reactions. Examples of vapor phase reactions include reactions between organic vapors with a chlorine atmosphere, vapor phase decomposition of materials such as ethylene oxide or organic peroxide, and reactions involving methane or hydrogen with air.
8. Interaction between reactants and other hardware materials used in the plant such as gaskets, pipe linings, fittings, pumps, seals, etc. For example, in an oxidation reactor, solids were known to be present, but nobody knew what they were. It turned out that the solids were pyrophoric and caused a fire in the reactor. Gasket materials react with oil or other hydrocarbons and swell, leading to leakage of the contents. At times, raw materials or solvents may not be compatible with oil used in vacuum pumps. Gaskets may crack at low or cryogenic temperatures, leading to leakage of chemicals. Water contamination in oil could lead to violent, superheated steam evolution during heating.
9. Develop a chemical interaction matrix. Compatibility of (1) raw materials with other raw materials, (2) intermediates/side products/waste between each other, (3) with wrong grade of raw materials, (4) with material of construction, and (5) with contaminants should be understood. Techniques such as the chemical interaction matrix can help in identifying possible adverse interactions (see Table 8.2). This is a systematic

TABLE 8.2
Interaction Matrix (✓ = compatible; ✗ = not compatible)

	Raw Material-2	Solvent-1	MOC of Reactor	MOC of Gasket	Intermediate	Product
Raw material-1	✓			✗		
Raw material-2	✓					
Solvent-1	✓					
MOC of reactor	✓					
MOC of gasket	✓					
Intermediate	✓					

method to identify chemical interaction hazards. This technique can be applied at any stage in the process life cycle, either at the research or at the plant stage.
10. Flammable liquids and solvents pose high probability of hazard. It has been realized that the probability of ignition of flammable liquids may be reduced by the presence of antistatic additives. This increases the conductivity of the solvent and hence minimizes the charge generation and accumulation and the chance of ignition. Of course, this solution would only be useful for liquids with a low conductivity.

In addition to the previously mentioned reaction hazard identification techniques, the following reaction process design considerations need to be considered for avoiding hazards:

1. Rapid reactions are desirable. In rapid chemical reactions the reactions occur immediately when the reactants come into contact and the reaction energy is quickly released, allowing one to control the reaction by controlling the contact (mode or time of contact) of the reactants. However, one must be certain that the reactor is capable of removing all of the heat and any gaseous products generated by the reaction immediately and that the mixing is perfect to avoid buildup of unreacted material.
2. Avoid batch processes in which all the potential chemical energy is present in the system at the start of the reaction step. The buildup is much less in a continuous process.
3. Use gradual addition or "semi-batch" processes for exothermic reactions. Addition could be stopped in case of temperature buildup. It is advantageous to operate a semi-batch reaction where one of the reactants is added gradually and slowly. The limiting reactant could be added using a metering pump, using a small feedline, or through a restriction orifice. The limiting reactant should react immediately when it is charged so that there is no buildup. The reactant feed can be stopped, if necessary, if there is a failure such as loss of cooling, power failure, or loss of agitation, and the reactor will contain little or no potential chemical energy from unreacted material.
4. Using control of reaction mixture temperature as the only means for limiting the reaction rate may not work. In autocatalytic reactions, an increase in temperature will result in a faster reaction and even more heat being released, causing a further increase in temperature and more rapid heat release. If

there is a large amount of potential chemical energy from reactive materials, a runaway reaction may ensue. Other strategies such as dumping the reaction mixture into a cold tank may need to be resorted to, as practiced in nitration reactions. Other questions that have to be answered are, "Will it be necessary to drown out the reaction mass into cold water or a coupling solution (for a diazotization process)?" or "Will it be necessary to vent the material?" Understanding the mechanism of reaction and the relation between operating temperature and reaction rate is important.

5. Account for the vessel size on heat generation and heat removal capabilities of a reactor. The heat balance in a chemical reactor is as follows:

> [heat accumulation in a vessel] = [heat addition]
> + [heat generation] − [heat removal by coolant]
> − [heat loss to surroundings].

Heat removal in a small laboratory reactor is very efficient, and even heat loss to the surrounding atmosphere is significant. Heat generation increases with the volume of the vessel, by the cube of the linear dimension. Heat removal capability increases with the square of the linear dimension (see Fig. 8.9). So the heat generated by a reactive system will increase more rapidly than the capability of the system to remove heat when the process is operated in a larger vessel. If the reaction temperature is easily controlled in the laboratory, this does not mean that the temperature can be controlled in a plant-scale

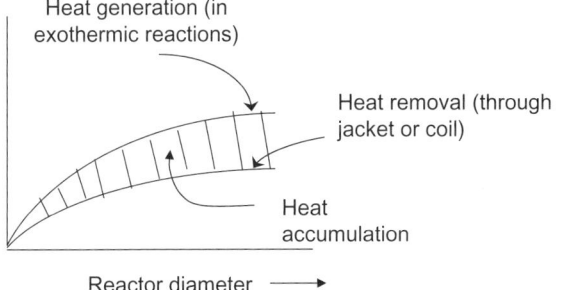

FIGURE 8.9. Effect of reactor diameter on heat accumulation.

reactor. Hence a large vessel will require more heat transfer area in the form of immersed coils. Heat generation due to agitation also becomes significant while mechanically agitating highly viscous fluids.

6. Use multiple sensors, in different locations in the reactor for rapid exothermic reactions. If the process is monitored and controlled based on a specific parameter such as temperature, pressure, or pH, then it is imperative to have multiple sensors for that parameter located at different points in the reactor so that the measurement is read accurately, so that improper measurement due to failure of one single sensor does not lead to an unusual occurrence. If the reaction mixture contains solids or is very viscous, or if the reactor has coils or other internal elements that might inhibit good mixing, then one temperature sensor may not give a true picture of the temperature inside the vessel. Redundancy also helps if one sensor fails. Temperature of the content will be higher near the walls of jacketed reactors when compared to the central axis. Large-diameter reactors will have a larger gradient, and location of the thermocouples should be based on all these considerations. Turbine-type agitators achieve radial mixing, while propeller types bring in axial mixing; hence the former are more suitable with jacket-type heating. Pressure gauges may get blocked if the solution contains solids, meaning they may not give accurate readings.

7. Avoid feeding a material to a reactor at a higher temperature than the boiling point of the reactor contents. This can cause rapid boiling of the reactor contents and vapor generation, and at times the boiloff could be violent. Also, avoid adding a chemical whose boiling point is lower than the temperature of the reactor contents. The added solvent may flash as soon as it comes in contact with the bulk contents in the vessel.

8. Identify operability issues. Could hazardous accumulation of reactants occur if the agitation fails? Viscous solvents should be avoided since they are difficult to agitate, which hinders process control and reduces the rate at which heat is conducted away from the reacting species, allowing the accumulation of heat. Also, agitating a viscous medium uses more energy than agitating one that is less viscous. Sizing of vent, safety, and emergency discharge valves has to be estimated accurately; if not, discharges during emergency situations could be hampered.

9. Focus attention on heat transfer fluids and refrigerants. Sufficient thought has to be given to these fluids, such as their stability at higher temperature and in the presence of contaminants such as normal plant items as well as the raw material and side products. These fluids have a fixed lifetime and may slowly degrade, leading to undesirable side effects. Closed systems are desirable since they prevent contamination by air and moisture.

10. Beware of vents and drains since the emissions and leaks occur predominantly here, and poorly designed nonreturn valves or transport could lead to higher leaks of VOCs and other gases.

11. Raw material and finished products charging and discharging zones are also potential areas of risk that may arise due to leaks, spills, human errors, etc. The risk of human exposure is greater in these areas. Solid products or reactants could also lead to dust clouds and associated problems.

Issues Related to Transportation

The manufacture of a product by a particular route may involve the transport of hazardous materials on a regular basis to the site of manufacture or even between sites. This could pose a very significant risk to the public at large. The quantity of material, the frequency and mode of transport, as well as the safety health and environment (SHE) issues will make or break the process. It is worth bearing in mind even at an early stage of the process that a development decision has to be made to avoid such a process. Other considerations to be borne in mind are: Can the material be made near the factory site, thereby avoiding its transportation? Can the process be changed so that the use of such a chemical could be totally avoided?

Losses to Atmosphere

Vapor, solvent, and gas losses to atmosphere occur because of several reasons, including losses through leaky valves, vents, and gaskets; during charging, discharging, loading, and unloading; and due to human errors; etc. To minimize losses to atmosphere, the following points need to be considered. Atmospheric losses will also depend on the ambient temperature of the locality, the season, and the time of the day. For example, these problems may be more serious in tropical countries than in countries away from the equator.

1. Avoid organic solvents that could generate VOCs. VOC generation also depends on the average (or maximum) ambient temperature of the locality and the season.
2. Avoid highly volatile solvents, especially those whose boiling point is less than 80°C.
3. Avoid carrier gases in distillation (e.g., inert gas sparging), since gases will carry low boiling solvents.
4. Pump liquids and slurries rather than blowing with high pressure, which could lead to solvent carryover. Of course, hazards of pumping certain classes of materials and problems of sealing materials in pumps should be considered while weighing this option.
5. Atmospheric-pressure distillation is better than vacuum distillation since loss of solvents is higher at the exit of the vacuum pump. Of course, both options are better than using inert gas sparging.
6. Transfers of materials should be minimized since each operation results in atmospheric emissions.
7. Bubbling of nitrogen to deoxygenate reaction mass also leads to VOC emission.
8. Avoid through-pan draughting of process vessels, but beware of the problems associated with flammable, toxic, or explosive materials during charging.
9. Extract air around the manhole, not through the vessel, if it is required to be open, and only extract air when necessary due to safety reasons. Extraction should be minimized as it increases solvent losses and the vacuum level determines the amount of vapor lost.

The Purification Step and Associated Hazards

As stated earlier, the purity of the raw material, the efficiency of the reaction, and the molar quantities of material used all affect the yield of the reaction. As a result, they also affect further downstream purification steps. Higher impurities—as well as lower yields—translate into a more rigorous purification step. For example, the equilibrium conversion of nitrogen and hydrogen to produce ammonia is barely 9–12%. So the unreacted gases are refrigerated, recovered, recycled, recompressed, and returned to the reactor. These operations lead to a very large inventory of inflammable hydrogen gas in the plant. A process with a high one-pass conversion will have a smaller amount of recycle material.

Inventory

Many processes have large inventories of toxic or flammable materials because the percentage conversion is low or the rate of reaction is slow. These are a major source of hazards, and every effort should be made to reduce the inventory. This thought should be given at the research stage, as it may not be possible to rectify the problem during the process development or pilot plant stage. It is worth remembering that the best way of avoiding a leak of hazardous material is to use so little that leakage will not matter.

The temperature and pressure at which the reaction is carried out will affect the inventory; for example, increasing the pressure in a phosgene plant reduces the inventory to one tenth of the original level. The inventory of the reaction is affected by the concentration. Can the concentration be increased safely, bearing in mind potential temperature rise due to the heat of reaction, as the reaction is carried out at concentrated conditions. In a biochemical process, increasing concentration may decrease the rate of reaction if the biocatalyst or organism is inhibited by substrate. The number of stages in the reaction will impact the overall inventory especially if intermediates are isolated at every stage. Isolation at each stage increases hazards.

Reaction Temperature

Choosing the ideal reaction temperature involves several factors. Using a lower reaction temperature means that the reaction mass is away from the onset of any exothermic or gas-evolving decomposition, but a lower temperature decreases the rate of reaction (a 10°C decrease in temperature reduces the rate of reaction by half). So the reaction kinetics determines the minimum reaction temperature. A lowering of the reaction temperature may be achieved if a suitable catalyst can be found, but the catalyst may bring its own hazards, such as dust and possible fire if it is pyrophoric.

Reaction Pressure

Performing a reaction at high pressure would result in more material being discharged in the event of vessel or pipe rupture. However, increasing pressure could increase the rate of the reaction, leading to smaller reactor size. In some cases increasing reaction pressure could lead to the gas being liquefied to a reduced reactor size. Increasing pressure leads to the potential for explosion and a flying missile hazard.

Concentration

The more concentrated a reaction mass is, the less solvent per ton of product will be used, leading to a reduction in both solvent inventory and reactor size. However, this results in a lower heat sink to absorb the heat of reaction. If the process is highly exothermic, this could be a concern. Selecting higher-heat-capacity solvents could minimize this problem. Solid–solid reactions suspended in water medium under violent agitation have been found to be a good answer to such problems, where the excess amount of water acts as a heat sink, and since the reactants are solid and neat, the rate of reaction is also high.

pH

pH-dependent decompositions are quite common. If gases are evolved during decomposition, the process could become violent. For example, the decomposition of hydrogen peroxide to water and oxygen gas occurs more rapidly in alkali, so acid is often added as a stabilizer. Decomposition of Na azide to produce nitrogen gas is another example. Generation of HCN gas from NaCN solution is also dependent on pH. Under alkaline conditions it is safe to handle the cyanide solution.

Thermodynamics

The heat of reaction for the hydrogenation of an aromatic nitro group to the corresponding amine is around 130 kcal/mol. The heat of reaction coupled with the mass and heat capacity of the batch can be used to calculate the maximum temperature rise that would occur in the absence of any cooling. This enables one to predict whether (1) the reaction is inherently safe, in that any temperature rise would be small or (2) it is inherently hazardous in that the potential temperature rise could reach a region of thermal instability or cause the batch to boil. In the latter case, the process control becomes very important and stringent procedures may need to be put in place to allow the process to be operated safely. In the extreme, it may not be possible to operate the process to the required degree of control, necessitating the selection of an alternative process.

Kinetics

If a reaction is rapid, there is the potential for very rapid heat release if all the reactant was added rapidly. To control the rate of heat evolution, the rate of addition of the reactant would need to be controlled. If a reaction is slow, this may be due to it being inherently slow or due to poor mixing. Problems can arise if cooling is lost while a large accumulation of material has occurred. This will lead to loss of control over the rate of heat and/or gas evolution.

Loss of agitation could also lead to accumulation of unreacted raw material. Sudden mixing of such a large amount of unreacted mass could lead to rapid reaction. For example, during nitration reaction, loss of agitation could isolate and build up the amounts of the two reactants, namely, the acids and the benzene. Restarting the agitator could force rapid mixing of large amounts of the two reactants, leading to a tremendous amount of heat generation and explosion. Always beware of restarting an agitator before the reaction mass has cooled below the reaction temperature.

Gas Evolution

Gas evolution could be a problem, for many reasons. First, the type of gas may be a problem. It may be toxic or flammable, requiring scrubbing or flaring. Second, the total volume and rate of gas evolution are important in terms of preventing the reactor from becoming overpressurized. The auxiliary units that handle the gas should also be able to handle this excess amount. Another issue to keep in mind during equipment design is that solvents and VOC may be carried over by the evolved gas into the atmosphere.

The Number of Stages and Telescoping Processes

Evaporation, distillation, and extraction processes are carried out in several stages. Even reactions are carried out in several vessels. Since each stage requires its own auxiliary equipment, telescoping the stages could be highly beneficial. Performing all the reactions in one pot followed by isolation is the most ideal situation. Fermentation is an example of this philosophy, where complex products are formed from simple chemicals.

Multiple stages would also lead to more chances for human error and generation of more effluents. Telescoping a process can be beneficial here, too. Using fewer reactors means less cleaning, which means less effluent. Less isolation of materials means less filtration and/or drying, which means less solvent usage (and need for recovery) and hence less atmospheric emissions. Reducing the number of isolated intermediates will reduce the requirement for in-plant storage and will result in improved yield, fewer chances of intermediates escaping through the effluent stream, and lower inventory of materials.

Effect of Operating Costs on Safety

Through his logic, Dalzell (1993) points out that safety and plant operating costs are interrelated. The risk of an event happening depends on the likelihood of the occurrence of the event and the consequences and can be mathematically represented as

$$risk = likelihood \times consequence.$$

The likelihood is a function of several factors, but the human cause plays an important role. So it is a function of human error or omission. The number of humans in a plant will depend on the activity level needed to run the plant, which in turn helps determine the operating costs. The consequence of an accident is either death or injury to the person. The consequence depends on the number of people in the plant and is determined by the activity level. Activity level, in turn, is a function of the operating cost. Therefore,

$$risk = operating\ cost \times operating\ cost.$$

The more personnel running the plant means the greater is the risk. Fully automated plants with minimum personnel have minimum human exposure in the case of an accident.

Inherent Security

Hazardous substances released from plants could cause serious problems to plant employees as well as to the nearby public. The release could be due to accidents (human error), instrument failure, control failure, and natural calamities or through deliberate acts of vandalism, terrorism, or sabotage. The application of inherent security can help in two ways: (1) by reducing the likelihood that a facility will be targeted and (2) by minimizing the severity of an incident should an attack occur. Similar to inherent safety principles, one can consider a set of guidelines for inherent security:

Perception. Plants should minimize the attention they attract. Vessels and tanks that can be seen from the outside could be easy targets.

Information. The less information publicly available on the manufacturing facility, the better it is from the viewpoint of security. One needs to balance the right-to-know against the need-to-know for the local communities.

Plant layout. The sensitive area should be well guarded and kept at a low profile. Control rooms should not be easily identifiable to outsiders.

Design. The design of tanks and large storage vessels should take into account protection against airborne attacks.

Protection to the safeguards or safety devices. Adequate precaution should be taken so that outsider/intruders cannot easily disable safety devices.

Computer systems. Process control computers should not be connected to the Internet, since being connected online opens the doors for possible "cyber attacks."

Buffer zone. A buffer zone should be set up between a sensitive area, whose access is restricted, and the general area that the majority of plant personnel use.

If it cannot be avoided, minimize impact. This involves maintaining low inventories, which, even if sabotaged, would lead to minimum damage.

Security may not have been as prevalent an issue 20 years ago, but at current times it has become a part of design strategy.

Waste and Effluent

Minimizing waste generation is prudent since it would also decrease efforts required for its treatment and disposal. Points to be kept in mind with respect to waste are as follows:

1. Solid wastes should be nonleachable and of low toxicity so that they can safely be land-filled.
2. Minimum gaseous effluent production would minimize its treatment. As far as possible, incineration as a treatment strategy should be avoided unless it has to be carried out to minimize NO_x and VOCs. Incineration requires spending energy and greater capital expenditure.
3. Liquid effluents will impact aquatic life and also groundwater.
4. Avoid "sunset" chemicals, i.e., those substances that have become undesirable because of safety, health, and environment (SHE) reasons, including

carcinogens (asbestos, benzene, etc.),
PCBs and PCTs,
the "drins" (aldrin, dieldrin, entrin, and isodrin),
chlorinated dioxins and furans,
CFCs,
halons,
carbon tetrachloride,
methyl chloroform.

Dust/Particle Handling

Dust and fine particles lead to explosion and health hazards. Suggestions for dust and particle handling include the following:

1. Use nitrogen as conveying gas as well as sealing gas instead of air.
2. Fill silos using cyclones to reduce dust cloud formation.
3. Reduce electrostatic problems by proper use of earthing.
4. Use lower mass flow rate to avoid static buildup.
5. Measure and control the electric field.
6. Keep the dust concentration below the explosive limits.
7. Design pipes to be explosion-proof.
8. The explosion limits of all compounds should be known, in order to prevent explosions.

Cost of Inherent Safety and the Profit–Risk Model

While safety features in a chemical plant are being designed, two frequently asked questions are, "How safe is safe?" and "How many safety features and redundancies must the company incorporate in the process to make it safe?" (such as those in Fig. 8.1). Each new safety feature to a process plant leads to capital expenditure as well as extra maintenance costs. For the public exposed to the risk of fatal injury as a result of a chemical-related accident arising from the activities of the company, the risk should not be significant when compared to other normal, "everyday" risks to which the person is exposed (e.g.,vehicle accident, lightning strike, or domestic gas explosion). For example, in Great Britain the probability of death from a lightning strike is 0.1 per person in one million years (1985 data)—a risk almost everyone would consider as insignificant or trivial and would thus not show any concern over. The same 1985 data set showed that the probability of death due to a domestic gas explosion is 1.0 per person in one million years; that for an employee working in an office, shop, or warehouse is 4.5 per person in one million years; and that by road accident is 100 per person in one million years. Based on these data, it can be extrapolated that an individual should not be exposed to more than 10 major accidents in 1 million years of exposure, which works out to an allowable frequency as follows:

f(major accidents per person) = 10 in 1 million years of exposure,

where a major accident is defined as an accident resulting from a chemical-related hazard that has the potential to cause multiple fatalities (about 10 deaths).

The reliability of a plant is a sum of reliabilities of all the equipment, instruments, and other hardware. The reliability of each item of equipment is estimated based on its past operational performance (failure rates, repair durations, etc.) and on current maintenance practices (preventive maintenance intervals and durations, test intervals, etc.). Once the economic impacts of various scenarios in which units or major equipment items fail has been established, the required safety features can be designed (e.g., profit–risk scenario and minimal cutoff could be established) (see Fig. 8.10). To the left of the dotted line in the figure, the litigation cost is higher than the cost of safety features; the reverse trend exists on the right of the dotted line.

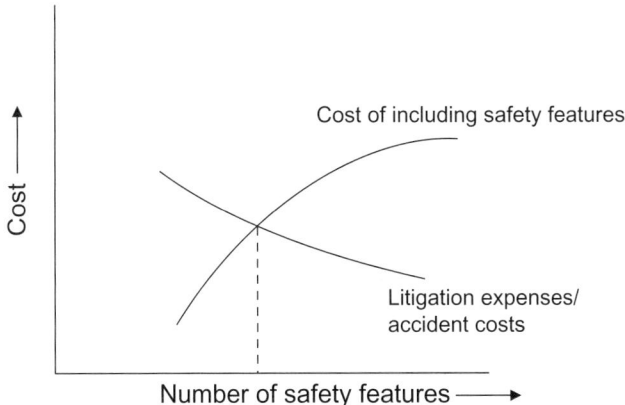

FIGURE 8.10. Profit–risk scenario.

The Weibull model can be a tool used for modeling the component's aging. It is based on the assumption that a component's failure rate changes with the component's age in accordance with the relationship

$$\text{failure rate} = (\beta/\eta)(\tau/\eta)^{\beta-1}, \tag{8.5}$$

where τ is the component's age (or time from last overhaul); β is the *shape factor*, which determines how the failure rate changes with the equipment's age (if $\beta < 1$, the failure rate decreases with age; if $\beta = 1$, the failure rate is independent of age; and if $\beta > 1$, the failure rate increases with age); and η is a characteristic time interval. The time-independent Poisson model is the specific case when $\beta = 1$.

The profit–risk modeling approach involves development of a mathematical equation relating to the profit, the probability of an equipment failure, and the probability that a safety device fails as follows:

$$\text{PR} = \text{profit} \times P \times P_{\text{safe}}, \tag{8.6}$$

where

PR = profit risk contributed by the equipment item,
profit = decrease in profit = profitability of optimized facility configuration when equipment item is available – profitability of optimized facility configuration when equipment item is not available,
P = probability that the equipment item is not available,
P_{safe} = probability that the relevant safeguards fail (such as a backup to the equipment).

The probability of not being available or of failure will, in turn, depend on the component's age, as shown in Eq. (8.5).

Conflicts Due to Inherently Safe Designs

A design may be inherently safer for one particular hazard while not being inherently safer for another hazard. For example, speed breakers are built near road crossings or near road junctions so that drivers slow down when seeing one; this helps pedestrians cross the road safely. But if a driver does not see the speed breaker and rides over it very fast, he or she (driver) could be hurt. Fatalities are reported because of speed breakers. Acrylic latex water-based paint is inherently safer than a solvent-based paint with respect to flammability, toxicity, etc., but due to their water base, the growth of microorganisms is a problem in water-based paints. This may make the paint useless for application and may also present a hazard to users. This hazard can be overcome by the addition of a biocide, which is an extra "add-on" and may pose its own hazard in terms of human handling and exposure (see Table 8.3). Also, the hazard the biocide may cause to the external environment is another issue to be addressed.

Another example relates to air bags in automobiles (see Table 8.4). Air bags are active safety devices that protect the occupants of the front seat of the automobile from injury in case of a collision. Air bags became a compulsory addition in all automobiles (U.S., 1997; Ref 49 CFR ChV § 571.208). But it was found that, when activated during collisions, these bags were sometimes blamed for serious injuries, even deaths, for women and children. The accidents in these cases were not severe and would not have created serious injury to the front-seat passengers, but the impact of the air bags' openings was so severe that it did result in serious injury to the front-seat passengers. Several alternatives were thought of to overcome this problem. Two of them were (1) children under 10 years of age always ride in the back seat, and (2) allow people to disarm the air bags. However, the second approach

TABLE 8.3
Conflicts in Inherent Safety (Replacement of Solvent-based Paint)

Starting Product	Hazards	Action	New Product
Solvent-based paint	Hazards due to handling of solvent	Remove solvent	Water-based paint
Water-based paint	Due to bacterial growth	Use biocide	Water-based paint with biocide
Water-based paint with biocide	Handling of biocide (biocide poison)	Protect people who handle this paint	

TABLE 8.4
Conflicts in Inherent Safety (Air Bags)

Starting Product	Hazards	Action	New Product
Automobiles	Injury to front seat passengers in case of accident	Provide air bags	Automobiles with air bags
Automobiles with air bags	Opening of air bags with high impact, leading to injury to front passenger	Human intervention to disengage air bag	Automobiles with air bags with option to disengage
Automobiles with air bags with option to disengage	Serious injury to front seat passengers in case of accident since air bag is disengaged		
New type of air bags with less energy while opening			

exposed the automobile occupants once again to the risk arising during serious collision. Newer generations of automobile air bags have been designed that provide adequate protection in case of a serious collision but do not have sufficient energy to seriously injure front-seat occupants.

In order to prevent pressure buildup and explosion in nitration reactors, the contents were dumped into a batch of cold water when the temperature rise was uncontrollable. Although this action prevented reactor explosion, it could led to release of toxic chemicals into the atmosphere (Chen et al, 1998).

Safety and Risk Assessment

Several risk indices have been practiced by the chemical and manufacturing industries over the past 50 years (see Table 8.5) (Heikkilä, 1999). The important ones among these indices are the Dow Fire and Explosion Index, the Dow Chemical Exposure Index, the HAZAN (Hazard Analysis), and the Prototype Index of Inherent Safety.

TABLE 8.5
Safety and Risk Assessment Methodologies

Stages	Safety Review	Checklist	Relative Ranking	Preliminary Hazard Analysis	What-if	Checklist	HAZOP	FMEA	Cause–Consequence Analysis	Human Reliability Analysis	Fault Tree	Event Tree
R&D			■	■	■							
Conceptual design	■	■	■	■	■	■	■					
Pilot plant		■		■	■	■	■	■	■	■	■	■
Detailed engineering	■	■		■	■	■	■	■	■	■	■	
Construction/startup	■	■			■	■	■			■		■
Operation	■	■			■	■	■	■	■	■		■
Expansion/modification	■	■	■	■	■	■	■	■	■	■	■	■
Accident investigation	■				■		■	■	■	■	■	■
Decommissioning	■	■			■	■						

The Dow Fire and Explosion Index (FEI) (Dow, 1994; Van den Braken, 2002) and the Dow Chemical Exposure Index (CEI) have been developed and practiced by Dow Chemicals for several decades. These tools measure process inherent safety characteristics, help to quantify the expected damage of potential fire and explosion incidents, and identify equipment that would likely contribute to the creation of the incident. The Mond Index was developed by ICI (UK) from the Dow Fire and Explosion Index. The Mond Index includes toxicity and covers a wider range of processes and storage installations than the FEI. The various aspects considered in the FEI are material factor (flammability and reactivity), general process hazards such as exothermic/endothermic reactions, and special process hazards such as toxic nature of the chemicals and dust explosion. For example, Table 8.6 shows the results of the FEI for various inventory levels of storage of ethyl acrylate.

Table 8.7 shows the difference in the CEI for the failure of a 2-in. pipe for a number of different chemicals present in the pipe in liquid or gaseous forms. The CEI value is higher when the chemical is in the liquid form than in the gaseous form, since the amount of the former is higher. In addition, the chemical's corrosivity and toxicity also determine the CEI value. For example, the CEI is very high for chlorine and hydrogen chloride than for organic chemicals. Using these indices, one could consider alternate storage, processing, and operational scenarios and select the best one based on minimum FEI and CEI values.

In addition to the chemical exposure, fire, and explosion, the various factors considered in the Mond Index are the material factor and its associated hazards, process hazard, quantity hazard, layout hazard, and toxicity hazard. The material factor relates to the type of materials, their flash point, explosive limits, etc., the nature of the process, and the process's inherent safety. The

TABLE 8.6
Fire and Explosion Index for Storage of Ethyl Acrylate

Million lbs of Inventory	FEI
2.0	151
0.2	130
0.05	120

TABLE 8.7
Chemical Exposure Index for the Failure of a
2-Inch Pipe

Chemical	CEI
Hydrogen chloride (liquid)	1000
Hydrogen chloride (gas)	610
Chlorine (liquid)	1000
Chlorine (gas)	490
Ammonia (liquid)	380
Ammonia (gas)	100
Butadiene (liquid)	290
Butadiene (gas)	100
Vinyl acetate (liquid)	40
Vinyl acetate (gas)	10

quantity of material used in the process has an effect as well, since a hazard's impact depends on the quantity of material stored or handled and also depends on the plant layout, whether or not buffer zones are placed between hazardous zones, the toxicity of the raw materials, and intermediated products that affect the CEI.

Hazard Identification and Assessment

Several methods are available for identifying and assessing hazards (Kletz, 1990). Hazards can be identified through checklists, failure mode effect analysis (FMEA), fault tree analysis, event tree analysis, what-if analysis, and hazard and operability studies (HAZOP). Assessing hazards can be done through hazard analysis (HAZAN), codes of practice, the Dow Explosion Index, and prototype index of inherent safety (PIIS).

The what-if analysis consists of imagining possible scenarios in the everyday plant operation, such as failure of a valve, failure of a controller, human error, addition of excess raw material, and addition of an impurity, and identifying the hazards that could arise due to these failures.

Safety audits and safety checks are generally used during the plant operation and identify wrong practices or wrong designs. The audit consists of a team of safety and design engineers walking through the plant on a typical working day and identifying wrong practices, unsafe situations, unsafe location of hardware

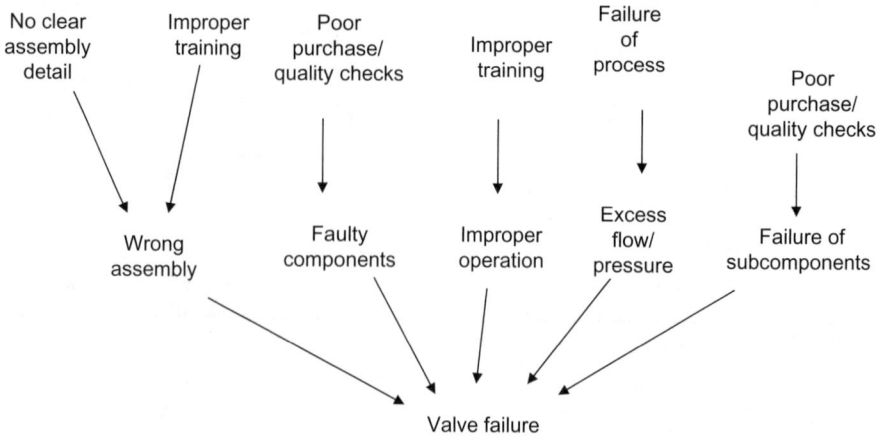

FIGURE 8.11. Fault tree for a valve failure.

components, and so on. At the end of this walk-through, the audit team may prepare a report that lists the unsafe situations and possible solutions.

Fault tree analysis consists of developing a fault tree connecting various negative scenarios to the hazard. A fault tree for a failure of a valve could be caused by several reasons, and each of these, in turn, could be caused by several other reasons, as shown in Fig. 8.11.

FMEA consists of imagining possible failures that could occur in a process and identifying various reasons for these failures. By attaching a probability to these reasons, one could arrive at the probability of the occurrence of the failure. A failure mode is defined as a manner in which a component or process could potentially fail to meet the design intent. A failure mode in one system could be the cause for a failure in another system. Once the possible cause for the failure has been listed, it may be addressed. A risk priority number (RPN) is estimated for each of the failures; the RPN is a product of the severity of the failure, the probability of its occurrence, and the probability of its detection. This RPN is used to prioritize the risks, and suitable action is taken, beginning with the most serious risk.

HAZOP is the most widely used technique by ICI developed several decades ago and involves vessel-to-vessel and pipe-to-pipe review. The entire flow sheet is systematically reviewed by design, safety, and process engineers using a set of guidewords to identify any deviation. Various process operations are also reviewed for potential hazards. Guidewords are meant to help the team identify potential hazards and process disturbances. The list of guidewords and their descriptions appear in Table 8.8.

For example, when guidewords are tested to check for hazards during the addition of a material into a vessel, possible deviations could arise that are listed in Table 8.9. Some of the deviations

TABLE 8.8
Guidewords Used in HAZOP

Guideword	Explanation
No	Negation of the intent
Less	Below the intended value
More	Above the intended value
As well as	Perform the intended and not-intended operations
Part of	Partially perform the intended action
Reverse	In the opposite direction
Other than	Something else instead of the intended action
Sooner than	Earlier time
Later than	Later time
In addition to	Something more than the intention

TABLE 8.9
Guideword When Applied to an Addition Operation

Guideword	Deviation	Possibility of Any Hazard
No	No material added	
Less	Less material added	
More	More material added	
As well as	In addition to intended material, some other material is also added	
Part of	Part of the material is added	
Reverse	—	
Other than	Some other material added	
Sooner than	Material added at an earlier time	
Later than	Material added at a later time	
In addition to	In addition to intended material, some other material is also added	

could lead to a potential hazard, while some may not lead to any hazard. The operations that may lead to hazards are addressed by the safety team. HAZOP studies are carried out at each stage of the project, starting from the development stage, to pilot plant, plant design, commissioning, manufacturing, and modification. These are called HAZOP Study 1 to 6, respectively.

HAZAN is a quantitative hazard analysis approach that can be used to estimate the probability of occurrence of a hazard given the probability of failure of various control systems and the probability of occurrence of various events that can lead to the hazard. The FMEA and HAZAN approaches are similar. The following example explains the principle behind HAZAN: Consider a packed bed, countercurrent absorber used to absorb chlorine gas from an exit stream using alkali solution in a countercurrent manner. Chlorine enters the bottom of the bed and alkali is sprayed from the top. Chlorine could escape from the top exit of this absorber for several reasons: a decrease in alkali concentration, insufficient alkali, excess chlorine in the inlet, leakage in the chlorine pipeline, and failure of the absorber column. If the probabilities of these possible failures are known, one could estimate the overall probability of chlorine leak.

The prototype index of inherent safety (PIIS) (developed by Edwards and Lawrence, 1993) is a numerical score based on the nature of chemicals used and the type of process. The PIIS was developed to compare processes based on raw materials and the sequence of reaction steps only. This method does not consider the other parts of the process. PIIS is calculated as a total score and consists of two parts: the chemical score and the process score. The first one is made up of inventory, flammability, explosiveness, and toxicity. The second one is made up of temperature, pressure, and yield.

The opportunity for installing inherent safety features decreases as the design progresses from the lab stage to the pilot stage and then to the production stage. For example, several inherent safety principles, such as (1) intensification, substitution, and attenuation, (2) limitation of effect, (3) simplification, (4) avoiding domino effects, and (5) ease of control, could be discussed during the conceptual and flow sheet stages. On the contrary, the principles of limit effects, prevent incorrect assembly, and clarify equipment assembly could be discussed during the line diagram stage.

The inherent safety parameters Edwards and Lawrence (1993) selected are inventory, temperature, pressure, yield, flammability, explosiveness, and toxicity. In addition to these parameters,

Heikkilä et al. (1996) selected heat of main reaction, unwanted side reactions, corrosiveness, chemical interaction, type of equipment, and safety of process structure. The unwanted side reactions could be polymerization, heat formation, toxic gas release, and so forth. From the safety point of view, it is important to know how exothermic the reaction is. King (1990) classified the heat as extremely exothermic ($\geq 3000\,J/g$), strongly exothermic ($<3000\,J/g$), moderately exothermic ($<1200\,J/g$), mildly exothermic ($<600\,J/g$), thermally neutral ($\leq 200\,J/g$), or endothermic. The hazardous substances present in the process are identified on the basis of their flammability, explosiveness, and toxicity. They give the following reasons for selecting only these three parameters:

1. The flammability of gases and vapors of flammable liquids is a great concern in the chemical process industries. The result of an ignition can be a fire, heat, or an explosion, or the sum of all, with varying magnitude. The flammability of liquids depends on the lower flammability limit of the material and its vapor pressure at that temperature. One needs to consider ambient and reaction temperatures. The boiling point may be taken as an indication of the volatility of a material.
2. Explosiveness is the tendency of chemicals to form an explosive mixture in air. When an explosion occurs in atmosphere, energy is released in a short time and in a small volume, so that a pressure wave is generated, which travels in all directions, releasing energy violently. The energy released in an explosion derives from either physical energy or chemical energy. The explosiveness of a vapor cloud depends especially on the lower explosion limit, which is the concentration of the vapor at which the vapor cloud is possible to ignite.
3. The toxic hazard can mean harmful physiological effects, ranging from irritation to skin to chronic effect to reactions leading to death. The ingestion of the chemical could be oral, inhaled, or through skin absorption. It is a measure of the likelihood that such damage can occur. The threshold limit value (TLV) is defined as the concentration in air that may be breathed in without harmful effects for five consecutive eight-hour working days. TLVs are based on different effects, ranging from irritation to physiological damage. In an industrial context especially, TLVs are the most usable toxicity values. The toxicity of microorganisms has not been fully understood, unlike chemical toxicity. While a linear relation may exist between chemical concentration and its toxicity, the relationship could be nonlinear, as in the case of microorganisms. Immunity could

be developed to microorganisms, which is not possible with chemical toxicity.

Chemical substances present in a process plant may react with each other, with air, or with water, causing safety problems. This chemical interaction is based on the chemical reactivity of each substance and with other substances present in the plant. As a potential process hazard, the chemical reactivity of any substance should be considered in the following contexts:

1. with elements and compounds with which it is required to react in the process,
2. with atmospheric oxygen,
3. with water,
4. with itself (to polymerize, condense, decompose, and explode).

In this context, using Table 8.2 would be the best approach.

In general, large inventories in one place are unfavorable in the case of fire or rupture of a vessel. Potential severity can be reduced by keeping inventories low, by minimizing the reactor size, and by avoiding storage of potentially hazardous materials in the synthesis train. For instance, large quantities of very toxic gases and volatile liquids were one of the major mistakes in the Bhopal accident. Temperature is a direct measure of the heat energy available at release (Edwards and Lawrence, 1993). Temperature (T) is the most important factor influencing the rate of reaction, as shown in the Arrhenius equation:

$$\text{rate of reaction } \alpha \approx e^{-E/RT}. \tag{8.7}$$

For example, Eq. (8.7) shows that a 10°C increase in temperature will increase a specific reaction rate by 200–400%, depending on the energy of activation (US EPA, 1999). In Eq. (8.7), E is the activation energy; the lower its value, the more facile the reaction is. The use of high temperatures in combination with high pressures greatly increases the amount of energy stored in the plant. There are severe problems with materials of construction in high-temperature plants. The use of high temperatures implies that the plant is put under undue thermal stresses, particularly during start-up and shutdown, since during start-up the material is heated up from ambient temperature to very high temperature while the reverse happens during shutdown. With high-pressure operations, the problem of leaks becomes much more serious. The amount of fluid, which can leak out through a given hole, is greater on account of the pressure difference (flow will be proportional to the square root of the pressure difference). If the fluid is a liquid, it will flash off as the pressure is reduced. Thermal stresses are high in low-temperature plants as well. If, in low-temperature plants, large amounts of fluids are kept in a liquid state only by pressure and temperature, then if for any reason it is not possible to keep the plant cold, the liquids will begin to vaporize. The ambient temperature of the locale will play an important role in such situations. Another hazard in low-temperature plants is the presence of impurities in the fluids, which are liable to precipitate out of a solution as solids. Deposited solids may cause not only blockage but also explosions.

The Inherent *Safety Index* developed by Hurme and Heikkilä (1998) is a sum of two indices, the Chemical Inherent Safety Index and the Process Inherent Safety Index (see Fig. 8.12). The Chemical Inherent Safety Index contains two subindices, the Reaction

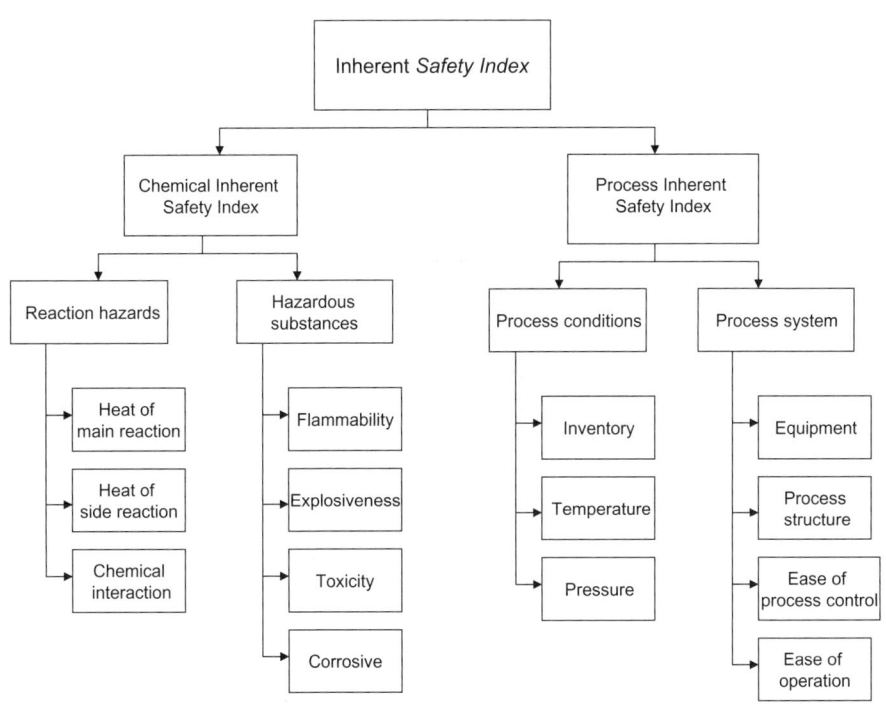

FIGURE 8.12. Components of Inherent *Safety Index*.

Hazards and the Hazardous Substances Handled indices. The former is once again divided into the Heat of the Main Reaction, Heat of the Side Reaction, and Chemical Interaction indices. The second one consists of flammability, explosiveness, toxicity, and corrosive indices. The Process Inherent Safety Index contains two subindices: Process Conditions and Process System. The Process Conditions Index is further divided into inventory, temperature, and pressure indices. The Process System Index consists of equipment, process structure, ease of control, and ease of operation indices. The ease of control and ease of operation indices should also include human interaction, but Hurme and Heikkilä (1998) did not include it. The Inherent Safety Index can assist the designers in choosing inherently safe alternatives, beginning with the conceptual design stage. This analysis can be carried out concurrently during the process development stage, unlike the traditional approach of incorporating safety features after the process has been scaled up to the pilot plant or semitechnical stage.

The INSIDE Project Toolkit

The INSIDE (Inherent SHE In Design) Project is a European government/industry project sponsored by the Commission of the European Community to encourage and promote inherently safer chemical processes and plants (INSIDE Project, 1997). One of the best-known measures for safety is risk, which is defined as (1) the potential for loss or the probability of a specific undesired event happening in a particular period of time and (2) the consequences of the undesired event:

$$\text{risk} = \text{function of (probability of undesired event,} \atop \text{consequences of undesired event).} \qquad (8.8)$$

An inherently safe chemical process is one that avoids hazards instead of controlling them, particularly by removing or reducing the amount of hazardous material in the plant or the number of hazardous operations.

Human Factors

Human factors play an important role in safety and need to be considered during the design process as well as during various stages of hazard analysis. Human factors can be grouped as

Situation-based—those related to the immediate work environment, complicated or dangerous workplace, and complicated
operating procedures.
Management-based—failures in communication, in rewards, in
training personnel.
Human-based—emotional stage, motivation, family problems, and
employee morale.

Giving a probability number to these factors is difficult. The first
two factors could be addressed through the use of various hazard
analytical methods, while the third factor requires Human Resource
Department support.

Corrosion of Hardware

Corrosion is another hazard that proceeds slowly and affects a
plant's integrity. It reduces the strength of materials and causes
leaks. Corrosion products affect process materials and moving
parts and cause fouling. The safety problems caused by the corrosive properties of the process streams can be prevented by selecting
proper materials of construction. Improper selection of materials
of construction or contaminants in the raw materials may lead to
corrosion. Presence of oxygen in water and moisture in combination with certain gases or salts in water both lead to corrosion. A
seemingly safe hardware will fail over a prolonged period of time
due to corrosion; if this issue is not addressed during the design
stage, it may appear as a surprise later. Extremely high or low
temperatures also may lead to failure of material(s). In addition to
the materials of the main equipment, one needs to consider the
materials of small items, such as gaskets, rivets, bolts, nuts, and
thermocouples. A gasket failure in Apollo 13 led to a major
accident. The disaster of space shuttle Challenger in 1986 also
resulted due to failure of O rings.

Molecular-Level Design for Safer Chemicals

Understanding the mechanism of action of a hazardous chemical
will help in designing new molecules that are stable and pose
fewer hazards. The concept of quantitative structure–activity relationship (QSAR) has been used effectively in drug design to suggest
new structures that are more active than the original structure. It
allows an estimation of the potential activity of an untested chemical by comparing its structure with a series of chemicals with

similar structural properties. This concept of structure–activity relationship can be extended for designing new chemical structures that will exhibit lower toxicity, hazards, and degradation, leading to quantitative structure toxicity relation (QSTR), quantitative structure hazard relation (QSHR), and quantitative structure degradation relation (QSDR).

Understanding the disposition of chemicals in organisms and how the body deals with a chemical once exposure has occurred should be described using the principles of toxicokinetics and toxicodynamics. The former includes the study of the time course of the disposition primarily of xenobiotics and examines the role of absorption, distribution, metabolism, and excretion (ADME) on toxicity. Each one of these components can be influenced through structural changes. Toxicodynamics includes the rate of absorption and excretion. This time-dependent phenomenon represents how fast certain chemicals can enter and leave the human system. Pesticides create problems to human and animals because they persist and remain in the body after ingestion. Pesticides are also time-dependent phenomena and represent how fast certain chemicals can enter and leave the human system.

For most chemicals, exposure occurs primarily through ingestion, inhalation, or dermal contact. The entire amount of chemical exposure will not cross the membranes with 100% efficiency. *Bioavailability* is defined as the portion of the total quantity of concentration of a chemical that is available for biological action. If the availability of the molecule can be decreased, the amount of chemical at the site of action is thereby decreased, which leads to decreased toxicity. Bioavailability depends on the hydrophilic–lipophylic balance, its stability on exposure to alkali or acidic conditions, its stability in the presence of body enzymes, and so on. So knowledge of biochemistry and pharmacokinetics is essential in determining safety measures.

Conclusions

It is estimated that 43% of today's high-production-volume chemicals do not have complete toxicity data—and 7% of these chemicals have no data at all. Inherent safety implies that the process is inherently safe and that it is not kept safe with the help of several safety checks and alarms. Several available hazard analysis techniques can be used at various stages of the process development. Developing inherently safe process is more advantageous than

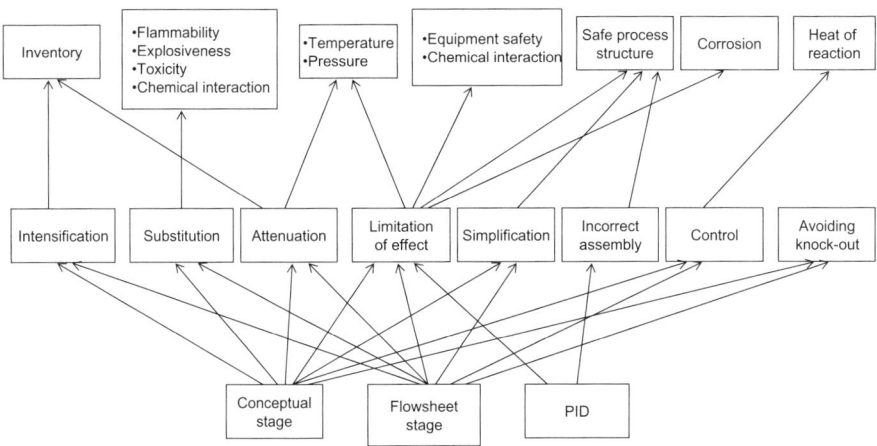

FIGURE 8.13. Principles of green chemistry and inherent safety.

adding layers of safety features. Industries would always like to weigh the cost vs. benefits of adding a safety feature. Figure 8.13 relates the various principles of green chemistry with the inherent safety parameters at various stages of the project. Safety is not only the domain of the engineers but also encompasses the work of biologists, toxicologists, and biochemists.

References

Anastas, N. D. and Warner, J. C., The incorporation of hazard reduction as a chemical design criterion in green chemistry. *Journal of Chemical Health & Safety*, (March/April): 9–13, 2005.

Chen, C. Y., Wu, C. W., Duh, Y. S., and Yu, S. W., An experimental study of worst case scenarios of nitric acid decomposition in a toluene nitration process, *Process Safely & Environment Protection*, **76**(B3), 211–216, 1998.

Dalzell, G., Is operating cost a direct measure of safety? In Barton, J. (ed.), *Chemical Reaction Hazards*, 1993.

Dalzell, G., Is operating cost a direct measure of safety? *Trans. IChemE*, Vol. 81, Part B, Nov. 2003.

Dow's Fire & Explosion Index Hazard Classification Guide. 7th Edition, AIChE, 1994.

Edwards, D. W. and Lawrence, D., Assessing the inherent safety of chemical process routes, *Loss Prevention & Safety Promotion in the Process Industries, 8th Symposium*, pp, 473–482, 1995.

Edwards, D. W. and Lansence, D., Assessing the inherent safety of chemical process routes: Is there a relation between plant costs & inherent safety? *Trans IChem*, Vol. 71, Part B 252–258, 1993.

Heikkilä, A.-M., Inherent safety in process plant design: An index-based approach, Ph.D. thesis, Helsinki University of Technology, Espoo, Finland, 1999.

Heikkilä, A.-M., Hurme, M., and Jänveäinen, M. Safety Considerations in Process Synthesis Computers, *Chem. Engng.* Vol. 20, Suppl. A, S115–S120, 1996.

Hendershot, D. C., An overview of inherently safer design, *Proc. CCPS 20th Ann. Intl. Conf.—Risk Management: The Path Forward*, American Institute of Chemical Engineers, New York, pp. 232–252, 2006.

Hendershot, D. C., Inherent safety strategies for process chemistry, *Chemical Health & Safety*, 5(4): 18–22, 1995.

Hendershot, D. C., Sussman, J. A., Winkler, G. E., and Dill, G. L., Implementing inherently safer design in an existing plant, *Proc. 7th Process Plant Safety Symp.*, K. Pearson and J. R. Thompson (eds.), American Institute of Chemical Engineers, New York, pp. 419–428, 2005.

Hurme and Heikkilä, Synthesis of inherently safe chemical process by using genetic optimization and case-based reasoning, *Human and Artificial Information Processing*, Finnish Artificial Intelligence Society. Espoo pp. 134–143, 1998.

Kletz, T. A., *Critical Aspects of Safety and Loss Prevention*, Butterworths, London, 1990.

Orrell, W. and Cryan, J., Getting rid of the hazard, *The Chemical Engineer*, (August): 14–15, 1987.

Osterwalder, U., Continuous process to fit batch operation: Safe phosgene production on demand, *Symp. Inst. Chem. Eng.*, IChemE, Warwickshire, UK, 1996.

U.S. Chemical Safety and Hazard Investigations Board, Investigation report: Chemical manufacturing incident, Report no. 1998-06-I-NJ, 2000.

U.S. EPA, Chemical safety case study: How to prevent runaway reactions, EPA, Chemical Emergency Preparedness and Prevention Office, Report no. EPA 550-F99-004, 1999.

Van den Braken, A., Process safety and industrial explosion protection, *Proc. Intl. ESMG Symp.*, Dow Chemicals, 2002; *AIChE Technical Manual*, 1994.

CHAPTER 9

Industrial Examples

According to the U.S. EPA, the introduction of green chemistry has resulted in about 140 million pounds of hazardous substances *not* being produced in the United States each year, saving more than 55 million gallons of process water, preventing 57 million pounds of carbon dioxide emissions, and preventing formation of 3 billion pounds of hazardous waste per year (Brundtland, 1987; *Chemical Week*, 2003; *C&E News*, 2006). McDonough and Braungart (1998) cite two examples that relate to eco-efficiency: 3M's saving over $750 million on pollution-prevention measures and Du Pont's continuous efforts that have led to a 75% reduction in airborne cancer-causing emissions since 1987. Since 1996, Bayer has cut the amount of waste it produces from about 5 kg to 2.3 kg/100 kg of product in 2002, which is more than a 50% reduction. Environmental releases from Bayer's U.S. plants dropped from 12 million pounds in 1996 to 2.2 million pounds in 2000 (EPA, 2001; Fraser, 2001).

Nike has adopted green chemistry techniques in several of its operations. These changes include (1) a technology to generate waste from outsole molds into Regrind® for premium-performance outsoles, (2) substitution of water-based solvents in adhesives, primers, degreasers, and mold release agents instead of petroleum-based solvents, and (3) substitution of Nike Air®, which contains sulfur hexafluoride (SF_6), a well-known ozone-depleting chemical, with a harmless gas. It should be noted that 1 g SF_6 has the global warming potential of 23,900 g CO_2. Dow manufactures styrene sheet foam packaging products totaling 700 million lb/year. It has developed a process that leads to total replacement of chlorofluorocarbons (CFCs) and volatile organic compounds (VOCs) by CO_2 as blowing agent in the manufacture of these foams. By implementing this process, Dow has removed 3.5 million pounds of CFCs from the atmosphere every year.

Patagonia (USA) has adopted green chemistry through recycle, replacement, and reduction. A few of Patagonia's examples of using a renewable feedstock in a more sustainable way include the use of PET from recycled 2 L soda bottles that are used as the base material for Synchilla®, a signature Patagonia fleece; the elimination of chlorine as a bleaching agent in fabric treatment; reduced formaldehyde use; the elimination of azo dyes; the elimination of PVC in luggage fabrics, plastic components, and clothing labels; and the introduction of organic cotton.

Because carpet manufacturers are unable to fully recycle carpet tiles, most used carpeting is sent to a landfill or incinerated. Shaw Industries (USA) has designed safer chemicals for its EcoWorx® polyolefin-based carpet tiles, which are more easily recycled than traditional carpet tiles. The company has transitioned to polyolefin from polyvinyl chloride and phthalate ester plasticizers. Through a joint agreement with Dow Chemical, it has developed an "interpolymer" of polyethylene and a long-chain α-olefin, such as 1-octene, which is less toxic and recyclable. Polyamide-6 (PA-6) and polyamide-6,6 (PA-6,6) belong to the group of polycondensation plastics. Rhodia recycles about 30,000 tons annually of PA-6 wastes to caprolactam in three different places in Europe. PA-6 production waste as well as used pure PA-6 wastes (fish nets and pure PA-6 fabrics) are used as starting products. PA-6,6 is also recycled. DSM and AlliedSignal opened a pilot plant in Richmond, Virginia, in 1997, where PA-6 carpets are depolymerized. The technology involves chemical processing of complete carpets without an expensive mechanical separation of fibers from the other carpet components.

3M, one of the world's largest specialty and consumer product companies, has phased out its Scotchgard products because they contain perfluorooctanyl sulfonate (PFOS), which is persistent in nature and has been found at levels of 10 to 100 parts per billion (ppb) in blood banks from several countries (some 3M employees were found to have levels of 2 parts per million). The phasing has affected its annual turnover by $320 million. Guilford of Maine (USA) (an Interface carpet maker) reduced phosphorus levels in effluent water at its manufacturing facility from 10 parts per million to trace amounts through product substitution. In collaboration with Sandia National Laboratories, the University of Minnesota, and Reaction Engineering International (USA), Dow Chemicals developed an oxidative cracking process for producing ethylene that may save up to 13 trillion BTUs of natural gas over conventional hydrocarbon cracking processes and reduce carbon dioxide emissions by 4 tons per annum.

GE Plastics manufactures Ultem®, a thermoplastic resin, through a catalytic process. A new catalyst that has been developed uses 25% less energy per pound resin, generates 90% less organic waste for off-site disposal, and produces 75% less waste in the manufacture of the catalyst itself when compared to the old catalytic process. All these examples are based on reuse and replace and have led to reduction in liquid, gaseous, and solid toxic effluents. The examples that follow deal with achieving a reduction in effluent while adding considerably to the bottom line (due to reduction in operating costs) (Ritter, 2002, 2003).

The ISO developed a series of standards (ISO 14040 Series) that discuss life cycle analysis (LCA) (www.ansi.org). The standards include Principles and Framework (ISO 14040), Goal and Scope Definition and Inventory Analysis (14041), and Impact Assessment (14042). Several approaches could be adopted at each stage of the product life, starting from the raw material to the disposal stage (see Fig. 9.1) to achieve the goal of green chemistry. The methodology shown here is like the cradle-to-grave concept.

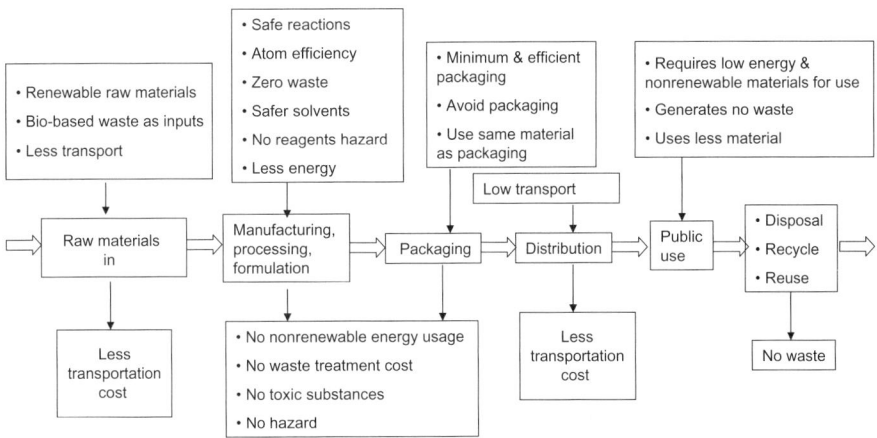

FIGURE 9.1. Green product manufacturing and usage methodology (cradle-to-grave or life-cycle analysis).

The Kalundborg Model

Integration of several manufacturing facilities so that the waste material or energy or byproduct produced in one facility can be used in another facility has several advantages. This principle is adopted in the industrial park at Kalundborg, Denmark. A coal-fired electrical power-generating station sends its waste steam to operate a pharmaceutical plant and an oil refinery. The waste heat is also piped to houses in the nearby town. The waste water from the oil refinery goes to the power plant, reducing that plant's fresh-water requirements. The waste gas from the refinery runs a factory making gypsum wallboard. The factory makes gypsum extracted from the power plant's wastes. Sulfur, a byproduct of the oil refinery, is converted into sulfuric acid at another plant. Fly ash, left over from the power plant, is converted into cement. The pharmaceutical plant has a yeast-based process, and the waste sludge produced in the process is used by the farmers to fertilize their fields (see Fig. 9.2). A complete integration of all these manufacturing facilities has led to a savings in heat, a reduction in waste, and a reduction in transportation cost of raw materials.

Ibuprofen: A Green Manufacturing Alternative

Analgesics are a group of drugs that include aspirin, acetaminophen (commercial name Tylenol), and ibuprofen (commercial

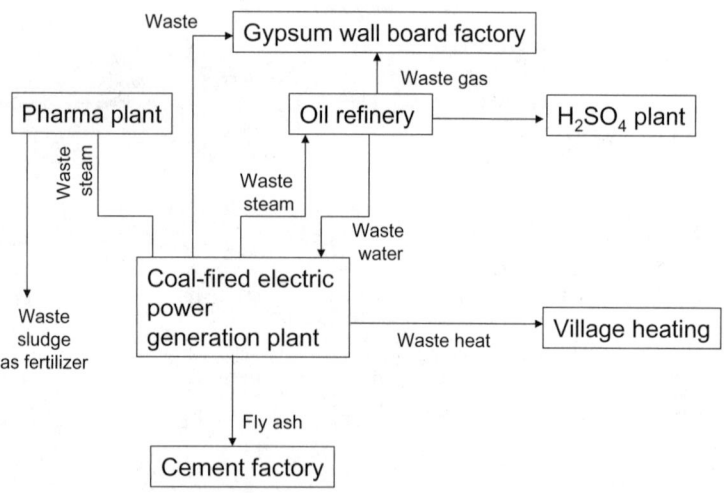

FIGURE 9.2. Kalundborg (Denmark) model.

names Motrin, Advil, and Medipren) and are used as painkillers. Aspirin and ibuprofen are also known as nonsteroidal anti-inflammatory drugs (NSAID). The original industrial synthesis of ibuprofen was developed and patented by the Boots Company (UK) in the 1960s (U.S. Patent 3,385,886). This is a six-step process and results in large quantities of waste and byproducts (see Fig. 9.3). The waste is generated due to poor atom efficiency, which is about 40%. In the 1980s, the BHC Company (U.S.) developed a new, three-step, green industrial process for the synthesis of ibuprofen

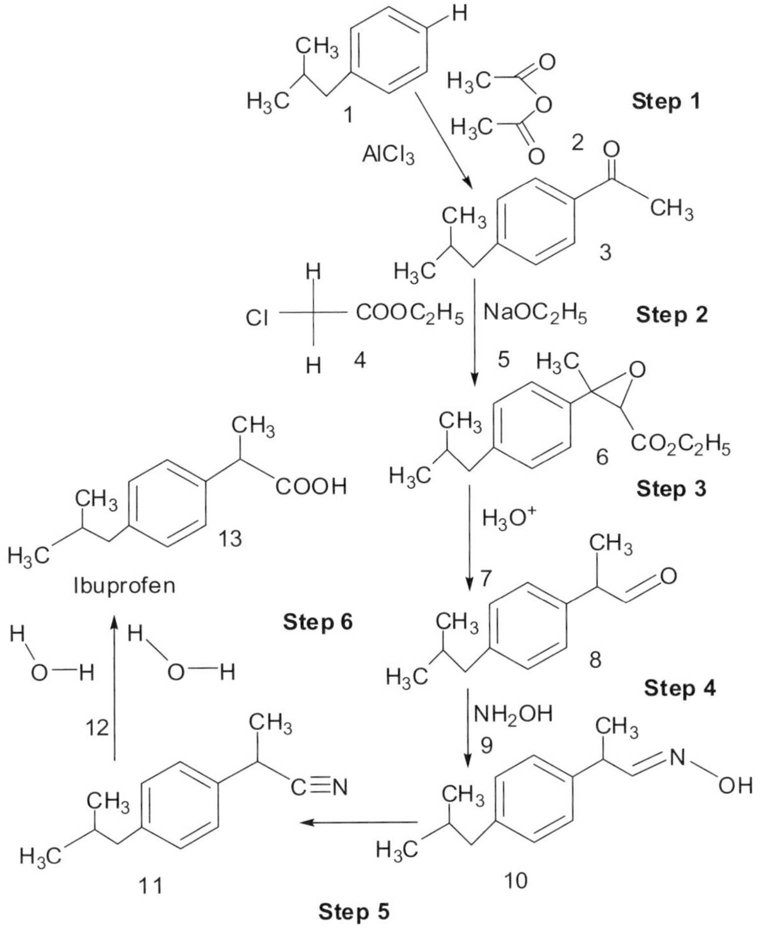

FIGURE 9.3. Ibuprofen manufacture (Books process).

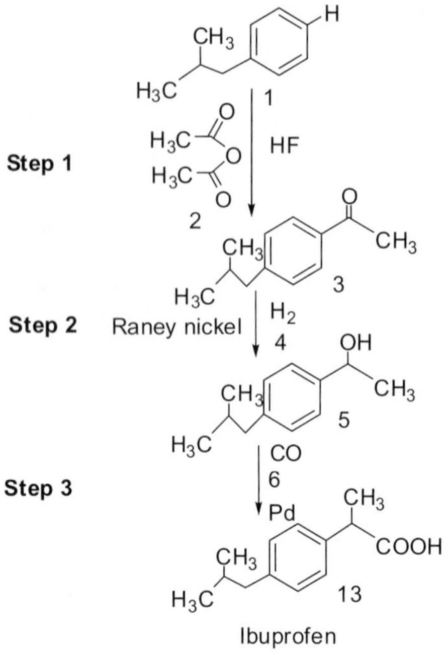

FIGURE 9.4. Ibuprofen manufacture (new BHC process).

starting from the same raw material (U.S. Patents 4,981,995 and 5,068,448, 1991). This process has a high atom efficiency (77%) when compared to the old one. This results in only small amounts of unwanted byproducts and waste (see Fig. 9.4; Cann and Connelly, 2000). Since about 30 million pounds of ibuprofen are produced each year in the U.S., this leads to more than 35 million pounds of waste. If the acetic acid generated in the process is recovered, then the atom efficiency increases to almost 99%.

The old Boots process uses aluminum trichloride in a stoichiometric amount, which is then converted into aluminum trichloride hydrate waste. In the BHC process, hydrogen fluoride, Raney nickel, and palladium are used as catalysts, which are recovered and reused. The batch cycle times and capital expenditure in the BHC process are much less when compared to the Boots process. This results in significant economic benefits to the company. The process won the Presidential Green Chemistry Challenge Award in 1997 and the Kirpatrick Chemical Engineering Achievement Award in 1993.

The BHC Company is a joint venture of the Hoechst Celanese Corporation and the Boots Company. BASF purchased the Boots Company and Celanese sold its interest in the BHC Company to BASF. Celanese operates the new ibuprofen manufacturing facility in Bishop, Texas, for BASF. Ibuprofen manufactured via the BHC process is marketed under the brand names Advil and Motrin. The industrial-scale facility created in Bishop, Texas, in 1992 is the world's largest ibuprofen manufacturing plant, is operated by the Celanese Corporation for BASF, and currently produces approximately 20–25% (more than 7 million pounds) of the world's yearly supply of ibuprofen.

Supercritical Carbon Dioxide

One solvent that is finding newer applications is supercritical carbon dioxide (sc-CO_2), which is nontoxic, nonflammable, and inexpensive (see Fig. 9.5; Sarbu et al., 2000) (Fukuoka et al., 2003; Licence et al., 2003). The first commercial chemical plant using carbon dioxide was set up in 2001 and operated by Thomas Swan & Company. A variety of reactions, including hydrogenation, Friedel Crafts alkylations, and etherifications, are carried out here. Heterogeneous catalytic hydrogenation of isophorene using H_2 gas and supported palladium as catalyst is one of the processes being carried out in sc-CO_2 (see Fig. 9.6). Partially oxidizing alcohols

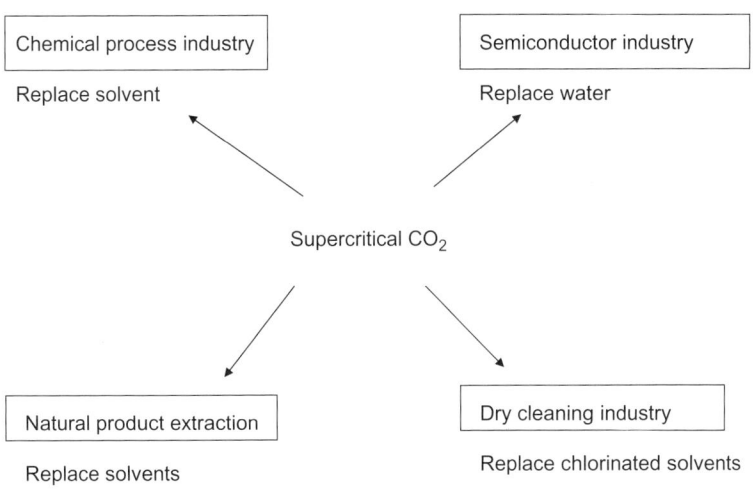

FIGURE 9.5. Uses of supercritical CO_2.

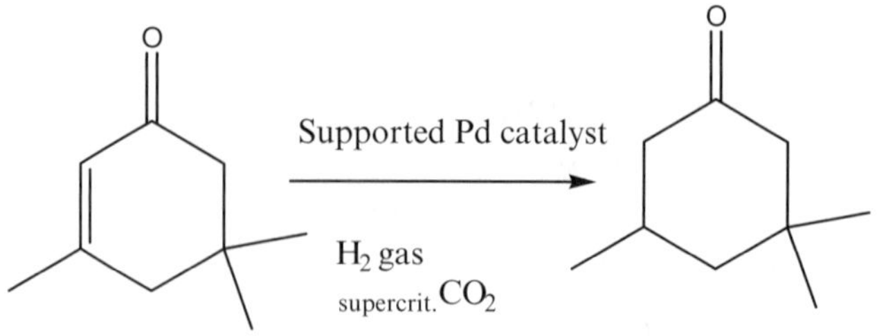

FIGURE 9.6. Hydrogenation of isophorene in supercritical CO_2.

such as methanol to ethers, aldehydes, esters, or acids are also being practiced in this solvent. sc-CO_2 is also used to decaffeinate coffee.

The manufacture of integrated circuits consumes large quantities of organic solvents and ultrapure water (5–7 L of water/cm^2 of wafer). In addition, it is an energy-intensive process. Removal of photoresist during this process employs acids, bases, organic solvents, and radio-frequency plasmas. Research at Los Alamos National Lab has led to a process that involves the use of sc-CO_2 as a better solvent for photoresist removal. This process replaces water in the final rinse step. Lowering the pressure at the end of the process returns the sc-CO_2 to vapor phase, leaving the chip dry and eliminating the use of isopropyl alcohol as a drying agent. The low surface tension of sc-CO_2 enables it to clean the fine features on the integrated circuit better than conventional solvents. The CO_2 is compressed and recycled again.

Supercritical CO_2 replaces hazardous dry-cleaning chemicals such as perchloroethylene. Through collaboration with Los Alamos National Lab (operated by Bechtel National, University of California BWX Technologies & Washington Group International for the Department of Energy's National Nuclear Security Administration), Los Alamos has developed dense-phase carbon dioxide. It is used instead of solvents in the food industry to extract dyes from red chilies and caffeine from coffee. The extracted chemicals are recovered when the pressure is reduced.

The Pharmaceutical Industries and Green Chemistry

The *E-factor* measures the efficiency of a chemical process and is defined as kg waste products/kg product (Sheldon, 1994). Bulk chemicals have an E-factor of less than 1 to 5; for fine

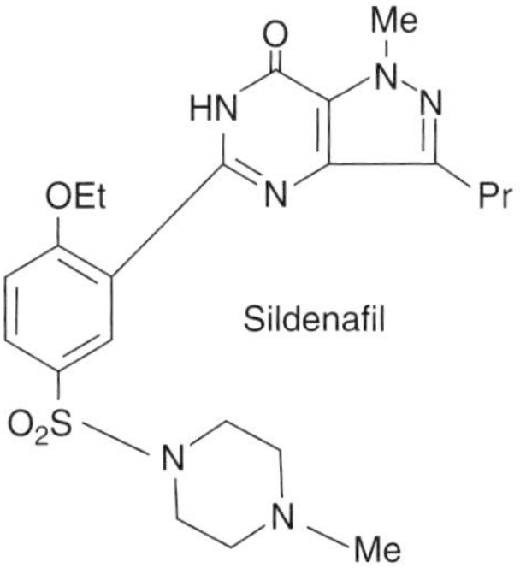

FIGURE 9.7. Sildenafil citrate.

chemicals it is typically between 5 and 50; and for certain pharmaceutical products it reaches a value of 100. Adopting green
chemistry strategies has led to a considerable reduction in waste
and use of solvents, an improvement in yield, an increase in
carbon balance, and a savings in operating costs (Kirchhoff,
2005).

Sildenafil Citrate

Pfizer has redesigned the synthesis of several of its pharmaceutical products to reduce generation of hazardous waste. Changes
were made in the synthetic route to sildenafil citrate (see Fig.
9.7), the active ingredient in Viagra® (Dunn et al., 2004), which
resulted in a more efficient process that required no extraction
and solvent recovery steps (see Fig. 9.8). The E-factor (Sheldon,
1992) for the process is 6 kg waste/kg product, which is substantially lower than an E-factor of 25–100, which is typical of pharmaceutical processes. Furthermore, all chlorinated solvents had
been eliminated from the commercial process. During the medicinal chemistry stage in 1990, the solvent usage was 1816 L/kg, and
the optimized process used 139 L/kg solvent, which was reduced
to 31 L/kg during commercial production in 1997 and to 10 L/kg
with solvent recoveries. Pfizer plans to replace t-butanol/t-butox-
ide cyclization with an ethanol/ethoxide cyclization. Combined
with other proposed improvements, this is expected to increase
the overall yield from 76–80% and further reduce solvent usage
and organic waste.

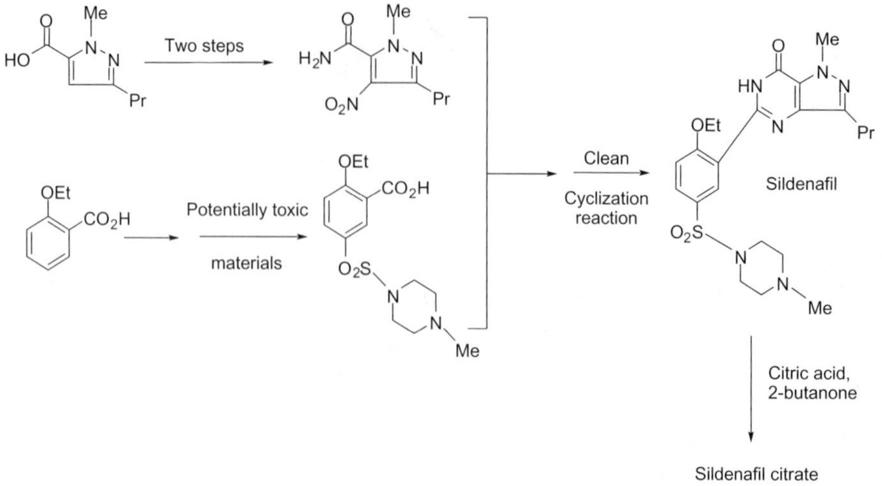

FIGURE 9.8. Sildenafil citrate manufacture.

Sertraline Manufacture

Sertraline is the active ingredient in Pfizer's top-selling antidepressant called Zoloft® in the United States. In the year 2000, 115 million prescriptions were issued in the United States alone (2002 Greener Synthetic Pathways Award, Pfizer Inc.; Green Chemistry in the Redesign of the Sertraline Process, U.S. EPA). The original seven-step manufacturing process was telescoped into a single step, thereby decreasing the raw materials usage by 20–60% (see Fig. 9.9). Additional process optimization eliminated the need for the use of four toxic solvents and hence the recovery steps. There was a reduction in the generation of the caustic waste by 100 metric tons, acidic waste by 150 metric tons waste, and solid titanium dioxide wastes by 440 metric tons per year, leading to a dramatic impact on the environment. At the discovery stage the amount of solvent used was 232 L/kg, which reduced to 98 L/kg during the first commercialization, to 81 L/kg at the second commercialization, and finally to 26 L/kg. Solvents such as methylene chloride were completely eliminated.

Ganciclovir

The Roche Colorado Corporation manufactures ganciclovir, which is an active ingredient in Cytovene® used to treat eye infections. The original commercial process was optimized to achieve air

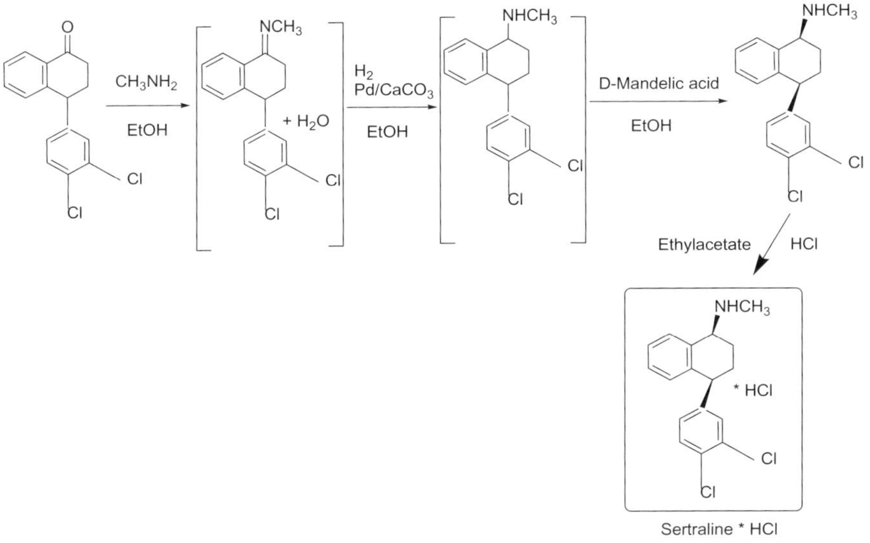

FIGURE 9.9. New sertraline process.

emission reduction by 66%, liquid and solid wastes by 89%, and solvent usage by 94%. This achievement was recognized in the year 2000 with an EPA Green Chemistry Award.

Benzodiazepines

Synthesis of 5H-2,3-benzodiazepine, one of the forms of a psychotherapeutic drug that is used to fight anxiety and relax muscles, among other uses, leads to the generation of a large amount of waste in the form of reagents, solvents, and separation agents. Approximately 340 L solvent and 3 kg chromium waste are produced per 1 kg of the drug. The new process not only improves the efficiency of the process, but also eliminates the waste generated.

Riboflavin

The conventional process for the production of 4000 tons/year of vitamin B_2 (riboflavin) consists of eight chemical and biochemical steps. The first step is a fermentation using *Bacillus* bacteria. A biotechnological synthesis of riboflavin in a single fermentation step with the help of bacteria, yeast, or fungi (Roche, ADM, and BASF, respectively) has reduced the production cost by 40%. The product crystallizes from the fermentation. The chemical steps have been completely eliminated in the manufacture of this vitamin by all three companies.

ACE Inhibitors

Captopril[®] is an ACE inhibitor used to treat high blood pressure. It is manufactured from two chemicals, D-β-hydroxy-isobutyric acid and L-proline. These building blocks are both manufactured by fermentation, with the yeast *Candida rugosa* and the bacterium *Corynebacterium* sp., respectively. Then both raw materials are joined by conventional chemistry in a reactor directly producing captopril. The fermentation process requires milder conditions and less toxic chemicals (see Fig. 9.10).

Vitamin B₃

Lonza, a fine chemical manufacturer, has developed a biotechnological route, starting with 3-cyanopyridine to nicotinamide (also known as niacin or vitamin B₃) (see Fig. 9.11). Conversions are based on enzymatic hydrolysis with nitrile hydratase from *Rhodococcus* bacteria or by bioconversion with living bacterial cells. The reactions are very specific, and the yields are quantitative. Novozyme has introduced an extremely thermostable lipase from the yeast *Candida (Pseudozyma) antarctica* (Novozyme 435), which is extremely suitable for carrying out specific esterifications in organic solvents.

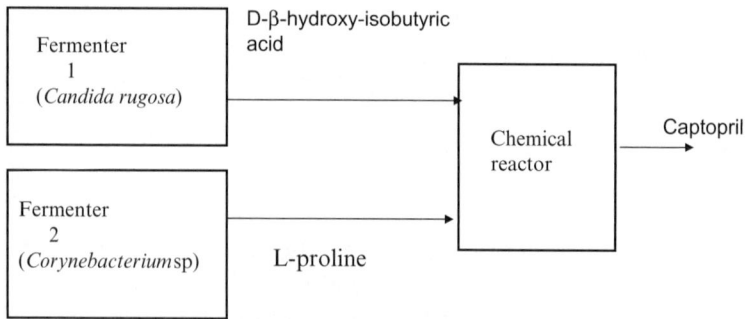

FIGURE 9.10. Biotechnology approach for the manufacture of captopril.

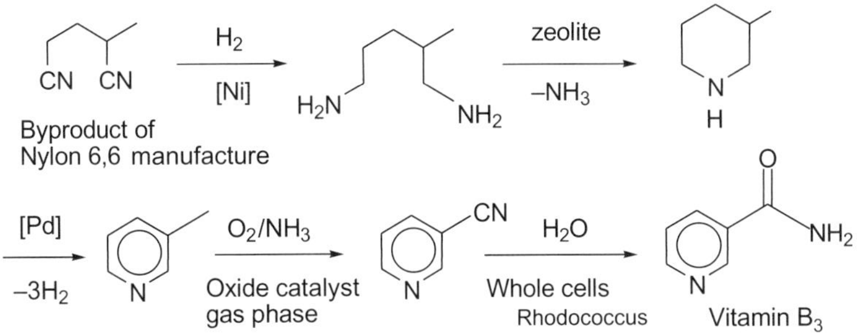

FIGURE 9.11. Process integration: Lonza nicotinamide process.

3-Cyanopyridine

3-Cyanopyridine (used to derive the vitamin Niacin) is manufactured in a three-step process starting from a byproduct of the Nylon 6,6 manufacturing process. All the steps are heterogeneous catalytic reactions involving Ni, Zeolite, Pd, and an oxide catalyst.

7-Aminocephalosporanic Acid

Biochemie (Austria) manufactures 7-aminocephalosporanic acid (7-ACA), a very important intermediate in the synthesis of semi-synthetic cephalosporin antibiotics from cephalosporin-C. The multistep process uses toxic reagents as well as chlorinated solvents. Most of the waste streams are burned because they poison normal biological waste water treatment systems, damaging the bacteria. The process is operated at very low temperatures and consumes high energy. In 1995, Biochemie replaced the chemical synthesis with a biotechnological route, which uses two naturally existing enzymes and takes place at room temperature under solvent-free conditions. Aqueous waste products are generated from the fermentor output and are processed in the biological waste water treatment system without problems. The waste mycelium is used as a fertilizer. This fermentation process has reduced the toxic waste from 31 kg to 0.3 kg/kg of 7-ACA, which is a 100× reduction.

6-APA

Penicillins belong to the beta-lactum class of antibiotics, and a number of different penicillins have been in use for treating infectious diseases since penicillin was discovered in the late 1920s. Of these, penicillin G and V are the most commercially important. 6-Aminopenicillinic acid (6-APA) is made from Penicillin G. The current fermentation approach gives a relatively pure aqueous solution containing a high concentration of Penicillin G. Solvent extraction is used to recover the product. However, ion exchange resins could be used to recover the product as well. The chemical method involves a very low temperature operation (−40°C) and the use of several solvents (see Fig. 9.12). The waste stream will contain chlorinated solvents and organics. The biotechnology process is operated at room temperature and does not produce waste. The immobilization of the enzyme is still a research area.

7-Aminodeacetoxy Cephalosporanic Acid

From 1975 to 1985, DSM (the Netherlands) produced 7-ADCA (7-aminodeacetoxy cephalosporanic acid), an intermediate for the synthesis of the semisynthetic cephalosporin antibiotic Cephalexin in a 10-step process. The waste stream was 30–40 kg/kg final product. In 1985, as a result of process optimization of the chemical process and the introduction of recycling, the waste generated per kg product was reduced to a value of 15. DSM-Chemferm introduced a biocatalytic process, comparable with the one used

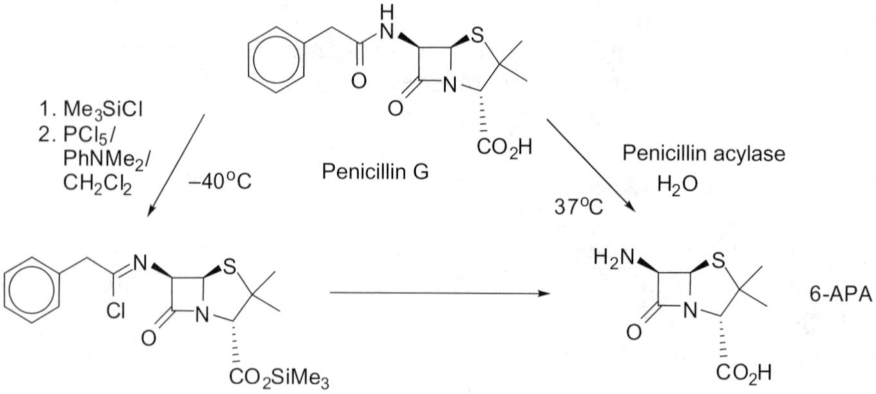

FIGURE 9.12. Enzymatic vs. chemical process for the manufacture of 6-APA.

at Biochemie, that resulted in reduction in waste of less than 10 kg/kg final product. Later DSM developed a fermentation process for the direct production of 7-ADCA in a single step, reducing the original 10-step process for cephalexin manufacture to a 4-step process, leading to a 50% reduction in cost and a 65% reduction in energy consumption. The quantity of waste produced and its toxicity were significantly reduced. The waste had only harmless inorganic salts. Using modern metabolic engineering techniques, it is believed that it may be possible to reduce the number of steps even further. A new organism may produce cephalexin or a direct precursor by means of fermentation, possibly reducing the waste to 2–5 kg/kg antibiotic (see Fig. 9.13).

L-Aspartame

L-Phenylalanine is an important amino acid used for the synthesis of L-aspartame, the artificial sweetener that is 200 times sweeter than sugar and used by diabetic people. It is used in many foods and beverages, and its worldwide production is 15,000 tons/year

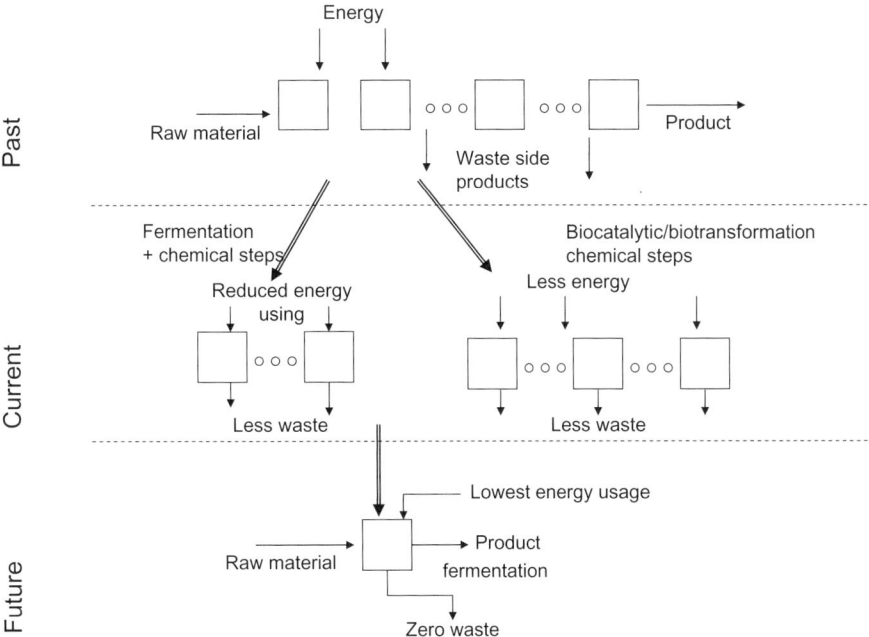

FIGURE 9.13. Philosophy of waste reduction and yield improvement.

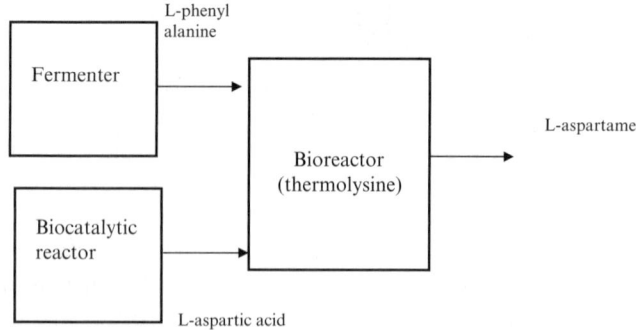

FIGURE 9.14. Biotechnological approach for the manufacture of L-aspartame.

at an approximate world market price of 35 €/kg (2001) (Royal Belgian Academy Council of Applied Science, 2004). The initial process for the manufacture of aspartame was based on chemical synthesis, but now it is based on a biochemical route. The two most important building blocks for L-aspartame, L-phenylalanine and L-aspartic acid, are produced by fermentation and biocatalysis, respectively (see Fig. 9.14). The Holland Sweetener Company uses a bacterial enzyme thermolysine to specifically link these two amino acids.

The Anticonvulsant Drug Candidate, LY300164

Lilly Research Laboratories has an anticonvulsant drug, LY300164 (see Fig. 9.15), also known as talampanol, for the treatment of epilepsy and neurodegenerative disorders. The drug is a noncompetitive AMPA receptor antagonist. The original solvent-based process has been redesigned using the principles of green chemistry. The original process consisted of seven steps with three intermediate isolations. The first step in the new synthesis utilizes the yeast *Zygosaccharomyces rouxii* in a three-phase reaction system, allowing for the removal of organic solvents from the aqueous waste stream. The yeast was deactivated with an increase in organic concentration; hence, a three-phase system was chosen, where the resin-supported biocatalyst was in the aqueous phase while the reactants were in the organic phase. The original process used approximately 340 L/kg product. A second key step in the process was the oxidation step, which used chromium. The improved process is carried out using compressed air. This eliminated the

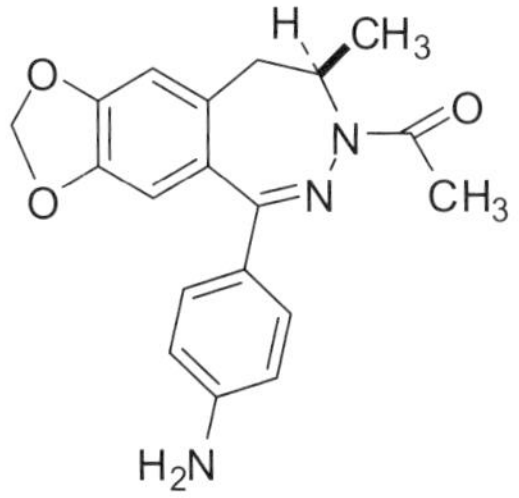

FIGURE 9.15. LY300164—an anticonvulsant drug.

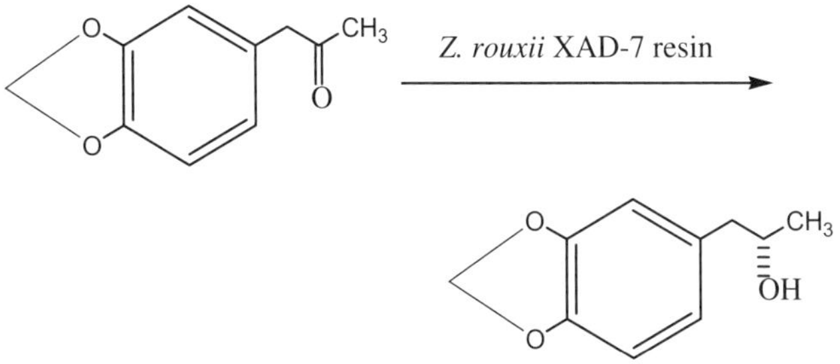

FIGURE 9.16. Biocatalytic reduction to produce optically pure alcohol.

use of chromium oxide (3 kg chromium/kg product), a possible carcinogen, and prevented the generation of chromium waste. In addition to environmental improvements, the process yield improved from 16% to 55%. Figure 9.16 shows biocatalytic reduction of a ketone to produce optically pure alcohol using *Z. rouxii*.

L-DOPA

The naturally occurring amino acid L-3,4-dihydroxyphenylalanine (L-dopa) was first synthesized as D,L-racemate. D,L-dopa exerted significant effects on glucose metabolism and arterial blood pressure. It has also been found to be effective against Parkinson's disease.

The 2001 Nobel Prize in Chemistry was awarded to Karl Barry Sharpless of the Scripps Research Institute in La Jolla, California, to Ryoji Noyori of Nagoya University in Chika, Japan, and to William S. Knowles, formerly at Monsanto Company in St. Louis, Missouri. The award recognized their contribution in the area of asymmetric catalytic oxidation reaction and asymmetric catalytic hydrogenation reaction. A hydrogenation process was developed

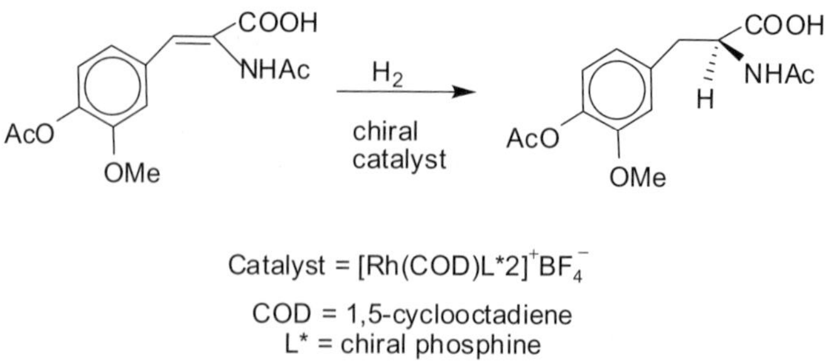

FIGURE 9.17. Monsanto L-DOPA process.

at Monsanto (called the Monsanto Process) for the selective synthesis of L-dopa using the asymmetric homogeneous catalyst (see Fig. 9.17) at 100% yield and at 95% enantiomeric excess. The Monsanto Process, which was the first commercialized catalytic asymmetric synthesis using a chiral transition metal complex, has been in operation since 1974.

Ascorbic Acid

Synthesis of vitamin C (ascorbic acid) is conventionally performed via the Reichstein–Grüssner procedure, which involves the fermentation of glucose followed by five chemical steps. Cerestar/BASF recently developed a new process that consists of one fermentation step and two simple chemical steps (via 2-keto-L-gluconic acid). It is predicted that soon a fermentation process will be developed to convert glucose into vitamin C in a single step, eliminating several recovery steps and reducing extraction solvents.

Hyperlipidimia-Controlling Drugs

Derivatives of clofibric acid were used in earlier times as hyperlipidemia-controlling drugs. Currently more interest is shown in the synthesis of 2-methyl-2-aryloxypropanoic acids because these classes of compounds are being considered as possible remedies for type II diabetes. The earlier GSK process involves reacting 2-bromo-2-methylpropanoic acid with a phenolic compound at 50°C, where both compounds are suspended in 2-butanone solvent. The acid is expensive, and large volumes of the organic solvent are also required.

An alternative synthesis of clofibric acid and analogues are based on the Biginelli reaction (published in 1906). This process involves reacting phenol with acetone and chloroform in the presence of sodium hydroxide, under reflux conditions for several hours, followed by a long workup to produce low yields of clofibric acid and analogues. Water-based biphasic reaction media are reported for exothermic reactions where water-immiscible organic reagents are made to react without the presence of any organic solvent. The reagents are mixed intimately using intense agitation and an excess amount of water. Water in excess helps in quenching the intense heat produced during exothermic reactions.

3,4-Dihydropyrimidin-2(1H)-ones are produced by the Biginelli reaction. The process involves reacting an aldehyde, α, β-ketoester, and urea or thiourea in the presence of an acid catalyst to produce dihydropyrimidinone. The reaction has been demonstrated in large scale using water as the medium under biphasic conditions at 50°C without any organic solvent. The product precipitates out of the reaction mixture. This water-based biphasic reaction strategy is adopted for the acetylation of amines as well.

Traditionally, acetylation of amines is conducted by the treatment of the solution of an amino compound with an acid anhydride or acid chloride in the presence of pyridine and a solvent such as ether or dichloromethane. In contrast, the new process has an aqueous mixture of a water-immiscible amine, stirred vigorously with acetic anhydride to produce in minutes the insoluble amide, which precipitates at high purity, produced in more than 90% yield.

Paclitaxel

The drug Taxol (paclitaxel), typically produced by extracting it from the yew tree using methylene chloride as a solvent, is used for treating ovarian cancer. In 1995, Hauser (USA) developed a lower-cost process that does not use methylene chloride to produce paclitaxel at a higher quality, yield, and concentration. Hauser's process uses cultivated yew trees that are renewable, more concentrated, and recyclable. The spent biomass can be recycled into compost rather than landfilled. Another renewable alternative is the use of the hazelnut tree. Researchers have recently isolated paclitaxel from fungi that grow on these trees, although at a lower concentration than from yew trees. Currently, Bristol-Myers Squibb has developed a plant-cell fermentation process to produce this anticancer drug. The drug was originally extracted from tree

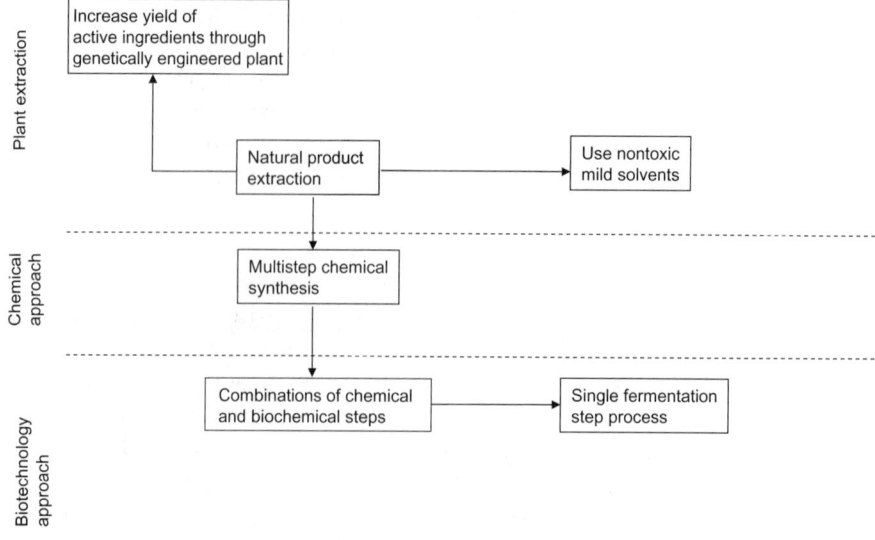

FIGURE 9.18. Drugs from natural sources.

bark, but now the current method involves a four-step synthesis (see Fig. 9.18).

Progesterone Synthesis

Oxidation of alcohols to carbonyl groups is traditionally carried out using a heavy-metal catalyst. Such a process generates a large amount of hazardous waste. Pharmacia and Upjohn have developed an alternative method that employs bleach and a catalyst cofactor system (such as sodium bicarbonate, potassium bromide, and 4-hydroxy-TEMPO) to convert bisnoralcohol to bisnoraldehyde, an intermediate in the synthesis of progesterone and corticosteroids. Soya sterol is converted to bisnoralcohol by fermentation. This produces a nontoxic aqueous waste stream and organics that could be recovered and avoids toxic reagents, such as organic peroxides and organoselenium compounds (see Fig. 9.19).

The Polymer Industry

The polymer industry is also adopting more sustainable practices. The polycarbonate (PC) Lexan is used in cars, appliances, CDs, and DVDs (Fukuoka et al., 2003). The polymer is termed an engineer-

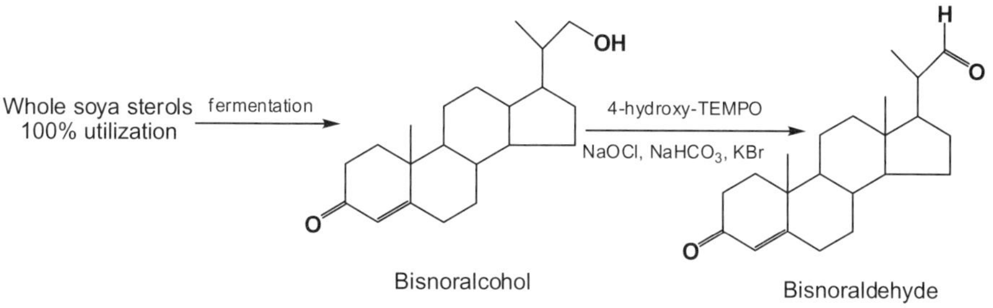

FIGURE 9.19. Synthesis of bisnoraldehyde from soya sterols.

ing plastic, is transparent, and has high strength. The Asahi Kasei Corporation (Japan) has developed a new process for the manufacture of PC. Originally PC was manufactured from bisphenol-A using large quantities of phosgene in methylene chloride solvent. This solvent is a suspected carcinogen, and phosgene is highly toxic. The new, less toxic process consists of bisphenol-A, ethylene oxide, and a byproduct carbon dioxide, obtained from Asahi's ethylene oxide plant under melt conditions. This process eliminates the use of methylene chloride as the polymerization solvent and reduces CO_2 emissions by 173 tons/thousand tons polycarbonate produced. GE also has produced another melt polymerization process, which involves reacting diphenyl carbonate with bisphenol-A at high temperature under melt conditions at very high vacuum. The phenol that is produced during the condensation is recovered and reused for the production of diphenyl carbonate directly to give low-molecular-weight "prepolymers" (MW 2000–20,000). These prepolymers are then converted to a higher-molecular-weight product by further polymerization. The amount of BPA left in the PC by the melt process is slightly higher than that left in the PC from the solvent process.

DuPont produces 1,3-propanediol (PDO), a monomer for producing its Sorona line of polymers (Ritter, 2003) using renewable resources. A genetically engineered *Escherichia coli* converts glucose derived from corn to PDO by fermentation. The fermentation process does not require an organic solvent or metal catalyst and is operated at room temperature, thereby saving energy. Other commercial routes to PDO use petroleum-based feedstock, high temperature and pressure, and heavy metal catalysts. Condensation of PDO and terephthalic acid yields this (Sorona® 3GT) polyester polymer, which is used in fiber and fabric manufacture. In a collaborative project between Genencor and DuPont, an *E. coli* production strain has been engineered with four foreign genes from other microorganisms to convert glucose to 1,3-propanediol. The natural *E. coli* does not produce this diol. The current commercial petroleum-based route to PDO involves hydroformylation of ethylene oxide to 3-hydroxypropionaldehyde, which is subsequently hydrogenated to PDO. The EPA lists ethylene oxide and acrolein as hazardous air pollutants.

Cargill Dow began using renewable resources, such as corn and sugar beet, in 2001 to manufacture (140,000 tons/year) its NatureWorks® polylactic acid (PLA) (Ritter, 2002). Glucose is converted to lactic acid by fermentation and subsequently polymerized to polylactic acid. No organic solvents are used in the production of PLA, and the product is recyclable, biodegradable, and compostable. The process requires 20–50% less fossil fuel resources than comparable petroleum-based plastics. The properties of the polymer are quite comparable to conventional polymers such as polyethylene or polypropylene. A number of consumer products are made from PLA, including plastic cups, clothing, and food packaging. It is also used in medical applications, such as for making biodegradable sutures and patches. PLA is also being used for the manufacture of biodegradable stents. Currently, the majority of the stents used in angioplasty are metal stents, which are not biodegradable and remain in the body forever. In addition, they lead to adverse reactions within patients, and once a metal stent has been implanted, the patient cannot undergo another angioplasty in the same artery. PLA is also used as the polymer in slow-release, drug-coated stents. These drug-eluting stents release their drug slowly, preventing immediate stenosis after the operation. The only negative side of PLA is its cost.

Disposing waste in landfills or by incineration is an unsustainable practice since it is not advisable if the material is toxic. This problem is being addressed through the design of consumer products that can be reused or recycled. For example, most carpet tiles contain polyvinyl chloride (PVC)-based backing material, and it is well known that disposal of PVC is a concern. Burning a PVC-containing product in an incinerator releases toxic dioxins. PVC cannot be used in landfills since biodegradation of this polymer takes several hundred years. To avoid this problem, Shaw Industries (Ritter, 2003) developed a carpet backing using polyolefin resins and waste fly ash. The backing of the carpet is PVC-free, and the waste material is collected and recycled into new carpet,

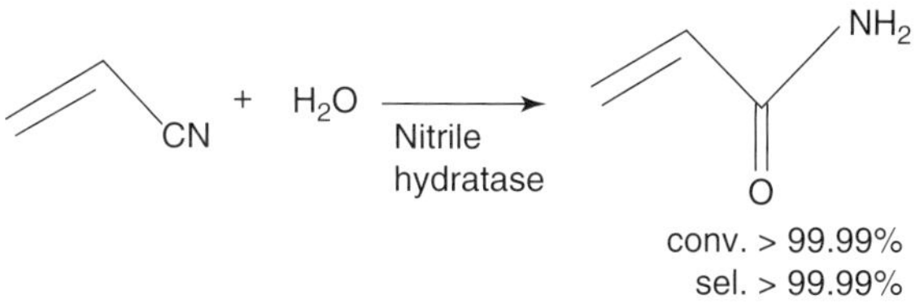

FIGURE 9.20. Biocatalytic production of acrylamide: Mitsubishi process.

thereby eliminating the need to landfill or incinerate the used product. [Pandel, Interface foam and backing manufacturer made carpet from corn-based polylactic acid polymers. In addition they also eliminated all lead, cadmium and barium stabilizers in their foam products.]

Mitsubishi Rayon produces acrylamide from acrylonitrile with the help of an immobilized bacterial enzyme, nitrile hydratase (see Fig. 9.20). This acrylamide is then polymerized to the conventional plastic polyacrylamide. This process was one of the first large-scale applications of enzymes in the bulk chemical industry and replaced the conventional process that used sulfuric acid and inorganic catalysts. The enzymatic process has several advantages over the chemical process. The efficiency of the enzymatic process is 100%, while that of the previous chemical process was only 30–45%. The energy consumption is only 0.4 MJ/kg product, compared to 1.9 MJ/kg product for the chemical route. The process generates less waste. The CO_2 production is only 0.3 kg/kg monomer, while the previous process produced 1.5 kg/kg. The reaction is carried out at ~15°C, which is milder than the original chemical route. About 100,000 tons of acrylamide are produced yearly now via this approach in Japan and other countries.

Baxenden Chemicals Ltd. (UK) has developed a large scale lipase-catalyzed process for the production of poly (hexane-1,6-diol adipate). The process consists of condensation of hexane-1,6-diol and adipic acid using *Candida antarctica* lipase. Constant removal of water from the enzyme-catalyzed polymerization process is crucial to shift the equilibrium to the right. This simple lipase-catalyzed polyester manufacture is operated at ambient temperature and does not require a solvent. In addition, the process is very economical when compared to the chemical process.

Polyurethanes are polymers that are widely used for a variety of household and industrial applications. Traditionally,

$$RNH_2 + COCl_2 \longrightarrow RNCO + 2HCl$$

(amine)   (phosgene)   (isocyanate) $\downarrow$ R' OH

$$RNHCO_2 R'$$
(urethane)

FIGURE 9.21. Synthesis of polyurethanes using phosgene.

$$RNH_2 + CO_2 \longrightarrow RNCO + H_2O \longrightarrow RNHCO_2R'$$
$$+ R'OH$$

FIGURE 9.22. Alternate synthesis to polyurethanes without using phosgene.

polyurethanes are manufactured through the use of phosgene (see Fig. 9.21). Phosgene is an extremely toxic and lethal gas. Monsanto has developed an alternate method of synthesizing polyurethanes and their precursors without the use of phosgene. The process uses CO_2, the gas that is known to cause global warming. This process not only decreases the greenhouse gas but also eliminates the use of dangerous and toxic chemicals (see Fig. 9.22).

Polyacrylic acid (PAC) is an anionic polymer used in many industrial applications including water treatment, but it is not biodegradable, so it ends up in waste water. An economically viable, effective, and biodegradable alternative is thermal polyaspartate (TPA). TPA is nontoxic and environmentally safe and is biodegradable. Donlar (USA) invented two highly efficient processes to manufacture TPA. The first process is a dry and solid polymerization technique involving converting aspartic acid to polysuccinimide without the use of any organic solvent. The only byproduct is condensed water, and the process is 97% efficient. The second step in this process is the base hydrolysis of polysuccinimide to polyaspartate. The second TPA production process involves using a catalyst during the polymerization step and performing the reaction at a lower temperature, resulting in a product with better performance characteristics and a mild color. The catalyst itself is recovered from the process.

Metabolix Inc. (USA) has developed a fermentation process to produce biodegradable polyhydroxyalkanoate (PHA) from renewable feedstock by incorporating genes from PHA-producing bacteria into a strain of *Escherichia coli*. The fermentation process

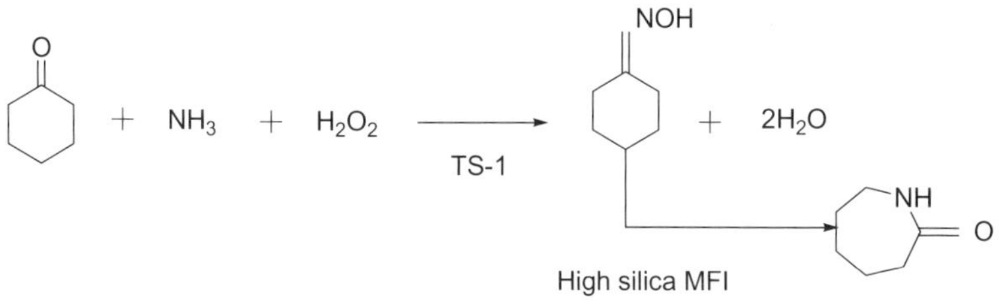

Atom economy = 75%, E-factor = 0.32

FIGURE 9.23. Sumitomo process: combined ammoximation and vapor phase Beckmann rearrangement for caprolactam synthesis.

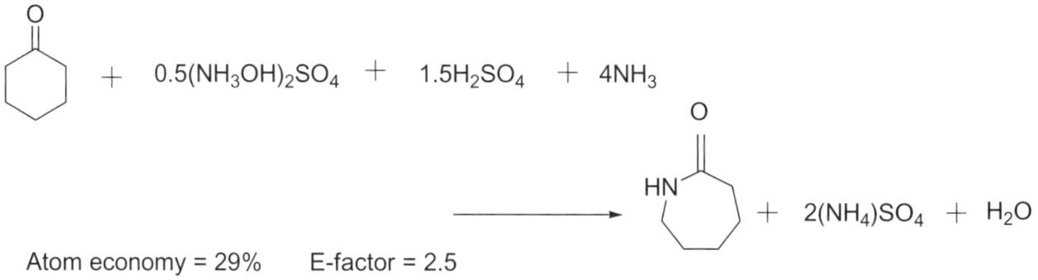

Atom economy = 29%   E-factor = 2.5

FIGURE 9.24. Overall reaction of current process for caprolactam synthesis.

converts sugars or oil into polymers. Cargill and DuPont are the other two companies that produce biodegradable polymers such as polylactic acid (NatureWorks®) and polypropylene terephthalate (Sorona®), respectively, from biomass.

Caprolactam is a cyclic amide (or lactam) and a monomer used industrially in the production of nylon. Most caprolactam is synthesized from cyclohexanoxime by Beckmann rearrangement (see Fig. 9.23). The two-step process uses titanium silicate (TS-1) as the catalyst in the first step and high silica for the second step. Rhodia (France) won the Potier Prize for Innovation related to the environment (a French award). The process eliminates the disadvantages of the original production process, which generates saline byproducts and solvents waste (see Fig. 9.24). This new innovation will generate no nitrogen oxides and will result in energy savings of 50%. The atom efficiency improved from 29% to 75%, and the E-factor decreased from 2.5 to 0.32. Another process for the

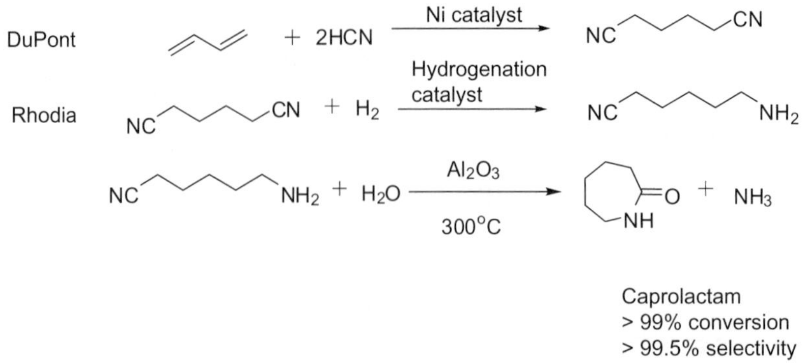

FIGURE 9.25. A new three-step salt-free caprolactam process developed by RHODIA.

production of caprolactam involves a three-step synthesis using butadiene as the starting material and several heterogeneous catalysts (see Fig. 9.25). This is also a Rhodia process; it is a salt-free process with both conversion and selectivity above 99%. All the catalysts can be filtered, recovered, and reused.

Pesticides, Antifoulants, and Herbicides

Agriculture is another area that needs green thinking, and it can benefit with the use of the principles of green chemistry. Agricultural runoff has led to eutrophication of water bodies, while the use of broad-spectrum pesticides harms beneficial organisms as well as pests. In addition, the pesticides get accumulated in cattle and other grazing animals and return to the public through drinking water, milk, diary products, and meat. Even soft drinks and bottled drinks end up having large quantities of pesticides, since these drinks use large quantities of groundwater. Dow Agro-Sciences has developed a number of pesticides that are more selective and less persistent. Instead of creating a chemical barrier around a structure to keep out termites, Dow's Sentricon® Termite Colony Elimination System (2002) is a targeted approach that employs traps baited with hexaflumuron, a substance that disrupts the molting cycle of termites. Unable to molt, the colony dies (Kirchhoff, 2005).

A bio-pesticide for sugarcane, called *Bio-Cane*®, has been launched in Australia. This product is based on a naturally occurring fungus that has been cultured on broken rice grains to provide a medium for distribution. Biocane granules are claimed to be particularly effective against greyback canegrub and can replace chemical pesticides, which are toxic, are persistent in soil, and are leached into the groundwater, causing environmental damage.

Traditionally, organotins [tributylin oxide (TBTO)] have been used as marine antifouling agents to prevent fouling of ships and boat parts that are underwater. These compounds prevent micro and macro fouling as well as deposition of large marine organisms such as barnacles and marine plants. Fouling of boats and ships leads to an increase in drag, reduction in speed, increase in fuel consumption, and corrosion of the ship body. This results in additional fuel consumption by the ships, totaling an extra $3.0 billion per year. Organotin compounds such as TBTO are very toxic, are persistent in water, cause plenty of damage to fauna and flora, and have the ability to increase shell thickness in shellfish. TBTO degrades more slowly, with a half-life of nine days in sea water and of six to nine months in sediment. These compounds have already been banned in the United States and Europe. The Organotin Antifoulant Paint Control Act of 1988 restricts the use of tin in the United States. A safer marine antifouling compound Seanine® (4,5-dichloro-2-n-octyl-4-isothiazolin-3-one), which degrades more rapidly and causes fewer pollution problems than organotins, has been introduced by Rohm and Haas. This antifoulant degrades rapidly, with a half-life of one day in sea water and one hour in sediment.

Rohm and Haas discovered a new class of chemicals, the diacylhydrazines, that offer a safer method for the control of insects in turfs and a variety of agronomic crops. The important product in that class is Confirm®, which is selective for the control of caterpillar pests in agriculture without posing significant risk to the applicator, the consumer, or the ecosystem. This insecticide mimics a natural substance found in the caterpillar's body, called 20-hydroxyecdysone, which is the natural "trigger" that induces molting and regulates development in the insects. It disrupts the molting process, causing the caterpillars to stop feeding within hours of exposure and die soon after. This action is safer than the action of other insecticides and does not harm the other insects.

Eden Bioscience (U.S.) developed a new class of nontoxic, naturally occurring proteins called harpins, which are produced in a water-based fermentation process that uses no harsh solvents or reagents, requires only modest energy inputs, and generates no hazardous chemical wastes. Harpins activate the proteins that trigger a plant's natural defense systems to protect against disease and pests and simultaneously enhance the plant growth systems without altering the plant's DNA. This is a better technique than using pesticides or growth hormones, which may persist in the environment, damaging ecology and finding their way into the food chain.

AgraQuest, Inc. (USA) has developed Serenade®, a biologically derived fungicide that can replace pesticides with heavy metals or chlorine. It has been tested on 30 crops in 20 countries and is registered for use on crops including grape vines, leafy vegetables, green beans, hops, potatoes, and peanuts. The product contains more than 30 lipopeptides produced by the *Bacillus subtilis* strain QST-713, which was discovered by AgraQuest scientists in soil samples. The *B. subtilis* lipopeptides are made up of a cyclic peptide portion and a fatty acid side chain (see Fig. 9.26). The pesticide is made by fermentation. The broth containing *B. subtilis* cells, spores, and lipopeptides is concentrated and then spray-dried to form a powder. Biopesticides accounted for about 1% of the $28 billion worldwide pesticide market in 2002, but as of that year market share was growing 20–30% annually (Ritter, 2003). Pheromone- and microbe-based pesticides have cropped up in recent years. These include several products derived from other *Bacillus* species, such as *B. thuringiensis* (Bt) strains. Natural products are inherently safer for agricultural workers and the environment since they reduce the reliance on conventional chemical pesticides and are not harmful to nontarget plants and animals. In addition, biopesticides take about three years and $6 million to develop, while a chemical product can take up to 10 years and $180 million to develop.

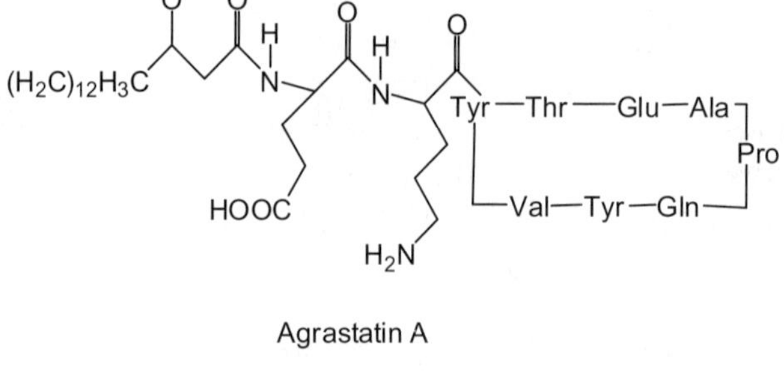

Agrastatin A

FIGURE 9.26. Structure of a lipopeptide.

Roundup®, manufactured by Monsanto (USA), is one of the most widely used herbicides in the world. A key intermediate for its manufacture is disodium iminodiacetate (DSIDA), which is traditionally made through Strecker's process. This process requires the use of ammonia, formaldehyde, and hydrogen cyanide. Hydrogen cyanide is extremely toxic, and its use requires special handling to minimize risk to workers, the community, and the environment. Ammonia and formaldehyde are also associated with human health problems and environmental hazards. The process produces 1 kg waste for every 7 kg product. The waste contains cyanide and formaldehyde, which need to be treated prior to safe disposal.

The new Monsanto process makes use of a copper catalyst, which is used to reduce diethanolamine to DSIDA, the intermediate to Roundup® (see Fig. 9.27). This process totally eliminates the use of ammonia, hydrogen cyanide, and formaldehyde, is free of contaminants and byproducts, and hence does not require further purification steps. The stream can be recycled after filtration of catalyst. Monsanto's process can also be used in the production of other amino acids such as glycine through reduction and is a general method for conversion of primary alcohols to carboxylic acid salts. The development of this technology for processes pertaining to the agricultural, commodity, specialty, and pharmaceutical sectors would have a pronounced impact on the environment.

DuPont practices biocatalytic hydrolysis of nitriles using immobilized whole cells of *P. chlororaphis* B23. The catalyst consumption is 0.006 kg/kg product. The conversion is higher and so

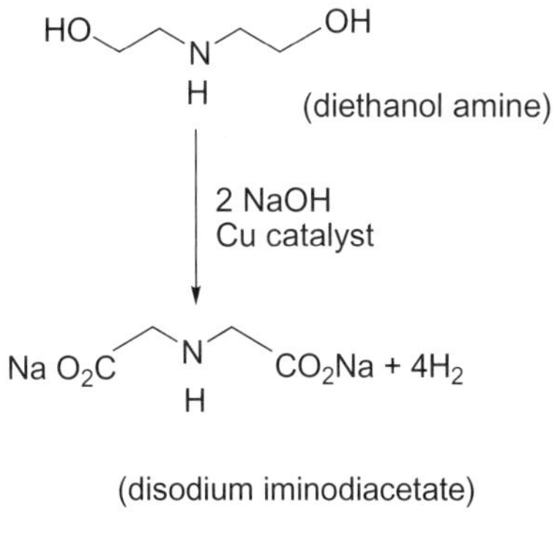

(diethanol amine)

2 NaOH
Cu catalyst

(disodium iminodiacetate)

FIGURE 9.27. DSIDA preparation.

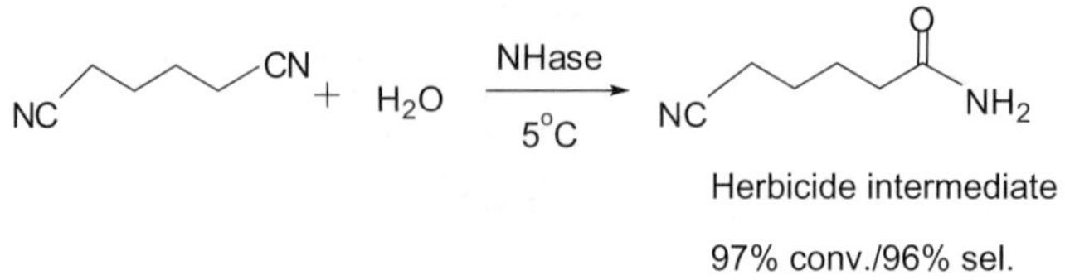

FIGURE 9.28. Hydrolysis of nitriles.

is the selectivity when compared to the chemical process, which is carried out with MnO_2 catalyst at a temperature of 130°C (see Fig. 9.28). The chemical process also produces effluent containing large quantities of manganese.

Solvents and Green Chemistry

Ethyl lactate is a nontoxic and biodegradable solvent derived from renewable carbohydrates. Its use and acceptance can be increased only if it is competitively priced to replace traditional organic solvents such as methylene chloride and chloroform. Argonne's National Lab developed a selective membrane technology for synthesizing ethyl lactate that eliminates the production of an equimolar amount of salt waste. The process uses pervaporation membranes and catalysts to directly convert the salts to esters, thereby eliminating the problem of salt waste. This approach reduces the product cost by half. The concept of pervaporation consists of the use of a membrane that selectively accepts one compound from a mixture of compounds; it is explained in Chapter 6.

The U.S. Bureau of Engraving has traditionally used a solvent mixture consisting of methylene chloride, toluene, and acetone for cleaning postage stamp and overprinting presses. The solvent mixture caused serious environmental issues in the District of Columbia. In 1999, the Bureau developed a replacement mixture that is a combination of isoparaffinic hydrocarbon, propylene glycol, monomethyl ether, and isopropyl alcohol (U.S. Bureau of Engraving, 1999). This mixture is less toxic and less polluting than the previous mixture and has now totally replaced the traditional solvent mixture in all uses in the printing process.

Adipic Acid

Traditional feedstock for the synthesis of adipic acid is benzene. Benzene is hydrogenated using $Ni\text{-}Al_2O_3$ catalyst at 370–800 psi to cyclohexane. This is oxidized using Co catalyst to a mixture of cyclohexanone and cyclohexanol. This is then converted to adipic acid using ammonium vanadate and nitric acid, with the help of a catalyst. Alternative feedstock in the synthesis of adipic acid is D glucose, which is converted to *cis*-muconic acid using *E. coli*, which is further hydrogenated to adipic acid using hydrogen gas. This approach eliminates several reaction and separation steps.

Fire Extinguishers and Flame Retardants

Traditionally, fire extinguishers have utilized halogens (CFCs), which today are well-known ozone-depleting agents. They also harm aquatic systems and contaminate water supplies. Pyrocool is a nontoxic, biodegradable, fire-extinguishing and cooling agent that can replace the traditional extinguishers, is just as effective in putting out fires, and does not deplete the ozone layer or persist in the environment, unlike CFCs.

Epoxy phenolic molding compounds (EMC) are mixtures of chemicals containing a base polymer resin matrix and various additives. Usually these additives are brominated epoxy resin and antimony oxide, which function as flame retardants. These compounds produce toxic fumes during flame and are also dangerous for the environment. In the last few years research has focused on developing polymers containing halogen-free flame-retardant additives, which lead to compounds containing P, Si, B, N, $Al(OH)_3$, and $Mg(OH)_2$. In addition to their low toxicity, their main advantages are that, in the case of fire, they do not produce dioxin and halogen acids, and they generate low amounts of smoke. Recently the microelectronics industry developed new halogen and antimony-free molding compounds based on phosphorous–based, organic, flame-retardant additives. These materials reduce the presence of toxic elements in the electronic package and the environment. It is known that halogens and other ionic impurities are responsible for metal corrosion under bias, humidity, and high temperatures. Elimination of these compounds in the formulation increases the life of these materials.

Catalyst

Current metal oxide catalyst synthesis involves oxidizing metal powder or chips with nitric acid at elevated temperatures under agitation. The resulting metal nitrate solution is treated with a base to precipitate the metal salt, which is washed with water to remove salts and ions. Finally, it is dried and then calcined to drive off excess water and NO_x, CO_2, or other species.

Süd-Chemie's (a global consortium) new route produces metal oxides by eliminating the nitrate formation step. The synthesis is carried out at room temperature by reacting the metal with a mild aqueous carboxylic acid as an activating agent and O_2 from air as an oxidizing agent. A minimum amount of water is used, which yields a slurry. Metal conversion to the metal oxide takes 24 to 48 hours. Any unreacted metal less than 1% is removed by a magnetic separator. This process eliminates the use of nitric acid and the handling of highly acidic waste and NO_x.

Hydrogen Peroxide

Titanium silicate catalyst (TS-1) is well suited for carrying out selective oxidation reactions such as the 4-hydroxylation of phenol to the commercially important hydroquinone. TS-1 has also been used in commercial epoxidations of small alkenes.

Hydrogen peroxide in combination with catalysts such as TS-1 acts as a good, "clean" epoxidation system. The reactions that could be carried with this catalyst include ammoximation of cyclohexanone, epoxidation of propene and other small alkenes, and hydroxylation of aromatics and linear alkanes (Chapter 4). The system produces little waste, avoids the use of hazardous chemicals such as alkyl hydroperoxide, and reduces process complexity. However, the key parameter for industrial development is the cost of H_2O_2. H_2O_2 is produced by only a few companies, and very large capital expenditure is required, because H_2O_2 synthesis (by alkylanthraquinone route) is economical only when large quantities are produced.

The current production of propene oxide (PO) from propene is based on the chlorohydrin process, which involves epoxidation using alkyl hydroperoxides. The process generates large quantities of waste water (40 tons waste water/ton PO produced) containing inorganic salts (tons $CaCl_2$/ton PO produced) and significant amounts of chlorinated organic byproducts (10%). The alternate route for the synthesis of PO is by epoxidation of propene with

H$_2$O$_2$ on a TS-1 (titanium silicate) catalyst, which reduces this environmental impact. This requires the production of H$_2$O$_2$ on site, to avoid its storage and transport, which are hazardous activities. A possible route for the synthesis of H$_2$O$_2$ could be from H$_2$ and O$_2$ mixtures using Pd-based catalysts. Proper reactor design to handle the explosive gas mixture has not been achieved so far.

Ferroelectric or Antiferroelectric Liquid Crystals

Optically active fluorine-containing molecules have been recognized as an important class of materials because of their possible application in optical devices such as ferroelectric or antiferroelectric liquid crystals (FLCs). In the manufacture of FLCs, fluorinated 6-deoxy sugars are key chiral dopants. The synthetic route to a furanol with a fluoroalkyl group is based on enzymatic resolution in organic or water media. These processes generate waste containing solvent media and enzymes. In the water medium, excellent resolutions of both the esters are obtained with *Pseudomonas cepacia* to give chiral alcohols >99% enantiomeric excess at 48% conversion. However, in this process, the reaction mixture has to be maintained at a pH of 7 by adding 1 N aqueous NaOH. Further, when the conversion reaches about 50%, flocculant is added and the reaction mixture is filtered through Celite-545 to remove the enzyme. But the separation of enzyme and products is not easy in the large scale, and also the enzyme is not reusable as it loses its activity. This process generates at least 30 L waste/1 kg product, and it is a mixture of water and organic solvents. In contrast, the enzymatic esterification in organic solvent with *C. antarctica* can give chiral ester of 93% (in CH$_3$Cl) and 97% (in CCl4) enantiomeric excess at 48 and 49% conversions, respectively. The reaction is free from water, and hence this process does not generate aqueous effluent stream.

The Food and Flavor Industry

Esterification and transesterification reactions using lipases are well-established processes in the area of foods and flavors. Value-added products are produced by these processes. The lipases are tolerant in organic media.

Vanillin

Vanillin is a flavoring agent used in syrups, ice cream, and other edible products. Xuebao Fine Chemicals Co. Ltd. (China) used to manufacture vanillin from *o*-nitro chlorobenzene. The process produced toxic chemicals, three to five different tars, high CODs, high VOCs, high health and safety risks, and unacceptable standards for a flavoring product. The plant dumped untreated effluents into a nearby river, and toxic tars were stockpiled in unmonitored landfills. Rhodia Chemicals purchased Xuebao in 2000. The process was modified so that it is now based on the catechol route (see Fig. 9.29). This process does not produce any waste and uses several heterogeneous catalysts.

Menthol

Menthol is the chief flavoring agent in mint and peppermint and is used in candy, tobacco, oral care items, pharmaceuticals, and various other products. Takasago developed the synthesis of (−)-menthol (in the 1980s) from myrcene, which is converted to diethylgeranylamine by the lithium-catalyzed addition of diethylamine. This is then catalytically isomerized to the chiral 3R-citronella enamine with 96–99% enantiomeric excess. Hydrolysis of this intermediate gave 3R-(+)-citronella a higher chiral purity than citronella from citronella oil. This is the second major com-

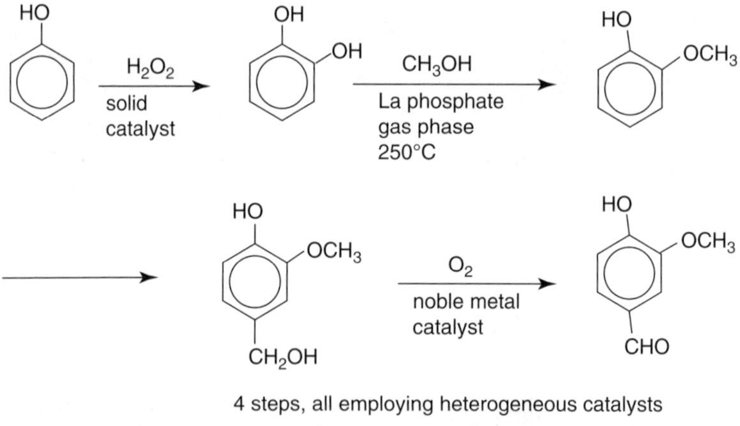

4 steps, all employing heterogeneous catalysts

Overall: $C_6H_6O + H_2O_2 + CH_3OH + H_2CO + \frac{1}{2}O_2 \longrightarrow C_6H_8O_3 + 3H_2O$

FIGURE 9.29. Catalytic vanillin synthesis: Rhodia process.

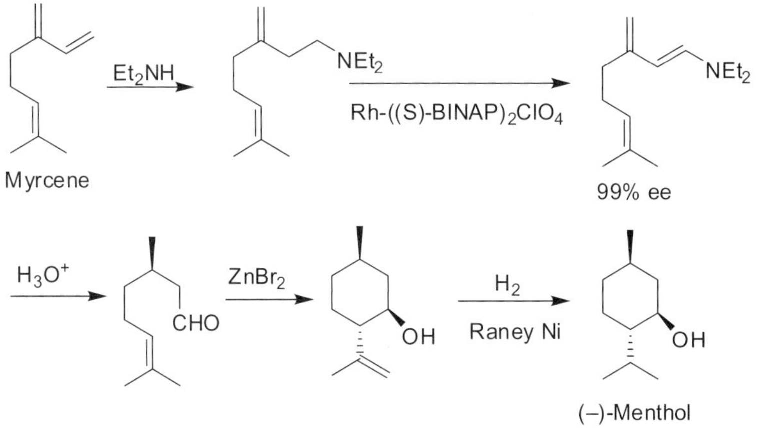

FIGURE 9.30. Menthol process by Takasago.

mercial route to (–)-menthol (see Fig. 9.30). The plant is operated at a scale of 400,000 tons/year. Takasago International Corp. has developed optically active substances using asymmetric synthesis catalysts technology. The company also uses asymmetric synthesis catalysts for the manufacture of several biodegradable polymers.

Bioethanol

Bioethanol has been identified as a clean fuel that is mixed with petrol to run automobiles without modifying the engine design. Bioethanol does not produce SO_2 or NO_x. Novozymes collaborated with NREL to improve cellulases enzymes, which are used for the production of bioethanol. These improvements in the enzyme production have reduced the cost of ethanol production from $5.00 to less than $0.18 per gallon. The raw material for the ethanol is corn stover. In 2004, Iogen (Canada) became the first company to produce 1 million gal/year of cellulosic ethanol from wheat straw at a plant in Ottawa. Dyadic (USA) sells nearly 50 liquid and dry enzyme products to fermentation companies and uses fungal strains to produce them. Its most important fungus are a C-1 protein-expression system based on a genetically modified *Chrysosporium lucknowense* and C-1, a soil fungus. Los Alamos and Motorola have jointly developed a fuel cell for their cellular phones that would last the lifetime of the phone and runs on methanol.

ADM uses an immobilized enzyme particle developed by Novozymes (Illinios, U.S.) to catalyze interesterification of fully hydrogenated oil and unhydrogenated oil to produce low-trans-fat vegetable oil. The enzyme replaces the conventional sodium methoxide-based, high-temperature process. The enzyme process is greener and saves capital and operating costs.

The Maleic Anhydride Manufacturing Process

The traditional maleic anhydride manufacturing process involves reacting benzene with excess air. A low benzene concentration is used in order not to exceed the flammability limit of the mixture. The reaction gas mixture is passed over a multitubular, fixed bed catalyst reactor at a pressure of 0.15–0.25 MPa. In addition to maleic anhydride, the reaction produces two moles of CO_2 and water. The reaction is highly exothermic, causing "hotspots" of 340–500°C. For each ton of benzene that is reacted, 27 MJ of heat are generated. Molten eutectic salts circulate outside the reactor tubes to dissipate the heat. Steam is generated when the molten salts are cooled, which is used to drive air compressors (see Fig. 9.31).

The vapor mixture leaving the reactor is cooled down to 55°C by heat exchangers, and 40–60% of the maleic anhydride is recovered as liquid. Contact of maleic anhydride with water results in the undesired formation of maleic acid. The maleic anhydride

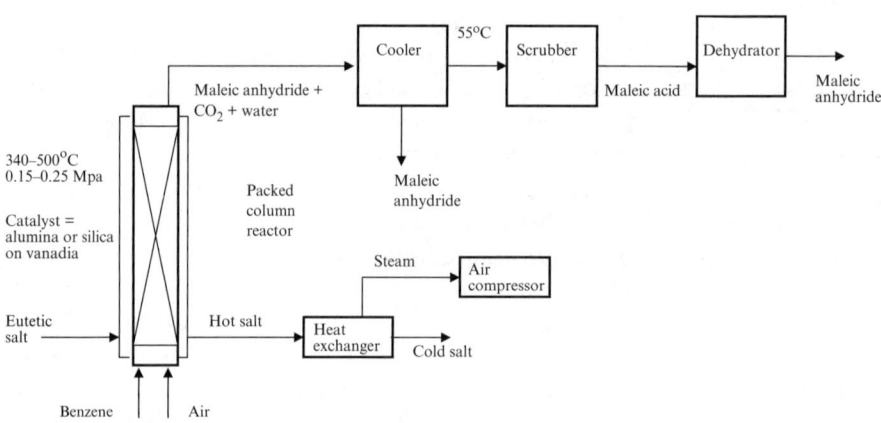

FIGURE 9.31. Maleic anhydride manufacturing process based on benzene.

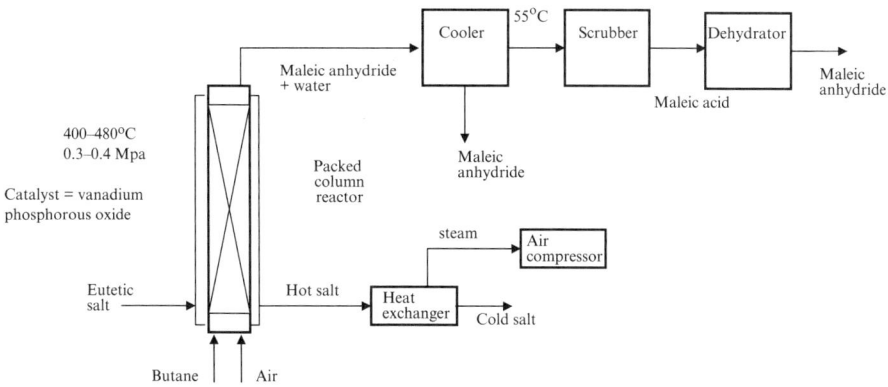

FIGURE 9.32. Maleic anhydride manufacturing process based on C₄.

that cannot be recovered is eventually washed out with water as maleic acid. Water scrubbing and subsequent dehydration of the maleic acid stream are required to purify and reform the remaining maleic anhydride. The commercial catalyst for benzene oxidation is alumina- or silica-supported V_2O_5, MoO_3, and a small amount of Na_2O. Molybdenum is the most popular element, but phosphorus and the alkali and alkaline earth metals tin, boron, and silver are also used.

The modified process is based on the oxidation of butane (see Fig. 9.32). A major advantage of C₄ over benzene is that no carbon is lost in the reaction. The yield from butane process is 30% greater than that from the benzene process. Also, C₄ is much cheaper than benzene. Benzene is a known carcinogen. So, the fixed bed process with *n*-butane as the raw material has been the only MA route used commercially since 1985 in the United States. In the fixed bed process, a low concentration of butane is passed over the catalyst at 400–480°C and 0.3–0.4 MPa. The process generates more water than the benzene process. Unsupported vanadium phosphorus oxide (VPO) catalysts with promoters such as lithium, zinc, and molybdenum are commonly used.

Chelants

A *chelant* is an organic compound that forms a chemical bond with metals through two or more of its atoms, which are used to clean up metal-contaminated groundwater. The contamination in the water could be heavy metals, agricultural nutrients, ions that increase the hardness of water such as calcium or magnesium effluent from the electroplating industry, and waste from photographic film processing units. Chelating agents are poorly biodegradable. Because of their high water solubility, they are readily released into the environment through runoff. Chelating agents are produced from amines, formaldehyde, sodium hydroxide, and hydrogen cyanide. The Bayer Corporation developed a cost-effective process to manufacture iminodisuccinate, which is an environmentally friendly and biodegradable chelating agent. Bayer's Industrial Chemicals Division markets the product under the name Baypure CX. Baypure is produced from maleic anhydride, sodium hydroxide, ammonia, and water and is used as solvent.

Conventional methylation reactions use methyl halides or methyl sulfate. These compounds are toxic and cause severe environmental damage. Also, the methylation of active methylene compounds often involves uncontrollable multiple alkylations. A method to methylate selectively using dimethylcarbonate has been developed by reacting arylacetonitriles with dimethylcarbonate at 18–22°C in the presence of potassium carbonate (see Fig. 9.33). The product is 2-arylpropionitriles with high selectivity (>99%). This process does not produce inorganic salts. This reaction can be carried out in continuous-flow and batch modes of operation.

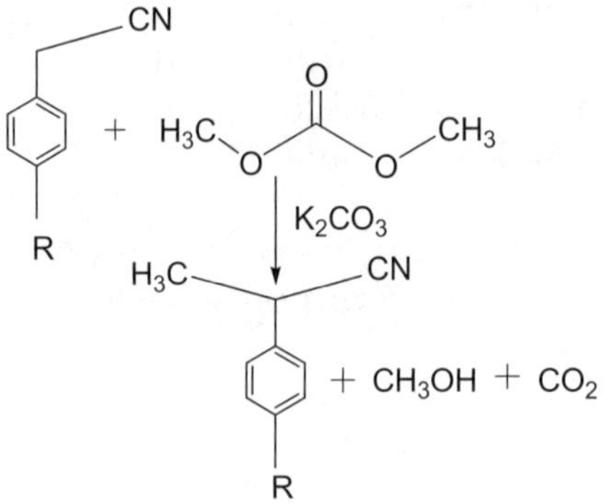

FIGURE 9.33. Selective methylation of active methylene group.

The Surfactant Industry

Surfactants, soaps, and detergents are used both for industrial and domestic purposes. The usage of these products over the past decade has increased by severalfold. Most of these compounds are nonbiodegradable and hence persist in the water bodies. The demand for non-ionic surfactants for use in health care and beauty products also has grown. Alkylglycoside, which is made from saccharide, can be a replacement for alkylaryl sulphonate anionic surfactants used in shampoos. A more environmentally benign replacement for phosphorus-containing additives in washing powder is sodium silicate. Coconut oil soap bases are mild alternatives to harsh anionic and cationic surfactants in liquid cleansing applications.

Rhamnolipids are natural glycolipids produced by bacteria such as *Pseudomonas aeruginosa* in soil and in plants. Jeneil Biosurfactant Co. (USA) has developed a series of rhamnolipid biosurfactants that can be substituted for or used in conjunction with synthetic surfactants such as nonyl-phenol ethoxylate that have environmental toxicity and environmental persistence. These products can be used in the petroleum, agriculture, and personal care industries. The biosurfactants contain a lipid portion and a hydrophilic region. Sucrose-based surfactants are very good emulsifiers and are biodegradable. Transesterification reactions involving sucrose and long chain fatty acid esters are carried out using *C. antarctica* lipase at 80°C (see Fig. 9.34). It is impossible to synthesize such surfactants effectively using chemical methods without generating waste and plenty of degraded side products.

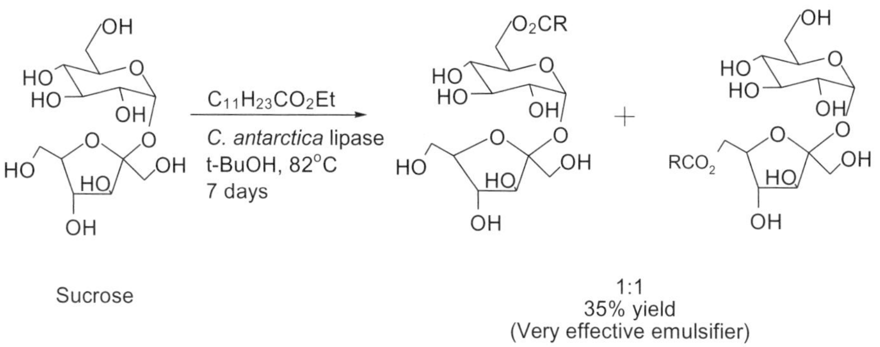

FIGURE 9.34. Synthesis of sucrose fatty acid esters.

Industries in Need of Support to Go Green

Companies such as BASF, DSM, and Degussa not only develop biocatalysts for their own manufacturing processes but also lend their expertise to other industries, which include paper and pulp, textile, natural fiber, and animal feed. These industries are generally plagued with old chemical-based processes that generate large quantities of effluent, toxic metals that need expensive treatment (Anastas et al., 1998; Binns et al., 1999).

Several industries currently operate at very poor efficiency, use toxic chemicals, use large quantities of water, are very energy inefficient, and generate very toxic effluents. These industries need a major research thrust to become green. The following section describes the various issues these industries are facing and opportunities available for applying the principles of green chemistry.

The Semiconductor Manufacture Industry

The semiconductor industry has had phenomenal growth in the past 25 years. It is a $150 billion dollar industry (Chepesiuk, 1999). Due to its tremendous growth, it now faces several environmental issues. Water usage in integrated circuit manufacture is among the highest in any industrial sector. Currently, semiconductor manufacturers use several toxic chemicals and large quantities of water and generate very toxic effluents. This industry requires large quantities of deionized water. Recycling of process water is not encouraged since it has to be very pure—hence, the amount of waste water generated is high. Many semiconductor fabs using chemical/mechanical planarization/polishing (CMP) consume water at a rate of 4.2–12 gal/min, i.e., more than 4.25 million gal/year.

The manufacture of computer chips requires large quantities of chemicals, ultrapure water, and energy. Weight per weight, the amount of fuel and chemicals required is 630 times the weight of the chip, as compared to a 2:1 ratio needed to manufacture a car. Current use of ultrapure water (UPW) is 5–7 L/cm^2 silicon, and the goal is to reduce this level to 4–6 L/cm^2 by 2005 (which is still not achieved by all manufacturers) (Peters, 2004). UPW usage in a wet bench is 53 L/wafer (for a 300-mm wafer), which should be reduced to 43 L/wafer by 2005 (Peters, 2004). The target for chemical usage is to reduce the quantity (in L/cm^2/mask layer) by 5% per year via more efficient use, recycle, and reuse. Reuse of waste water (for cooling towers) should be increased from the current average levels of 65% to 70% in 2005 (Peters, 2004), 80% in 2010, and 90% in 2013. The current energy usage for all fab tools is 0.5–0.7 kWhr/cm^2, which should be brought down to 0.4–0.5 kWhr/cm^2 in 2005 (Peters, 2004) and to 0.3–0.4 kWhr/cm^2 in 2008. As the World Semiconductor Council agreed, by 2010, PFC emissions must be reduced by 10% from the 1995 baseline. Through process optimization, alternative chemistries, recycling, and abatement, the industry must continue to diminish the emissions of byproducts that have a high global warming potential. The estimated cost to the U.K. economy, for example, could be as much as US$761 million a year to comply with the Waste Electrical and Electronic Equipment Directive. A further US$334 million annually may be needed by the industry to meet the legislation restricting use of certain hazardous substances. Use of sc-CO$_2$ for cleaning as a replacement for water is being investigated to reduce water usage (Phelps et al.). Sulfur trioxide is being tried as a cleaning agent instead of wet chemicals for removing residual photoresist and organic polymers. This attempt could reduce the handling of large quantities of hazardous chemicals. Since sc-CO$_2$ has low viscosity, it is more efficient than water as a cleaning fluid as it can enter through small openings.

Recovery and reuse of water, acids, and other chemicals could solve many of this industry's waste disposal problems, but the need for high-purity water and chemicals makes the industry hesitant to reuse the recovered chemicals. Biofilters and biotrickling filters appear to be good technologies for treating vapors and gaseous effluents from the semiconductor plant. Coagulation followed by settling and filtration of the liquid effluent is effective and cheap in removing the hazardous material from the effluent, but the disposal of the toxic sludge generated is a serious, as-yet unsolved problem. Bioremediation for liquid effluent appears to be very limited except for biosorption for extracting metals. Phytoremediation also appears to be a good technique available for treating contaminated soil and solid wastes.

The Dye Industry

Approximately 10,000 different dyes and pigments are manufactured worldwide, with a total annual market of more than 7×10^5 tons/year. There are several structural varieties of dyes, such as acidic, reactive, basic, disperse, azo, diazo, anthraquinone-based, and metal-complex dyes. All absorb light in the visible region.

Two percent of dyes that are produced are discharged directly in the effluent. In addition, 10% is lost during the textile coloration process. Studies have shown that many of the dyes are carcinogenic, mutagenic, and highly harmful to the environment. Untreated dye effluent is highly colored and hence reduces sunlight penetration, preventing photosynthesis. They are toxic to fish and mammalian life, inhibit growth of microorganisms, and affect flora and fauna. They are carcinogenic in nature and hence can cause intestinal cancer and cerebral abnormalities in fetuses.

Several chemical and physical methods are available for the removal of color from the textile dye effluent. But the disposal of the sludge or precipitate is another problem that has not been solved. Addition of adsorbent provides several advantages, including adsorption of toxic compounds—and hence reduction of toxic effects on the microorganisms—and better sludge-settling characteristics.

The Textile Industry

The textile manufacturing process consumes a considerable amount of water. Principal pollutants in the textile effluent are recalcitrant organics, colors, toxicants and inhibitory compounds, surfactants, soaps, detergents, chlorinated compounds, and salts. Dye is the most difficult constituent of the textile waste water to treat. The type of dye in the effluent could vary daily or even hourly, depending upon the campaign. It is generally difficult to degrade waste water from the textile industry by conventional biological treatment processes, because the BOD/COD ratio is less than 0.3. In 2005, the worldwide annual production of wool was about 1.22 million tons (Japan Commodities Funds Association, 2005). The scouring (contaminant extraction) or washing of this wool removes an approximately equal weight of wool wax and uses around $10 m^3$ water/ton raw wool. The percentage COD and the volume contribution due to various treatment operations are shown in Figs. 9.35 and 9.36, respectively. The desizing and scouring are the two worst operations that contribute to the generation of toxic effluents.

A combination of biochemical and chemical/physical processes appears to be promising in degrading such an effluent. The presence of dyes in the effluent poses the biggest problem, since they are recalcitrant and toxic. Both aerobic and anaerobic pro-

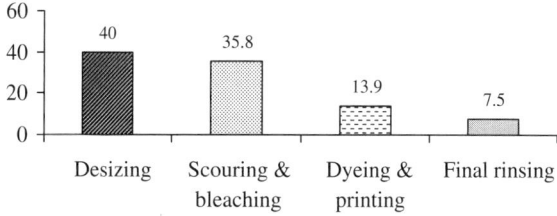

FIGURE 9.35. Percentage of COD contribution to the final effluent from various operations.

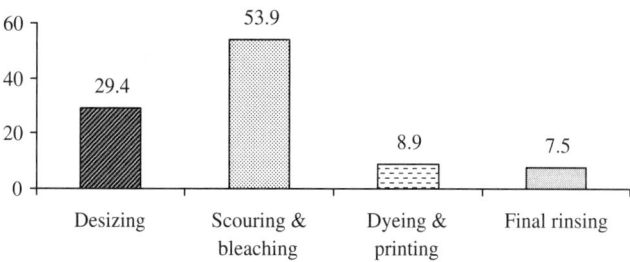

FIGURE 9.36. Percentage of volume contribution to the final effluent from various operations.

cesses have been successfully used for degrading the dyes, but the best appears to be a combination of both. Adsorption of dyes by dead cells appears to be a better alternative than treatment with live cells.

Novozymes developed an enzymatic process (BioPreparation) for treating cotton textiles that meet the performance characteristics of the original alkaline scour systems while conserving chemicals, water, energy, and time. Field trials indicated that textile mills could save 30–50% in water costs by replacing caustic scours or by combining separate scouring and dyeing steps into one. Typically, 162 knitting mills use 89 million m^3 water/year in processing the goods from scouring to finishing. The new technology would conserve 27–45 million m^3 water/year. In addition, the biological and chemical oxygen demands decrease by 25% and 40%, respectively, when compared to conventional sodium hydroxide process. The conversion would provide costs savings of 30% or more. Using biomass to manufacture chemicals has numerous advantages, including low or no net CO_2 emissions and conservation of nonrenewable petroleum resources.

The Tannery Industry

The global leather industry produced about 18 billion ft^2 leather in 2003, with an estimated value of about US$40 billion (World Leather Magazine). Developing countries now produce over 60% of the world's leather needs. Conversion of raw hide into leather requires several mechanical and chemical operations involving use of many chemicals in an aqueous medium, including acids, alkalis, chromium salts, tannins, solvents, auxiliaries, surfactants, acids, metallorganic dyes, natural or synthetic tanning agents, sulfonated oils, and salts. The quantity of effluent generated is about 30 L/kg hide or skin processed. The total quantity of effluent discharged by Indian tanneries is about 50,000 m^3/day and contains high concentrations of organic pollutants (Tare et al., 2003).

Chromium is an important heavy metal used in leather, electroplating, and metallurgical industries. More than 170,000 tons of Cr wastes are discharged into the environment worldwide each year. In India in 1999 700,000 tons of wet salted hides and skins were processed in about 3000 tanneries, which discharged a total of 30 billion liters of waste water with a concentration of suspended solids between 3000–5000 mg/L and chromium as Cr^6 between 100–200 mg/L (Rajamani).

Reduction and recycle of the various streams at source appear to be a good approach to dramatically decrease the present quality and quantity of effluent generated by this industry. Disposal of the sludge after biochemical treatment also has not been satisfactorily solved. Use of enzymes and microorganisms for dehairing and stabilizing could reduce the use of toxic metals and chemicals. However, the leather industry needs major breakthroughs to achieve nontoxic discharge.

The Sugar and Distillery Industries

Sugar cane is one of the most common raw materials used in sugar and ethanol production. More than 30 billion L of spent wash are generated annually by 254 cane molasses–based distilleries in India alone (0.2 to 1.8 m^3 waste water/ton sugar produced). The effluent has a pH between 4 to 7, a COD between 1800–3200 mg/L, a BOD between 720–1500 mg/L, total solids of 3500 mg/L, total nitrogen of 1700 mg/L, and total phosphorus of 100 mg/L. Several countries, including Thailand, Malaysia, Taiwan, and Brazil, also produce sugar from sugar cane. The waste water contains not only a high concentration of organic matter (BOD) but also a large amount of dark brown pigments called melanoidin.

In India, primary spent wash is generally utilized in an anaerobic digestion step to utilize its high COD load for methane production. The secondary spent wash produced by the anaerobically digested primary molasses spent wash (DMSW) effluent is darker in color, needing huge volumes of water to dilute it, and is currently used in irrigation, causing gradual soil darkening. Its disposal into natural water bodies also results in their eutrophication. The dark color leads to a reduction in penetration of sunlight in rivers, lakes, and lagoons, which in turn decreases both photosynthetic activity and reduction in dissolved oxygen concentrations, causing harm to aquatic life. Disposal on land is also hazardous since it reduces the alkalinity of the soil and manganese availability, thereby inhibiting seed germination and ruining vegetation. The decolorization of molasses spent wash by physical or chemical methods and subsequently being directly applied as a fertilizer have also been attempted and were found to be unsuitable.

The Paper and Pulp Industry

The pulp and paper-making industry is very water-intensive ($\sim$60 m^3 water used/ton paper produced) and, in terms of freshwater withdrawal, ranks third after the primary metals and the chemical industries. The major raw material used by the pulp and paper industry is wood, which is composed of cellulose fibers. The wood is broken down to separate the cellulose from the noncellulose material and then dissolved chemically to form a pulp. The pulp slurry is then vacuum-dried on a machine to produce a paper sheet. Dyes, coating materials, and preservatives are also added at some point of the process. Lignin is a complex aromatic polymer that is an integral cell wall constituent giving strength and rigidity to the tissues and resistance to microbial attack of vascular plants. The presence of residual lignin affects some properties of the manufactured pulp and paper products. Therefore, lignin is selectively removed during pulping without significantly degrading the cellulose fibers. The paper industry is composed of several sectors such as packaging board, newsprint, boxes, printing and writing, and tissue. In 1996, the worldwide production of paper and board was $\sim$320 million tons (Thompson et al., 2001). North America produces more than half, Western Europe about 20% (60% of this amount is produced by Sweden, Norway, and Finland), and Japan about 12% (Lacour, 2005). The consumption of water depends on the type of paper being produced. Manufacture of tissue, of printing and writing, of newsprint, and of packaging material require ~60, 35, 30, and 18 m^3 water/ton, respectively.

In the Kraft pulping process, sodium hydroxide and sodium sulfide are added to dissolve the nonfibrous material. The effluent generated (black liquor) in such a process is very alkaline and contains high dissolved solids, alkali, lignin, and polysaccharide degradation byproducts. The chemical oxygen demand (COD) of the effluent is as high as 11,000 mg/L. Dissolved small organic molecules in the effluent give a high BOD, while more complex lignin molecules do not cause BOD but create a high COD and dark color.

The paper and pulp process uses plenty of water, and the waste generated from this industry contains solvents, chlorinated compounds, resins, and most importantly lignin, which is highly resistant to degradation. Chlorinated compounds are also toxic to many of the microorganisms. Physicochemical treatment methods are expensive. Conventional biological methods such as activated sludge and aerated lagoons help in reducing the COD load and toxicity, but these methods cannot effectively remove color of bleach plant effluents. White rot fungus appears to be efficient in color removal. Anaerobic degradation is affected by the presence of sulfate. Destruction of absorbable organic halogen is another aspect that needs to be addressed. Carrying out biodegradation at thermophilic conditions would be advantageous since the waste stream from the paper mill is generally at around 50°C, but finding efficient microorganisms that can perform well and overcome other problems, such as those mentioned above, is still a research challenge. Use of enzymes and other innocuous chemicals instead of chlorine to bleach paper and paper products is a necessary research target to make this industry more acceptable.

Buckman Laboratories International (USA) developed a liquid esterase formulation (trade name Optimyze) using an enzyme produced by Novozymes to improve processing of recycled paper. When this formulation is added during the paper recycling process, it degrades and removes the sticky chemicals such as adhesives, plastics, and inks, which foul the processing equipment and lead to poor products. The enzyme formulation is inherently safer than the organic solvents and surfactants currently used in most paper recycling plants.

The Pharmaceutical Industry

Excess medication excreted by humans and animals and disposed unused or expired medicine both find their way into the municipal sewage effluent treatment plants. Since the 1980s, pharmaceuticals like clofibrate, various analgesics, cytostatic drugs, antibiotics, and others have been reported to be present in the surface water of many European countries (see Fig. 9.37). This has raised growing concerns since some of these persistent products find their way back into the drinking water. Genotoxic substances may represent a health hazard to humans as well as affect organisms in the environment. Since antibiotics mainly interfere with the bacterial metabolism, it can be assumed that bacterial communities are the primarily affected part of aquatic ecosystems due to discharge of effluents containing antibiotics. Also, resistance to certain antibiotics gives rise to infections that are difficult to treat. Antibiotics are consumed by humans and used in livestock and poultry production and in fish farming. The increasing use of these

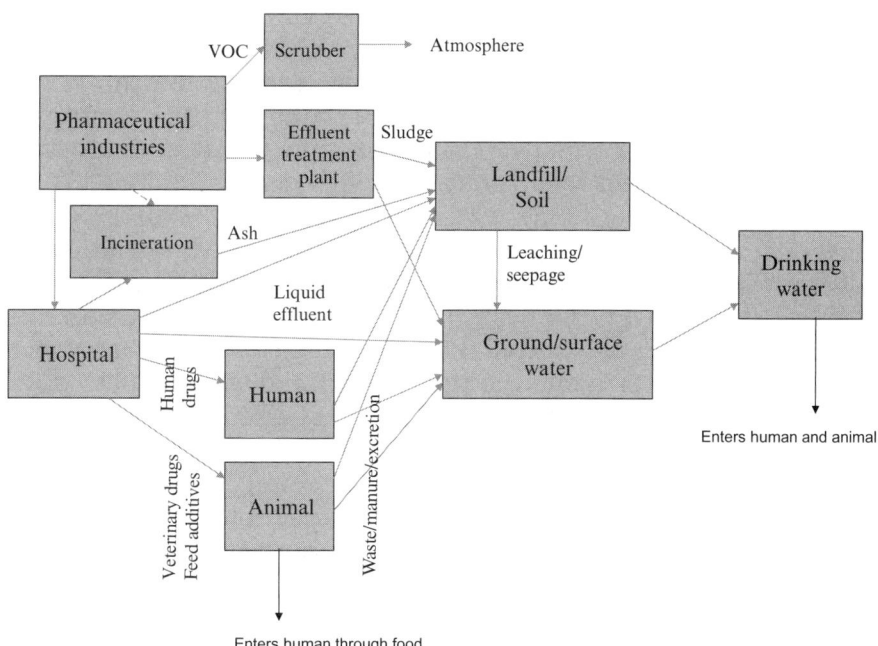

FIGURE 9.37. Life cycle of pharmaceutical drugs.

drugs during the last five decades has caused genetic selection of more harmful bacteria (reported veterinary drug usage in the EU was 1600 tons in 1999). When applied to agricultural fields as fertilizer or manure, animal excreta (which contain unmetabolized drugs) contaminate soil and also the groundwater, depending on their mobility, and affect the terrestrial and aquatic organisms as a result of leaching from the fields. Solid waste from industrial effluent treatment plants are disposed as landfill, which may lead to leaching of unmetabolized drugs into the groundwater. Excess drug usage during fish farming leads to 70% of the drugs administered to fish being released into the environment, specifically into the sediments near the fish farms. The modes of action of most pharmaceuticals in humans, animals, and fish are often poorly understood. The possible effects/side effects on nontarget receptor organisms and synergistic effects that might be produced due to the mixing of drugs are also not known. The growth promoters, antibiotics, and other veterinary drugs given to poultry and cattle also end up in humans as they consume their meat. Natural and synthetic estrogens produce deleterious effects in aquatic organisms, such as feminization and hermaphroditism (Fossi and Marsili, 2003; Gross-Sorokin et al., 2004). The persistence of a drug in a sediment or soil depends on its photostability, its binding and adsorption capability, its degradation rate, and its solubility in water. Strongly absorbing pharmaceuticals tend to accumulate in soil or sediment, and, in contrast, highly mobile pharmaceuticals tend to leach into groundwater and be transported with groundwater, drainage water, and surface runoff water.

Tetracycline and its derivatives chlortetracycline and oxytetracycline are antibiotics widely used in stockbreeding and aquaculture. The average concentration of oxytetracycline in German surface waters has been estimated at $0.01\,\mu g/L$. In Germany, $0.165\,mg/L$ of clofibric, a lipid-regulating agent, was found in river water, groundwater, and drinking water, and 80% of 32 drugs belonging to the classes of antiphlogistics (Ternes, 1998), lipid regulators, psychiatric drugs, antiepileptic drugs, beta blockers, and sympathomimetics were found in the sewage treatment plant effluent, with concentration levels of the order of $6\,\mu g/L$ (National Ground Water Association, 2001). The sewage treatment plant was treating household effluent and consisted of preliminary and final clarification and an aerator tank. The concentrations of tetracycline and its derivative oxytetracycline in the Lee River near London have been estimated at $9.5\,\mu g/L$, and tetracycline in British surface waters have been estimated at about $1\,\mu g/L$ (Waggot, 1981). The concentrations found in sediments near fish farms range from

0.1 to 10 mg sediment/kg. In the United Kingdom, drugs like diazepam, methaqualone, and penicilloyl antibiotics were found in both potable water and groundwater. A nationwide study carried out by the U.S. Geological Society in 2002 found pharmaceuticals, hormones, and other organic wastewater contaminants in surface water (Koplin et al., 2002). Apart from ceftriaxone and tilmicosin, several drugs, animal growth promoters, and antibiotics were found at the nanogram level in river sediment, river water, and drinking water in Italy. The concentrations found were several orders of magnitudes lower than the amount to produce any pharmacological effect. But possible effects of lifelong exposures of these pharmaceutics on humans are not known (Zuccato et al., 2000).

Conclusions

Industries have resorted to heterogeneous catalytic, biocatalytic, and fermentation routes to improve process efficiency, reduce energy usage, and decrease raw material usage (see Fig. 9.13) (Sheldon, 1987; Anastas et al., 1998). A one-step fermentation process is more attractive than a combination of biological and chemical routes. Developing atom-economical processes to help in sustainable development is a challenge to chemists and technologists (Rouhi, 1995). Currently, improvements in enzyme technology have become the focus of many pharmaceutical, food, and specialty industries. Altus Biologics has developed cross-linked enzyme crystals (CLECs) for use in organic reactions. Unlike free enzymes, CLECs can withstand extremes of temperature and pH and can be used in both aqueous media and organic solvents. The CLEC-mediated synthesis of the antibiotic cephalexin eliminates the need for N-protection of d-phenylglycine (see Fig. 9.38).

Replacement of the chemical step with a biocatalytic step is the first move toward green chemistry, which is followed by replacing the entire multiple-step process with a single fermentation process. Also, scientists have realized that biotechnology routes are inherently safe and mild and do not adversely impact the environment. Heterogeneous catalysis also appears to offer several advantages since the reactions are clean and selective. Also, the spent catalyst can be filtered and recovered for later use.

The semiconductor, tannery, dyes, paper and pulp, sugar and distillery, and textile industries are reeling from large amounts of toxic effluents. Pressures from government and NGOs are pushing these industries to the corner. They need to embrace the concepts

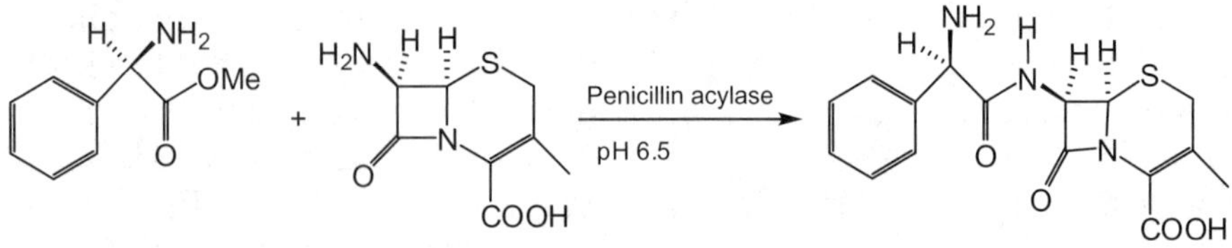

FIGURE 9.38. Use of CLEC for the synthesis of the antibiotic cephalexin.

of green chemistry and develop innovative processes that would eliminate use of toxic chemicals, use of large amounts of pure water, and generation of large quantities of VOC. While fine chemicals and pharmaceutical manufacturers have moved ahead with innovative alternatives, these just-mentioned industries are still lagging behind by using traditional, inefficient processes. Of course, the leather industry is developing processes that use lipase instead of harsh chemicals for dehairing, while the semiconductor industry is using SO_3 and sc-CO_2 for cleaning printed circuit boards.

References

Anastas, P. T. and Warner, J. C., *Green Chemistry: Theory and Practice*, Oxford University Press, New York, 1998.

Anastas, P. T., Heine, L. G., and Williamson, T. C., eds., *Green Chemical Syntheses and Processes*, American Chemical Society, Washington, DC, 2000, p. 1.

Anonymous, Promoting sustainability through green chemistry, *Chem. Week*, **165**(24): 14, 2003.

Argonne National Laboratory, 2004, http://www.techtransfer.anl.gov/techtour/ethyl-faqs.html.

Binns, F., Harffey, P., Roberts, S. M., and Taylor, A., Studies leading to the large-scale synthesis of polyesters using enzymes, *J. Chem. Soc., Perkin Trans.*, 1: 2671–2676, 1999.

Brundtland G., ed., *Our Common Future: The World Commission on Environment and Development*, Oxford University Press, Oxford, 1987.

Cann, M. C. and Connelly, M. E., The BHC Company, Synthesis of Ibuprofen; Real World Cases in Green Chemistry, ACS, Washington DC, 2000.

Chepesiuk, R., Where the chips fall: Environmental health in the semiconductor industry, *Environ. Health Perspect.*, **107**(9): A452–457, 1999.

Dow AgroSciences, LLC, Sentricon termite colony elimination system, a new paradigm for termite control, Presidential Green Chemistry Challenge Award, EPA744-K-02-002, U.S. Environmental Protection Agency, Washington, DC, pp. 30–31, 2002.

Dunn, P. J., Galvin, S., and Hettenbach, K., The development of an environmentally benign synthesis of sildenafil citrate (Viagra™) and its assessment by green chemistry metrics, *Green Chem.*, **6**: 43–48, 2004.

Fossi, M. C. and Marsili, L., Effects of endocrine disruptors in aquatic mammals, *Pure Appl. Chem.*, **75**(11–12): 2235–2247, 2003.

Fukuoka, S., Kawamura, M., Komiya, K., Tojo, M., Hachiya, H., Hasegawa, K., Aminaka, M., Okamoto, H., Fukawa, I., and Konno, S., A novel nonphosgene polycarbonate production process using byproduct CO_2 as starting material, *Green Chem.*, **5**: 497–507, 2003.

Gross-Sorokin M. Y., Roast S. D., and Bright G. C., Causes and consequences of feminization of male fish in English rivers; Science Report SC030285/SR, Environment Agency, July 2004.

Industrial Biotechnology and Sustainable Chemistry; Royal Belgian Academy Council of Applied Science, 2004.

Japan Commodities Funds Association, world wide chemical fibre production in 2005, www.jcfa.gr.jp/english.

Kirchhoff, M. M., Promoting sustainability through green chemistry, *Resources, Conservation and Recycling*, **44**(3), 237–243, 2005.

Koplin, D. W., Furlong, E. T., Meyer, M. T., Thruman, E. M., Zargy, S. D., Barber, L. B., and Buxton, H. T., Pharmaceuticals, hormones & other organic wastewater contaminants in US streams 1999–2000: A national reconnaissance, *Environ. Sci. Technol.*, **36**, 1202–1211, 2002.

Lacour, P. A., AFOCEL, Pulp & paper markets in Europe—2003 Study, UNECE/FAO, Sep. 2005.

Licence, P., Ke, J., Sokolova, M., Ross, S. K., and Poliakoff, M., Chemical reactions in supercritical carbon dioxide: From laboratory to commercial plant, *Green Chem.*, **5**(2): 99–104, 2003.

National Ground Water Association, Proceedings of the 2nd International Conference on Pharmaceuticals & Endocrine Disrupting Chemicals in Water, 2001.

Noyori, R., Sato, K., and Aoki, M. A., Green route to adipic acid: Direct oxidation of cyclohexenes with 30 percent hydrogen peroxide, *Science*, **281**: 1646–1647, 1998.

Perspective on leather—its place in the world, World Leather Magazine, http://www.tannerscouncilict.org/perspective.htm

Peters, L., The road map for ESH, *Semiconductor International*, **4**(1), 2004.

Phelps, M. R., Hogan, M. O., Snowden-Swen, L. J., Barton, J. C., Laintz, K. E., and Williams, S. B., Waste reduction using carbon dioxide: A solvent substitute for precision cleaning applications, 316, Precision Cleaning 195 Proceedings.

Rajamani, S., Generation of hazardous sluge from tannery effluent treatment plants & disposal problems in India, Central Leather Research Institute, India.

Ritter, S. K., Green challenge, *Chem. Eng. News*, **80**(26): 26–30, 2002.

Ritter, S. K., Green rewards, *Chem. Eng. News*, **81**(26): 30–35, 2003.

Rouhi, A. M., Atom-economical reactions help chemists eliminate waste, *Chem. Eng. News*, June 19 issue, pp. 32–35, 1995.

Royal Belgian Academy Council of Applied Science, Industrial Biotechnology and Sustainable Chemistry, 2004.

Sarbu, T., Styranec, T. J., and Beckman, E. J., Design and synthesis of low-cost, sustainable CO_2-philes, *Ind. Eng. Chem. Res.*, **39**: 4678–4683, 2000.

Scorr, F. L., Frazer, L., SCORR one for the environment, *Environ. Health Perspect.*, **109**(8): A382–A385, 2001.

Sheldon, R. A., Catalysis and pollution prevention, *Chem. Ind.*, Jan. 6 issue, pp. 12–15, 2000.

Sheldon, R. A., Consider the environmental quotient, *Chem. Tech.*, **24**(3): 38–47, 1994.

Sheldon, R. A., Organic synthesis—past, present and future, *Chem. Ind.*, pp. 903–906, 1992.

Tare, V., Gupta, S., and Bose, P., Case studies on biological treatment of tannery effluents in India, *J. of Air & Waste Management Assoc.*, **53**: 976–982, 2003.

Ternes, T. A., Occurrence of drugs in German sewage treatment plants and rivers, *Water Research*, **32**(11): 3245–3260, 1998.

Thompson, S., Swain, J., Kay, M., and Foster, C. F., Review Paper: The treatment of pulp and paper mill effluent. *Bioresource Tech.*, **77**: 275–286, 2001.

U.S. Bureau of Engraving, ISOMET: Development of an alternative solvent, The Presidential Green Chemistry Challenge Awards Program, summary of 1999 award entries and recipients, EPA744-R-00-001, 1999.

U.S. EPA, Toxics Release Inventory (TRI), 2001 public data release, executive summary, EPA 260-S-03-001, 2001.

Waggot, A., Trace organic substances in river Lee, In *Chemistry in Water Reuse*, Vol. 2, Ann Arbor Science Publications, Ann Arbor Mich., pp. 55–59, 1981.

Westheimer, T., The award goes to BHC, *Chem. Eng.*, **12**: 84–95, 1993.

Zuccato, E., Calamari, D., Notangelo, M., and Fanelli, R., Presence of therapeutic drugs in the environment. *Lancet*, **355**, 2000.

CHAPTER 10

Conclusions and Future Trends

Chemists, engineers, and biochemists who strive to achieve green and sustainable development face several challenges and focus areas. The magic word appears to be "reduce." Reductions of waste, raw material, energy, solvent, cost, risk, and product usage contribute to sustainable development (see Fig. 10.1).

Energy

Petroleum-based substances are in everything from lipstick to laundry detergents, clothes to computers to chocolate bars—even fertilizers and pharmaceuticals. In the 1950s and 1960s, humans accepted that petroleum would remain forever. But now the picture is very clear, and we know that we have to look for renewable and less polluting energy sources for our sustenance and growth. Peak petroleum production will happen in about 10 to 15 years, and the production will decrease unless new sources of petroleum are discovered. Designers of new energy sources should keep the following issues in mind:

1. How much pollution do the new energy sources create?
2. Are they renewable sources?
3. What is the cost of the energy produced?

All these points need to be addressed to attain sustainable energy. The European Union has set an objective to obtain 14% of its

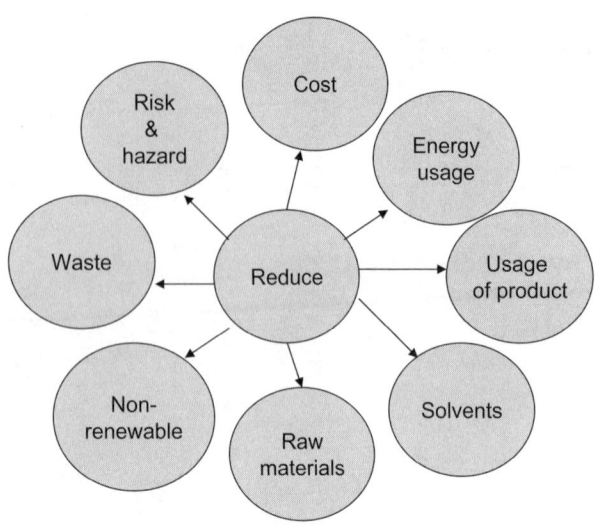

FIGURE 10.1. Reduction leads to sustainable development.

energy supply from sustainable sources by the year 2010. The need to design substances and materials that are (1) effective, (2) efficient, and (3) inexpensive to capture, store, and transport to different locations is a major challenge. Although there is plenty of talk about hydrogen being an effective renewable energy, the challenges of its storage and transport have yet to be solved. Compressing hydrogen to 10,000 psi requires a multistage compressor system that will lose 15% of the energy contained in the hydrogen. The more hydrogen is compressed, the smaller the tank. But as the pressure is increased, the thickness of the vessel's steel wall has to be increased—and hence the weight of the tank increases. Thus, the cost of storage and the weight of the storage unit increase as pressure increases. The polluting energy sources are ones that green chemistry must address in the future.

It has been predicted that by 2040 the yearly global worldwide energy consumption will have increased to 935 EJ for the estimated 10 billion people on earth (i.e., 3000 W energy/person). In the year 2002, 79% of the global energy consumption was met by fossil fuels, while the remaining 12% was met by renewable conventional sources such as wind and solar energy (Royal Belgian Academy Council of Applied Science, 2004). The global winds' technical potential is believed to be 5 times the current global energy consumption, or 40 times the current electricity demand. This requires 12.7% of *all* land area—at a height of 80 m. As of

2003, fossil fuels consisted of crude oil (35%), natural gas (23%), and coal (21%) (Royal Belgian Academy Council of Applied Science, 2004). In 2002, 9% of energy worldwide was provided by hydraulic power and nuclear energy, and 12% was met by renewable sources such as conventional firewood, wind, and solar energy. Natural gas is expected to be available at a constant amount for the next 50 years, after which this source is expected to dry up. At present, nuclear energy meets only a small percentage of our energy needs (2%) and will completely disappear by 2050 (Royal Belgian Academy Council of Applied Science, 2004). As of 2000, solar energy and biomass began to gain importance; by 2100, these two sources will meet more than half of our energy needs. Solar energy radiation is the fundamental source of all energy on the earth (except for nuclear energy), and the amount of this energy is 2.8×10^{24} J/year, which is 3000 times the predicted annual energy consumption in 2040 (Royal Belgian Academy Council of Applied Science, 2004). Hence, it appears to be the cheapest, cleanest, and most abundantly available source.

Process Intensification

As we've discussed in earlier chapters, process intensification has several benefits, including lower waste, higher yields, less energy requirements, less floor space, and inherently safer and smaller inventory.

Challenges to process intensification are many: chief among them is its acceptability. Design equations-correlations and mathematical models are available for conventional separation equipments and multiphase reactors, whereas such general-purpose equations and correlations are not yet available for high-G reactors, plate-heat exchanger-reactors, microchip reactors, or spinning disc reactors. Newer plants could be built using process intensification methodologies, but the current running plant cannot be scraped and modified to miniature scale since that will involve major investments by the industries. In addition, process intensification has limitations when handling slurries and solid reactants.

Product Replacement

The new nontoxic, safe, and biodegradable products designed to replace existing products have to be cost-competitive, otherwise the customers may not be willing to spend the extra money. Biodegradable polymers such as polylactic acid or polyhydroxyl alkonates are more expensive than the synthetic polymers and hence have not penetrated the market as expected. Also, their performance to date has not matched that of traditional materials. Few success stories in this area of polymers have been reported. For example, NatureWorks has been able to lower the price of its biorenewable packaging plastic from more than $1.00/lb to about $0.63/lb, yet the price of the equivalent petroleum-based plastic packaging currently costs as little as $0.40/lb.

Oxidation Reagents and Catalysts

The chemical industry currently relies predominantly on petroleum-based feedstock, namely the hydrocarbons such as alkanes, alkenes, alkynes, and benzenes, which are in their fully reduced state. So oxidation of these compounds is necessary for further functionality and for creating chemical building blocks. Traditional oxidation reagents and catalysts that have been used are toxic substances such as heavy metals (e.g., chromium), $KMnO_4$, $K_2Cr_2O_7$, and so forth. These substances, which are generally used in stoichiometric amounts, in turn release metals as well as oxides of Mn and Cr, for example, into the environment, all of which have substantial negative effects on human health and on the environment. On the contrary, if biomass were used as the starting material, it would not require too much oxidation since it already contains oxygen.

New oxidation chemistry needs to be developed based on the philosophy of green chemistry so that it will be environmentally benign, efficient, selective, and economical. The new oxidation chemistry will have to be catalytic processes rather than the current organic reagents of stoichiometric quantities. The processes have to be robust, with high turnover rates. The catalyst should be recoverable, reusable, and innocuous. Oxidations need to be carried out with molecular oxygen or hydrogen peroxide.

Design of Multifunctional Reagents

Synthetic catalysts and reagents generally carry out one discrete transformation such as reduction, oxidation, and methylation at a time. Biochemical systems, on the contrary, often carry out several manipulations with the same reagent. These manipulations may include activation, conformational adjustments, and one or several actual transformations and derivitizations. The *biomimetic* approach to the design of catalysts involves implementing some special features found in proteins and enzymes. A single catalyst that can perform the three tasks of reduction, isomerization, and methylation could be very useful rather than having three independent catalysts to carry out these three operations. Three catalysts need three different reactors for carrying out the reactions.

Combinatorial Chemistry and Green Chemistry

Combinatorial chemistry is based on the principle of making a large number of chemical compounds rapidly on a small scale in small reaction cells. This practice is widely adopted by the pharmaceutical sector for use during the drug designing and screening stages. If a "lead or promising compound" is identified by the drug design group, then a large number of derivatives of this lead are rapidly tested for their efficacy using the combinatorial approach. This philosophy has enabled a large number of substances to be made and their properties assessed without generating a sizable amount of waste and its later disposal. Green chemistry would benefit from the principles of combinatorial chemistry, since the latter approach produces practically very little waste.

Solventless Reactions

Generally, synthetic reactions are carried out in a solvent medium. This creates several problems such as VOCs, solvent loss in the effluent, need to recover and recycle solvents, and contamination of solvent in the product. Multiple reaction steps would require different solvents. The philosophy of green chemistry believes in solventless or neat reaction systems. Synthetic organic chemists have to move into this thinking mode. The ultimate dream is to conduct reactions in molten-state, dry-grind, plasma, or neat solid-supported reactions such as clay and zeolites. These techniques need utilization of nontraditional conditions such as microwave, ultrasound, and visible light as the activating agents. New separation and purification techniques that can be operated without the use of solvents also need to be developed. So this creates new innovative challenges to the chemists and chemical engineers.

Noncovalent Derivatization

Chemistry is based on bond making and bond breaking, which are efficient but nonselective processes, leading to side and waste products. Through the utilization of dynamic complexation, or through nonbonded interaction, temporary modified chemical structures should be formed that would modify the properties of a molecule for the necessary period of time to carry out a particular function without all of the waste that would be generated if full derivatization is carried out.

Biotechnology: The Solution to All Problems

A McKinsey report indicated that the market share of industrial biotechnology will strongly increase in all areas by 2010, but particularly in fine chemicals production (Bachmann, 2002). The estimated degree of penetration by biotechnology in 2010 is estimated to be between 30–60% for fine chemicals and between 6–12% for polymers and bulk chemicals. Considering the entire chemical industry, the penetration of biotechnology is presently estimated at 5%, but this is expected to increase to between 10–20% by 2010 and strongly increase even further after that year (Anastasio and Viglia, 2006). The rate of penetration will depend mainly on a number of factors such as the prices of crude oil and agricultural raw materials, technological developments, and the political will to support and structure this new technology.

The Kyoto Protocol obliges all ratifying countries to reduce greenhouse gas emissions between 2008 and 2012 by 8% relative to the 1990 reference level [CSI 011 Specification—*Projections of green house gas emission and removals*, EEA-Indicator Management Service. An Introduction to the Kyoto Protocal Compliance Mechanism, UNFCC (http://unfccc.int/kyoto_protocol/compliance/introduction); Kyoto protocol, from Wikipedia (http://en.wikipedia.org/wiki/kyoto_protocol)]. On the contrary, emissions have actually increased to 7% above the 1990 level, which means the challenge is to reduce greenhouse gas emissions by about 14% in order to meet the Kyoto target. Here biotechnology-based energy sources would help in achieving the target.

Interdisciplinary Approach

The clean technology pool (see Fig. 10.2) is made up several technology principles such as renewable energy, renewable raw materials, life-cycle assessment, principles of reaction cum separation, nonvolatile solvents, heterogeneous catalyst, biotechnology approach, supercritical fluids, air as oxidant, solventless reactions, and so forth. These technologies in turn are based on a good understanding of the underlying science, such as the fields of synthetic

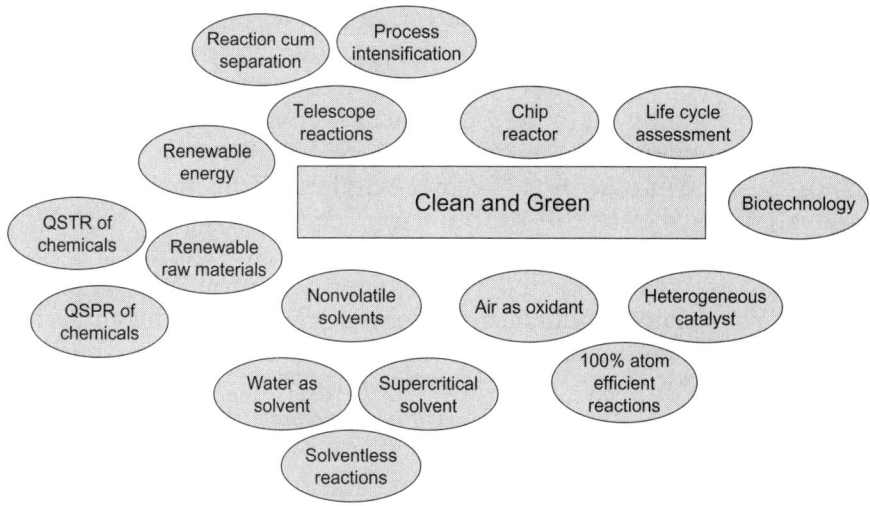

FIGURE 10.2. Clean technology pool.

organic chemistry, physical chemistry, catalysis, and chemical engineering. Hence, clean technology requires an interdisciplinary approach, with engineers and scientists from various disciplines joining together to create a clean environment, one that does not deplete the currently existing resources and does not destroy the flora and fauna through dumping of toxic chemicals. The green process design appears to encompass the green chemistry paradigm and several other engineering principles.

Biomass: The Future Raw Material

As of 1999, the total annual biomass production is estimated to be 170 billion tons, which is divided into 75% carbohydrates (sugars), 20% lignins, and 5% other substances such as oils and fats, proteins, terpenes, and alkaloids (Okkerse and Van Bekkum, 1999; Dale, 2003). Of this biomass produced, 6 billion tons (3.5%) is presently used for human needs. The breakup of this amount is 3.7 billion tons for human food use; 2 billion tons of wood for energy use, paper, and construction needs; and 300 million tons to meet the human needs for nonfood raw materials such as clothing, detergents, and chemicals. The remainder of biomass produced is used in the natural ecosystems, namely as food for the wild animals, and the rest is lost as a result of the natural mineralization processes. The renewable raw materials mentioned here are almost all from agriculture and forestry.

The estimated world production of renewable raw material in 2002 was cellulose, 320 millions tons/person/annum (Mtpa); sugar, 140; starch, 55; glucose, 30; bioethanol, 26; and glutamic acid, 1. The corresponding main petrochemical components are ethylene, 85 Mtpa; propylene, 45; benzene, 23; terephthalic acid, 12; isopropanol, 2; and caprolactam, 3. While ethylene, benzene, and propylene all cost about 350–400 Euro/ton, cellulose costs 500 E/ton while sugars and starch are cheaper, costing 180–250 E/ton. Products that are manufactured through fermentation are generally cheaper than those manufactured through chemical or biocatalytic routes. Also, large quantities of material could be processed in a fermenter. Table 10.1 gives current world production of various drugs and chemicals through fermentation and their approximate prices.

Currently, processing, separating, and refining of crude oil are carried out in a refinery. Refineries that process biomass, agricultural feedstock to produce commodity, and value-added products are called biorefineries. In the future, such refineries will be built. When the earth's crude oil reserves are totally exhausted, the oil refiners may have to close, but the biorefineries will be operational. In biorefineries, feedstock will be processed, fractionated into intermediate basic products, and converted into final products. Right now, it may appear very far-fetched, but already several

TABLE 10.1
World Production of Various Drugs and Chemicals in 2002

World Production (Ton/Year)	World Market Price (€/kg)
Bioethanol	26,000,000	0.40
L-glutamic acid (MSG)	1,000,000	1.50
Citric acid	1,000,000	0.80
L-lysine	350,000	2
Lactic acid	250,000	2
Vitamin C	80,000	8
Gluconic acid	50,000	1.50
Antibiotics (bulk products)	30,000	150
Antibiotics (specialties)	5,000	1,500
Xanthan	20,000	8
L-hydroxyphenylalanine	10,000	10
Dextran	200	80
Vitamin B_{12}	10	25,000

products are being made in large quantities starting from an agricultural product such as corn.

Some of the products already being industrially produced from a single feedstock such as corn are

- glucose (a natural sugar) as a raw material for foods,
- citric acid as a food additive,
- bioethanol as a motor fuel,
- bioplastic (polylactate), used as a packaging material and as a textile fiber,
- starch carboxylate as an ingredient of washing powders,
- lysine as an animal feed additive,
- antibiotics as a pharmaceutically active substance for drugs,
- vitamins for human food and animal feed use,
- biocolorants for the food industry,
- xanthane biopolymer as a viscosity control agent in numerous applications.

Life-Cycle Assessment

The FDA approves drugs for human use after it has been thoroughly accessed with animal models and human volunteers, whereas various agencies pay very little focus or interest to access the impact of new products on the ecosystem. Data on toxicity and biodegradability are not requested by the agencies. The public at large is also not aware of the impact.

The impact of a product on the ecosystem has to be understood and studied before it is introduced in the market. Life-cycle assessment (LCA, or ISO 14040 as it's called) uses comparative analyses to measure the impact a product has on the environment and on people at each stage of its life, starting from extraction of raw materials, processing, packaging, transportation, usage, and final disposal (see Fig. 10.3). In Europe, a different system is used to translate the total environmental impact of a product during its life cycle into a single number called an ecoindicator. Sustainable design requires the standard approved by the U.S. BEES (Building for Environmental and Economic Sustainability), which lags behind Europe. Industrial designers, product designers, and process designers must combine all the negative impact through LCA while designing new products or processes.

One of the fastest-growing industries is that of personal computers, with a product technology changeover of less than 6 months and a process technology cycle of 2–4 years, while the

FIGURE 10.3. Product life cycle.

semiconductor industries work on a 1–2-year and 3–10-year cycle, respectively. At the other end of the spectrum is the slowly changing petrochemical industry (Warner et al., 2004) with a new product technology cycle of 10–20 years and 20–40 years for major process changes. The pharmaceutical industry is in between these two, with a product cycle of 7–15 years and a process cycle of 5–10 years. This means that the semiconductor and personal computer industries should develop products that are recyclable, reusable, and/or biodegradable; otherwise, at each product design change, a huge amount of obsolete products could be dumped as waste. Although the current trend is to view the petrochemical industries as the largest polluters and the computer industry as the cleanest, the truth is far from it. The strategy followed by all the sophisticated instrument manufacturers is to make the instruments obsolete in 5–10 years, so that customers switch over to the new generation of products, thereby keeping their sales turnover high. But such product obsolescence leads to the generation of large amounts of waste. These high-technology industries need to be socially responsible and design products that can be recycled or reused after upgrades. The responsibilities of the manufacturer do not end at the factory gate but extend until the safe and ecofriendly disposal of the product at the end of its life.

Remediation

It was estimated in the year 1980 that industrialized countries devoted 1–2% of GDP to the prevention and reduction of pollution (Kates, 1986). In the 1990s, it was estimated that the social cost of coping with technological hazards in the United States was 7–12% of GDP, with about half devoted to hazard management and the remainder used to take care of damage to people, material, and the environment (Warner et al., 2004). It is estimated that the United States produces over 260 million tons of hazardous waste

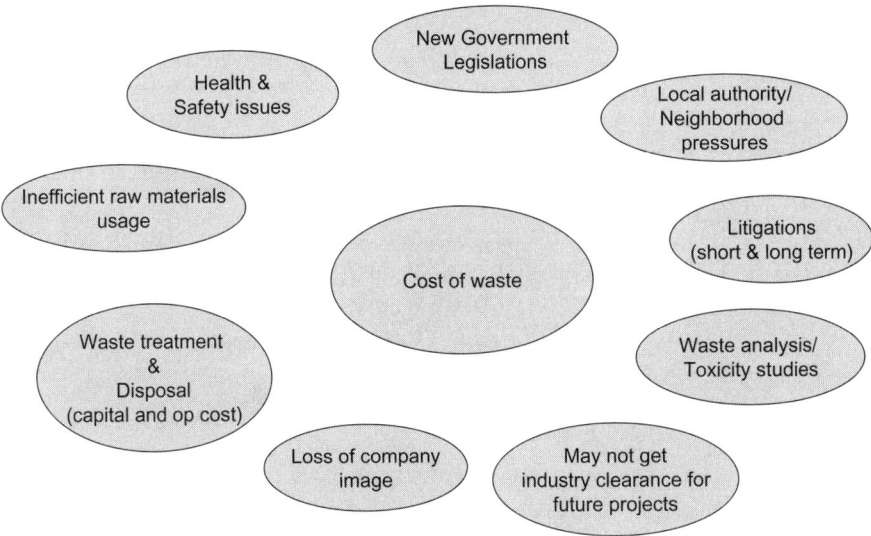

FIGURE 10.4. Various costs involved in remediation.

every year, which works out to be around 1 ton for every resident (see Fig. 10.4).

Fifty-five percent of the chemicals on the Toxic Release Inventory list have full testing data. Of the 3000 high-volume chemicals (where their production is around 1 million lb/year) on this list, 43% have no testing data or basic toxicity, and only 7% have a full set of basic test data. For 38,000 chemicals (66% listed by the EPA), fewer than 1000 have been tested for acute effects, and only about 500 have been tested for cancer-causing, reproductive, or mutagenic effects. It was estimated that it would cost only US$427 million to collect the basic data. So it is not the cost of data collection that is the bottleneck but the industries' collective lack of will that is the cause.

There is evidence that children (from USA & U.K.) who are breast-fed may be exposed to a significant number of known toxic chemicals in breast milk, including methylene chloride, styrene, perchloroethylene, toluene, trichloroethylene, 1,1,1-trichloroethane, xylene, dioxins, benzene, polychlorinated biphenyls (PCBs), chloroform, and polybrominated diphenyl ethers (PBDEs). Over the last 30 years, the total PBDE levels in breast milk have doubled, especially during the past five years. The chemical "body burden" of males has slowly increased over this time period (Lyons, 1999; Lakind et al., 2004).

Future Predictions

By 2025 to 2030 it should be possible to

- eliminate nearly 100% of emissions during polymer manufacturing and processing,
- replace all solvents and acid-based catalysts that have adverse environmental effects with solids or greener alternatives,
- achieve a 30–40% reduction in waste,
- reduce by more than 50% the quantity of plastics in landfills,
- achieve 100% raw material utilization,
- see a 50% reduction in energy consumption.

In the area of polymers synthesis and processing, it is believed that the following targets and performances will be achieved by the year 2020:

- a 40–50% improvement in energy efficiency,
- a 30–40% reduction in waste during processing,
- a 50% replacement of solvent-based processes by benign processes,
- a 50% reduction in energy usage,
- a 100% raw material utilization,
- a 50% reduction in plastics used in landfills,
- a 50% improvement in performance,
- a 75% rate of polymer recycle and reuse,
- 30% of raw material will be derived from biomass,
- an unspecified reduction in additives,
- thorough knowledge of LCA of all polymers.

A few examples of alternate chemicals to achieve a "clean environment" include

- using CO_2 instead of toxic chemicals in dry cleaning,
- taking chromium and arsenic, which are toxic, out of pressure-treated wood,
- using new less toxic chemicals for bleaching paper,
- substituting yttrium for lead in auto paint,
- using enzymes instead of a strong base for the treatment of cotton fibers,
- using sulphur trioxide for cleaning computer chips instead of water,

- using air as blowing agent in polymers instead of CFCs,
- using melt polymerization techniques instead of solvent process,
- using CO_2 as benign plasticizer for polymer melts and dispersion,
- using low-vapor and low-pressure organic liquids/ionic liquids,
- using solvent-free coatings,
- using supercritical fluids and high-pressure water instead of solvent-based processes,
- using only enzymes instead of toxic chemicals in the leather industry,
- gleaning a better understanding of polymerization mechanisms will lead to sophisticated mathematical models that can be used to predict durability, wearability, morphology, and miscibility of polymers.

Just a few of the future challenges for theoretical and physical chemists and scientists are

- What is the molecular basis of hazard and toxicology? Can QSTR and QSPR relations be developed between toxicity and structural parameters and property and structural parameters?
- Can we use weak forces as a design tool for achieving performance as we have done with covalent forces (i.e., instead of using reactions that lead to the formation and breaking of covalent bonds, can we use hydrogen bonds, van der Waals forces, or electrostatic forces to form products selectively)?
- Can we design catalysts from first theoretical principles? (In the area of catalysis, the current understanding of the mechanism comes after the synthetic process has been established.) Is it possible to design an effective process based on the understanding of the mechanism?
- Can we use energy in the place of matter to effectively carry out transformations catalytically on a commercial scale?
- Are the reaction types and feedstock we currently use in chemical manufacturing the ones we should be using in the next 10 or 20 years? Alternate feedstock, and design new reaction schemes that would be optimal for the alternate feedstock.

The focus of the chemists, technologists, and engineers in the next decade should be toward

1. clean synthesis (e.g., new routes to important chemical intermediates),

2. enhanced atom utilization (e.g., more efficient methods of bromination),
3. replacement of stoichiometric reagents (e.g., catalytic oxidation using air as the only consumable source of oxygen),
4. new solvents and reaction media (e.g., use of supercritical fluids and reactions in ionic liquids),
5. water-based processes and products (e.g., organic reactions in high-temperature water),
6. replacement of hazardous reagents (e.g., the use of solid acids as replacement for traditional corrosive acids),
7. novel separation techniques (e.g., the use of novel biphasic systems such as those involving a fluorous phase),
8. alternative feedstock (e.g., the use of plant-derived products as raw materials for the chemical industry),
9. new, safer chemicals and materials (e.g., new natural product–derived pesticides and insecticides),
10. waste minimization and reduction (e.g., applying the principles of atom utilization and the use of selective catalysts),
11. renewable energy sources (e.g., microwave energy),
12. intensified processes with reduced mass and hear transfer resistances,
13. design products that can be recycled or biodegraded after use.

Conclusions

Worldwide public perception is that the chemical industry is the most polluting industry. Humans are universally concerned about the impact of chemical products on health, safety, and the environment and the damage the manufacturing industries have created on the ecology. This perception has become more significant since the mid-1980s, in spite of the fact that major improvements have been achieved in the reduction of emissions. There is thus a link between the public's attitude and their behavior as consumers. In addition, the public has easier access to data through NGOs, which explains why the awareness level has increased considerably. In addition, the data on the short- and long-term effects of large number of chemicals used in the industries are not fully known. Finding new uses to biomass or new uses to CO_2, for example, are also very important. Biotechnology appears to be the only answer to solve many of the problems mentioned in the book. It can help in reducing waste, making processes more benign, detoxifying waste, decreasing energy usage, and decreasing raw material usage. All industries should adopt the use of life-cycle analysis. A "responsible consumer" is necessary to achieve sustainable development. He or she is a person who reduces usage, recycles, and reuses the product. The economics of countries such as India are such that product recycle and reuse are practiced vigorously, whereas use and throw are commonly adopted in the United States and European countries. This latter practice has to be reviewed very seriously in order to achieve sustainable development. It should be very simple: if the value of a product's net energy release (i.e., energy output from a product to the energy input to the product) is positive, then that product is sustainable. If the value is negative, however, the product is not sustainable.

Introducing green chemistry teaching aids in the undergraduate and postgraduate curricula of various degree courses is an important step in bringing awareness of the importance of sustainability, atom efficiency, and waste minimization. Many teaching aids in the form of books and websites have already flooded the global market (Collins, 1995; Cann, 1999; Scott et al., 2000).

Chemistry is always a journey rather than a conclusion, and so is green chemistry. It is also expected to be based on continual improvement, discovery, and innovation. It should lead to the path toward the perfect goal of environmentally benign reactions, synthesis, processes, and products. Many areas of research pose a scientific challenge to chemists and also have the potential for large benefits. Such areas require major financial support as well as plenty of research input. The research is highly interdisciplinary, needing support from chemists, scientists, and engineers across disciplines. The funding has to be from industries, governments, and NGOs. The world has to be left untarnished by the present generation for the future generation to live and enjoy. Unless the governments of various nations take interest, green chemistry and greenhouse gas emission reduction will never be a reality, instead remain an academic exercise.

References

Note to readers: Several websites are listed here that represent worthwhile sites to introduce to classrooms.

American Chemical Society, green chemistry resources, ACS homepage: http://www.acs.org/education/greenchem/.

American Chemical Society, green chemistry teaching materials: http://center.acs.org/applications/greenchem/; http://www.acs.org/education/greenchem/cases.html.

Anastasio, M. and Viglia, A., Worldwide Biotech Process Overview on Industrial Biotechnology. Part 2, *The White Industry*, Workshop, Biotech Process Overview, Milan, Italy, Feb. 2006.

Bachmann, R., Industrial biotechnology—New value creation opportunities. McKinsey & Co. Study, 2002.

Cann, C. M., Bringing state-of-the-art, applied, novel, green chemistry to the classroom by employing the Presidential Green Chemistry Challenge Awards, *J. Chem. Educ.*, **76**: 1639, 1999.

Collins, T. J., Introducing green chemistry in teaching and research, *J. Chem. Educ.*, **72**: 96, 1995.

Dale, B. E., "Greening" the chemical industry: Research and development priorities for biobased industrial products, *J. Chem. Tech. & Biotech.*, **78**: 1093–1103, 2003.

Green Chemistry Institute: http://www.lanl.gov/greenchemistry/.

Green Chemistry Network: http://chemsoc.org/networks/gcn/.

Jansen, M. P., The cost of converting a gasoline-powered vehicle to propane: An excellent review problem for senior high school or introductory chemistry, *Chem. Educ.*, **77**: 1578, 2000.

Lakind, J. S., Wilkins, A. A., and Berlin Jr., C. M., Environmental chemicals in human milk: A review of levels, infant exposures, and health, and guidance for future research, *Toxicology and Appl. Pharmacology*, **198**(2), 184–208, 2004.

Lyons, G., Chemical Trespass: A Toxic Legacy, A WWF-UK Toxic Programme Report (July 1999), WWF-UK, Surrey, Gu71xR, UK.

Okkerse, H. and Van Bekkum, H., *From fossil to green*. Green Chemistry, 107–114, Apr. 1999.

Reed, S. M. and Hutchison, J. E., Resources for incorporating green chemistry into teaching green chemistry in the organic teaching laboratory: An environmentally benign synthesis of adipic acid, *J. Chem. Educ.*, **77**: 1627, 2000.

Royal Belgian Academy Council of Applied Science, Industrial Biotechnology and Sustainable Chemistry, Jan. 2004.

Royal Society of Chemistry, *Green Chemistry*: http://www.rsc.org/publishing/journals/gc/index.asp; http://www.rsc.org/publishing/journals/gc/reviewsarchive/asp

Singh, M. M., Szafran, Z., and Pike, R. M., Microscale chemistry and green chemistry: Complementary pedagogies, *J. Chem. Educ.*, **76**: 1684, 1999.

U.S. EPA's Green Chemistry Program: http://www.epa.gov/greenchemistry/.

Warner, C. John, S., Amy, Cannon, Kevin M. Dye, Green Chemistry. *Environmental impact Assessment Review*, **24**: 775–799, 2004

Index

Page numbers followed by "f" denote figures; those followed by "t" denote tables.